THE NEW NATURALIST LIBRARY

A SURVEY OF BRITISH NATURAL HISTORY

SHALLOW SEAS

OF NORTHWEST EUROPE

EDITORS
SARAH A. CORBET, ScD
DAVID STREETER, MBE, FIBiol
JIM FLEGG, OBE, FIHort
Prof. JONATHAN SILVERTOWN
Prof. BRIAN SHORT

*

The aim of this series is to interest the general reader in the wildlife of Britain by recapturing the enquiring spirit of the old naturalists. The editors believe that the natural pride of the British public in the native flora and fauna, to which must be added concern for their conservation, is best fostered by maintaining a high standard of accuracy combined with clarity of exposition in presenting the results of modern scientific research.

THE NEW NATURALIST LIBRARY

SHALLOW SEAS
OF NORTHWEST EUROPE

PETER J. HAYWARD

WILLIAM COLLINS

This edition published in 2016 by William Collins,
an imprint of HarperCollins Publishers

HarperCollins Publishers
1 London Bridge Street
London SE1 9GF

WilliamCollinsBooks.com

First published 2016

© Peter J. Hayward, 2016

All rights reserved. No part of this publication may be reproduced, stored in a retrieval system or transmitted in any form or by any means, electronic, mechanical, photocopying, recording or otherwise, without the prior written permission of the copyright owner.

A CIP catalogue record for this book is available from the British Library.

Set in FF Nexus

Edited and designed by
D & N Publishing
Baydon, Wiltshire

Printed in Hong Kong by Printing Express

Hardback
ISBN 978-0-00-730729-6

Paperback
ISBN 978-0-00-730730-2

All reasonable efforts have been made by the author to trace the copyright owners of the material quoted in this book and of any images reproduced in this book. In the event that the author or publishers are notified of any mistakes or omissions by copyright owners after publication of this book, the author and the publisher will endeavour to rectify the position accordingly for any subsequent printing.

Contents

Editors' Preface   vii
Author's Foreword and Acknowledgements   ix

1 The Shallow Seas   1

2 The Benthic Environment   17

3 Mud, Sand and Gravel   71

4 Hard Grounds   145

5 Some Key Habitats: Kelps and Seagrasses   181

6 More Key Habitats: Maerl Beds and Biogenic Reefs   233

7 Time and Change   279

References and Further Reading   363
Species Index   382
General Index   392

Dedication

For Sam, Ben, Isaac and George

Editors' Preface

The New Naturalist Library has produced few titles on the seas, and up to now the marine benthos has hardly been touched on. Two classic works produced more than half a century ago were C. M. Yonge's *The Sea Shore* (1949) (succeeded by Peter Hayward's *Seashore* (2004)) and Sir Alister Hardy's *The Open Sea*, published in two volumes. The second volume, *Fish and Fisheries* (1959), devoted 50 pages to the benthos because of its vital importance to fishes 'either as their prey or as voracious predators for limited supplies of food'.

Since then human activities have put the integrity of benthic communities at risk but have not diminished their significance for fisheries, and advances in research and techniques have greatly increased our understanding of happenings on the floor of the shallow seas that overlie the continental shelf. Now Peter Hayward extends and updates our understanding of this fascinating habitat, so that for the first time naturalists, even if they are not scuba divers, can begin to appreciate the variety of environments and the diversity of organisms found there. He introduces us to a hitherto mysterious and inaccessible system that is important not only as a source of fish for the table and a sink for rubbish from the land, and as a warning signalling the potential consequences of climate change, but also because it is for most naturalists a new and unfamiliar world that deserves to be widely appreciated and wisely managed.

Author's Foreword and Acknowledgements

The margins of the continents, especially broad in the North Atlantic region, are drowned by shallow seas, creating a sea-floor environment termed variously the sublittoral or shallow shelf, and part of the wider and deepening benthic realm. Continental-shelf seas are the most biologically rich and productive areas of the world ocean. This book addresses some aspects of the natural history of the benthic environment of the shelf seas of northwest Europe, and its biological communities, the benthos.

In northern Europe, naturalists began to investigate coastal marine habitats in the seventeenth century, and seashore studies reached an apogee in the second half of the nineteenth century, by when the development of small bottom dredges had initiated the exploration of the shallow sea floor. Marine biology began with the studies of enthusiastic amateur naturalists, and there are many who continue the amateur tradition today. Scuba, together with digital cameras and video recorders, provides the modern naturalist with opportunities for exploring the marine environment that Victorian enthusiasts would have wondered at. Thus equipped, the amateur naturalist is able to continue to make original contributions to marine biology, but only with regard to inshore habitats, and at the shallowest depths. Surveying the benthic environment offshore involves expensive, ship-borne sampling procedures. Many biological science students will have had the opportunity to see grabs, dredges and trawls deployed from research vessels operated by marine biology stations, and to examine and to identify their catches in seawater tanks. Some postgraduate students will have had the privilege of participating in research projects and programmes that involved extensive shipboard fieldwork, and long periods in the laboratory developing familiarity with benthic faunas. It is also likely that many amateur marine biologists will have become familiar with benthic biology and ecology through attending classes provided by field stations. My own experience of the benthos derives from more than enough hours aboard coastal research vessels, and many more, and more comfortable, hours in the laboratory, over a period of 28 years conducting practical classes in benthic marine biology.

It is a challenge to present a convincing account of the natural history of an environment that is rather difficult to envision. Away from rocky coastlines the sea floor is rather flat, often muddy, beneath turbid water with low or no visibility. Benthic faunas mostly live within the sediment of the sea floor, or are sparsely and patchily distributed upon it, and if at all motile are likely to withdraw into burrows or move quickly away on disturbance. Yet, dredges and grabs reveal an often extraordinary diversity and density of animals, suggestive of complex interacting communities. This is not a textbook of marine benthic ecology, nor is it a comprehensive review of the benthic communities of the northwest European shelf seas. Rather, it describes the natural history of some benthic habitats and associations characteristic of our region. It is based upon my joint experience, with friends, colleagues and students, of marine animals and communities, and further informed by the published research of many other fellow marine biologists. I wish to thank all whose work I have relied upon in exploring and explaining those aspects of the field that I find of particular interest. All sources consulted are cited, but I take responsibility for any errors of fact or interpretation.

Finally, it is a pleasure to offer sincere thanks and appreciation to all the friends and colleagues with whom I have spent so many pleasant hours. Special thanks are due in particular to Jill Ireland, John Lancaster, Helen Marshall, Keith Naylor, Joanne Porter, Emily Roberts, John Ryland, and the students and demonstrators of the Millport forays. Jim Ellis, John Lancaster, Joanne Porter and John Ryland kindly provided many of the fine photographs embellishing this book.

FIG 1. At Europe's rim: the Cliffs of Moher, Co. Clare, Ireland. (Liam Morrison)

CHAPTER 1

The Shallow Seas

At distant points on the northwestern edge of Europe the continent seems to end dramatically, dropping from vertiginous promontories such as the Cliffs of Moher, Co. Clare, Ireland (Fig. 1), into a forbidding greenish ocean that rolls west over an unimaginable abyss. At other places, such as the Wash, or the Solway Firth, it does not seem to end at all; high spring tides lapping at salt-marsh grazing recede to indistinct horizons with no discernible boundary between land and sea. Actually, the European landmass does not end at Land's End, the Blasket Islands, the coast of Clare or St Kilda, but 150 km further west where it finally peters out at the true ocean's margin, a sharp edge termed the continental break where the basaltic rocks underlying the continent plunge towards the great ocean depths. It is interesting to reflect that at this edge the sea's depth, at around 180 m, is less than half the height of some of the greatest cliffs of Britain and Ireland, and that over much of the sea floor between the coast and the continental break, Beachy Head would protrude comfortably above the sea surface. This area of sea floor lying between shore and break is termed the continental shelf, appropriately so as it is mostly flat and borders the whole of Europe's coastline. It is deepest to the west, shallowest to the east, with most of the southern half of the North Sea being shallower than 50 m. Its area is greater than the combined land area of the British Isles and Ireland. On low spring tides hundreds, perhaps thousands, of square kilometres of it are exposed to the inspection of the curious, and through the past 20,000 years substantial portions of it have alternated between dry land, coastal marsh and shallow sea, through cycles of centuries. The shallow sea floor of northwestern Europe constitutes the largest continuous environment in the region.

The environment of the European shelf sea floor, the benthic environment, and its fauna, the benthos, are the focus of this book. Benthic is used in contradistinction to pelagic, the environment of the water column, although the two terms provide only the coarsest division of the marine realm. Benthic and pelagic environments may both be subdivided by depth (Fig. 2). The sea floor of the continental shelf is sometimes termed the sublittoral zone, although shallow benthic is perhaps more descriptive. From the shelf break the steeply descending continental slope constitutes the bathyal zone; between 2,000 and 3,000 m depth the gradient flattens and from 3,000 to 6,000 m the sea floor is the zone of the vast abyssal plains that underlie 75% of the world ocean. The deep ocean trenches, dropping to beyond 10,000 m, are the hadal zone. The pelagic environment of the shelf seas is termed neritic. Beyond the shelf edge it is divided into an epipelagic zone, to around 200 m depth, a mesopelagic zone extending to 1,000 m, and below that, to 4,000 m, a bathypelagic zone. The next 2,000 m down is the abyssopelagic, and the water column of the trenches is the hadalpelagic.

Vertical zonation of benthic and pelagic environments reflects changes in physical and biological characteristics with depth. Light is one important physical correlate: 50% of the sunlight reaching the sea surface is reflected or scattered, or absorbed within the top few metres of the water column, but the remainder, light with wavelength between 400 and 700 nm, penetrates deeper. In the clearest coastal waters sunlight reaches to around 250 m depth, approximating the lower limit of the epipelagic zone, but in translucent ocean water it will define the boundary between the mesopelagic and bathypelagic zones. In the top 50 m of the epipelagic zone light intensity is sufficient for net photosynthetic production, and thus growth, in macroalgae and phytoplankton, and these critically important metres are termed the euphotic zone. This zone is

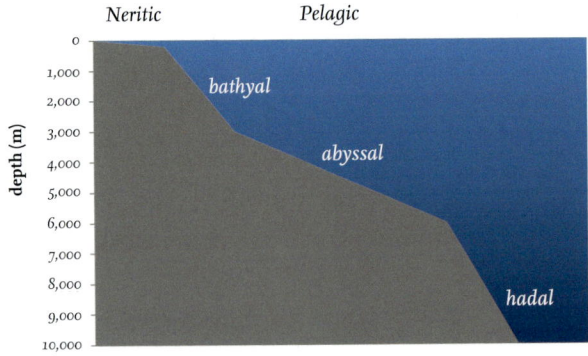

FIG 2. The vertical ecological divisions of the ocean. Not to scale.

at its maximum extent on the outer shelf, while in turbid inshore water it may be only a metre or less in vertical extent. The residual light that penetrates the mesopelagic is of insufficient intensity for photosynthesis but does allow for sight in specially adapted fish, crustaceans and cephalopods; beyond 1,000 m no light is detectable, or usable, by any biological process.

The benthic and pelagic environments of the northwest European continental shelf are, of course, interconnected, and biological and ecological processes within the water column and on the sea floor are often tightly coupled. However, marine biology is a science of many disciplines in which benthic ecology and pelagic ecology each constitute distinct components. Benthic ecology demands understanding of sediments and sedimentary processes, while pelagic ecology is intimately associated with oceanographic processes. Fisheries science has also developed as a distinct discipline that incorporates both benthic and pelagic perspectives, focusing upon the biology and ecology of both demersal (i.e. bottom-living) and pelagic fish. However, although benthic ecology encompasses autecological studies of larger or numerically dominant species, it is mostly concerned with the ecology of benthic communities, assemblages of species, while fisheries science has typically addressed life cycles and population dynamics of commercially important species of fish and shellfish, and until recently often without detailed reference to the wider ecosystem.

The benthic environment of the shelf is characterised by substratum patchiness and heterogeneity at all temporal and spatial scales; temperatures fluctuate diurnally and seasonally, inorganic and organic matter are abundant and productivity is high, often with huge seasonal peaks. With increasing depth habitat heterogeneity declines, temperatures decrease and fluctuate less, and inputs of organic matter are low, and often pulsed. The deep-sea environment, that of the abyssal plains that comprise three-quarters of the sea floor, was once thought to be generally inhospitable to life, and mostly barren. Deep-sea exploration, from the cruise of HMS *Challenger*, 1872–1876 (Fig. 3), to the advent of the *Alvin* generation of manned and unmanned submersibles, has shown that this is not so, and has revealed rich diversity and

FIG 3. A scientist (with boater and companion) sorting samples aboard HMS *Challenger*. (From *Natural Science*, July 1895)

complex life modes. Nonetheless, for their area, the shelf seas remain the most diverse and productive part of the marine environment.

While shelf seas comprise less than 10% of the world ocean they are overwhelmingly the most productive portion of the marine realm, and they have been exploited throughout human history. In Europe, the classical civilisations of the eastern Mediterranean had sophisticated fisheries for fin fish, shellfish, sponges and coral, and also nurtured the first marine biologist, Aristotle, whose researches and observations were not improved upon until the Renaissance. Scientific enquiry flowered later in the maritime nations of northern Europe, and in Britain the first naturalists to embrace marine biology began to appear only in the late seventeenth century. The naturalist John Ray (1627–1705) included marine sponges, hydroids and bryozoans in his catalogue of the British flora (*Synopsis Methodica Stirpium Britannicarum*, 1690), and the physician Martin Lister (1638?–1712) published the first comprehensive English malacological catalogue (*Historia sive Synopsis Methodica Conchyliorum*, 1685–1692). John Ellis (1710?–1776) was another early specialist, publishing in 1755 the elegant *Essay towards a Natural History of the Corallines*, which described and illustrated many British species of hydroids and bryozoans, then together classed as 'zoophytes' and newly confirmed to be animals rather than plants. Ellis was also an early enthusiast for fieldwork, personally collecting many of his specimens from the shorelines of southeast England. Daniel Solander (1736–1782), a student of Linnaeus, arrived in England in 1760, with a letter from Linnaeus recommending him to Ellis, and evidently worked closely with him, and was able to complete the jointly authored *Natural History of Zoophytes* (1786) after the death of the latter.

The study of natural history burgeoned throughout Europe in the eighteenth century, as European exploration of the rest of the world began to reveal its apparently limitless biological diversity, and reached an apogee in the nineteenth century as a focus for serious academic and scientific research. Marine natural history, especially malacology, was at the forefront of this focus. Exploration of the sea floor is difficult. Pioneer marine naturalists probably relied upon fishermen, whose various nets, grapples and grabs would have supplied samples of sea-floor habitats and specimens for study, but by the end of the eighteenth century a small iron-framed dredge had been designed especially for biological sampling. Britain's pioneer, Edward Forbes, began to employ this 'naturalists' dredge' in the Irish Sea, and then in 1839 embarked upon a dredging survey around the Shetland Isles. The results of this expedition were presented before a meeting of the British Association and stimulated such interest that a committee was formed to promote exploration of the British sea area through dredging, and in the ensuing three to four decades dredging surveys were conducted off much of the coastline of Britain and Ireland.

The results of the activities initiated by the dredging committee were published in the reports of the annual British Association meetings, as faunal lists, but the most significant outcome was a stream of often beautifully illustrated books that monographed much of the shallow-water marine fauna of northwest Europe. These included the familiar and popular, in *The Fishes of Great Britain and Ireland* (Day, 1880–1884) and *British Conchology* (Jeffreys, 1862–1869), as well as the more arcane, such as *A Monograph of the British Nudibranchiate Mollusca* (Alder & Hancock, 1845–1855) and *A History of the British Marine Polyzoa* (Hincks, 1880). The early shoots of marine biology in Britain and western Europe have been briefly reviewed in *The Sea Shore* (Yonge, 1949), *The Open Sea: the World of Plankton* (Hardy, 1956) and *Living Marine Molluscs* (Yonge & Thompson, 1976).

Through much of the nineteenth century, marine natural history study was promoted by a large community of amateur enthusiasts. There were comparatively few academic specialists. Edward Forbes was appointed Professor of Natural History at Edinburgh University just before his early death, in 1854, and Charles Wyville Thomson, who studied at Edinburgh in the 1850s, became Professor of Natural History at Queen's College (now University), Belfast, but it was not until the 1880s, with the beginnings of fisheries science, that marine biology became an academic discipline. The amateur community, in contrast, thrived from the first decades of the century. Many amongst it made a living from writing popular books, and of these Philip Henry Gosse is especially notable. Gosse became a professional writer following unsatisfying ventures into business, farming and teaching, during which he nurtured a talent for natural history, and developed deep religious convictions. His list of publications tops 400, encompassing popular natural history, religious tracts, textbooks on zoology and marine biology, and an impressive number of scientific papers based on his own original research (Freeman & Wertheimer, 1980). P. H. Gosse was an acknowledged authority on the biology and ecology of several groups of marine invertebrates, and his exquisitely illustrated monograph of British sea anemones, *Actinologia Britannica* (1858–1860), was the standard reference to the group for more than 50 years, and is still considered a useful source today.

Many of Gosse's contemporaries, enjoying private incomes, or supported by church benefices, and probably free of his financial exigencies, pursued their interests along strictly scientific courses. They collected, identified and classified marine plants and animals, naming numerous new species from the dredge surveys conducted around the entire coastline of northwest Europe, and many became taxonomic specialists on an international scale. They were also collectors, and the collections accumulated by such accomplished amateurs as Canon A. M. Norman, the reverends G. Busk, T. Hincks and A. W. Waters, and the lawyer J. G. Jeffreys, now form important core components of national collections, in the Natural History Museum, London, as well as in the larger provincial museums. Rich in original type specimens, these primary collections are important records of the taxonomic diversity of marine communities of European shelf seas, and of numerous other areas of the world from which these early specialists obtained their specimens.

Seashore holidays became popular with the emerging middle classes of Britain through the first half of the Victorian era, and the study of seashore natural history was encouraged as an improving distraction. Collecting excursions between the tides, often led by an informed naturalist, introduced visitors to the diverse plant and animal communities to be found there, private aquaria provided exhibitions of fish, decapods and other large invertebrates usually found only below low water mark, and the natural history cabinets of newly inspired amateurs often contained series of shells, pressed seaweeds and various dried tests, casts and egg cases.

Scientific enquiry into the marine environment began to increase from 1870 as both academic and amateur biologists turned their attention towards the physiology and biology of the many newly familiar animals and plants they were encountering, and was marked by the development of marine laboratories, especially equipped for scientific research. The first of these was that developed at Naples by the marine physiologist Anton Dohrn, through the 1870s. Small laboratories were then founded on the shore at St Andrews, and on a barge, *The Ark*, moored in a flooded quarry close to Edinburgh, in 1884, and the following year towed via the Forth and Clyde Canal to the Clyde Sea island of Cumbrae, where it became transmogrified eventually into the shore-based Millport Marine Biological Station, of the Scottish Marine Biological Association (SMBA). The Marine Biological Association of the United Kingdom (MBA), also inaugurated in 1884, established the then largest research facility, the Plymouth Marine Laboratory, in 1888 on a site at Citadel Hill. The MBA continues to operate from the PML, but the SMBA, now the Scottish Association for Marine Science (SAMS), is today located at a much larger site at Dunstaffnage, Oban.

Further marine laboratories were established as university facilities through the twentieth century: the University of Liverpool assumed responsibility for the marine station at Port Erin, Isle of Man, founded in 1892 by the Liverpool Marine Biology Committee; the University of Newcastle upon Tyne created the Dove Laboratory at Tynemouth; and the Millport Marine Biological Station became a joint teaching and research facility of the universities of Glasgow and London.

Marine research laboratories were also established by government agencies such as the Central Electricity Generating Board (CEGB), the Ministry of Agriculture, Fisheries and Food (MAFF) – now the Department for Environment, Food and Rural Affairs (DEFRA) – and the Natural Environment Research Council (NERC), while the Field Studies Council (FSC) created several field centres at coastal locations, specialising in teaching coastal and littoral ecology for secondary school and undergraduate students. Into the second decade of the twenty-first century, British marine biological and oceanographic institutions are largely concentrated at Lowestoft and Weymouth, for the Centre for Environment, Fisheries and Aquaculture Science (CEFAS, an executive arm of DEFRA), Plymouth (MBA and NERC), Southampton and Liverpool (the universities and the NERC National Oceanographic Centre), Menai Bridge, Anglesey (Bangor University, Centre for Applied Marine Sciences), and Dunstaffnage (SAMS), although smaller establishments continue to operate at the Gatty Laboratory, St Andrews, the Dove Laboratory, Tynemouth, and Millport.

Each of the maritime nations of northern Europe developed an equally diverse range of facilities devoted to marine environmental research and teaching, at centres such as Arcachon, Bordeaux, Concarneau and Roscoff, Brittany (France), Texel (the Netherlands), Sylt and Helgoland (Germany), Helsingør (Denmark), Drøbak, Oslo and Espeland, Bergen (Norway) and Kristineberg (Sweden). Cooperation began in 1902, at Copenhagen, when Denmark, Norway, Finland and Sweden, together with the United Kingdom, the Netherlands, Germany and Russia, founded the International Council for the Exploration of the Sea, now familiarly known as ICES or CIEM (Conseil International pour l'Exploration de la Mer). The Council determined upon three main objectives: to stimulate research into the marine environment of the northeast Atlantic region, particularly in relation to commercially important biological resources; to develop collaborative research programmes agreed by the member states of the Council; and to publish, or otherwise communicate, the scientific results of such programmes. At the centenary meeting, again in Copenhagen, the representatives of the member states, now comprising all of the European Atlantic maritime nations, the Baltic states, the Russian Federation, the United States of America and Canada, reaffirmed the original objectives of ICES, while recognising the need for wider partnerships aimed to conserve and sustain the world's ecosystems in accord with the 1992 Rio declaration. Scientific understanding of the physical, chemical and biological functioning of marine ecosystems, and of the effects and consequences of human impacts on the marine environment, are critical to the development of ecosystem-scale management policies that protect and conserve marine resources, while allowing sustainable exploitation.

The early focus of ICES, on pelagic production and fishery resources, has now expanded to encompass the whole of the marine environment and all marine scientific disciplines. It has especially valuable roles in the collection and storage of marine environmental data, and in providing large, long-term datasets for scientific analysis, and on which states and agencies may base environmental management strategies. For example, ICES data have been used to map nine 'ecoregions' for the northeast Atlantic and adjacent Arctic areas, and a further four for the Mediterranean and the Black Sea. An ecoregion can be defined as a large geographical area characterised by a distinct suite of environmental conditions, natural communities and species, more or less corresponding to a biogeographical province, or alternatively simply as a geographical area of most practical use in devising and implementing ecosystem management strategies. The nine Northeast Atlantic/Arctic ecoregions constitute the Food and Agriculture Organization (FAO) major fishing area 27. The northwest European shelf seas are encompassed by two ecoregions. The Greater North Sea ecoregion, covering 750,000 km^2, extends from 5° W eastwards as far as the Kattegat, and is bounded to the north by the 62° N parallel, while the western shelf of the British Isles and all Irish coasts fall within the Celtic Seas ecoregion; these have been subdivided by ICES for the purposes of statistical recording of data, particularly relating to fish stocks (Fig. 4). However, benthic data are more usefully recorded in relation to ICES statistical rectangles. Each ICES rectangle occupies an area of 30 × 30 nautical miles, equivalent to 0.5° of latitude and 1° of longitude, giving a total of 211 rectangles for the epicontinental North Sea; at 53° N the ICES rectangle has an area of 3,720 km^2.

Exploration of shallow benthic environments expanded through the second half of the nineteenth century, and continental shelf seas, worldwide, were sampled by dredging and trawling. Dredge designs were progressively refined and modified to enable sampling of soft muds, compacted sands and gravels, and rocky patches. The bottom trawls traditionally employed by inshore fishermen were similarly adapted for scientific sampling. The beam trawl, so called for the rigid beam that holds open the wide mouth of a triangular net, is principally used for commercial fishing for shrimps and prawns, scallops and flatfish. Commercial beam trawls used within the three-mile limit of territorial waters have a 4 m wide beam, while larger gear used in deeper waters of the outer shelf may be up to 12 m in width; smaller trawls, with a 2 m beam (Fig. 5), are conventionally used in scientific sampling. The Agassiz trawl (Fig. 6) has an iron-framed, rectangular mouth extending between symmetrical iron runners; it operates efficiently whichever side lands on the sea floor, catching sedentary and slow-moving animals, but it is unsuited to catching flatfish and is not used commercially.

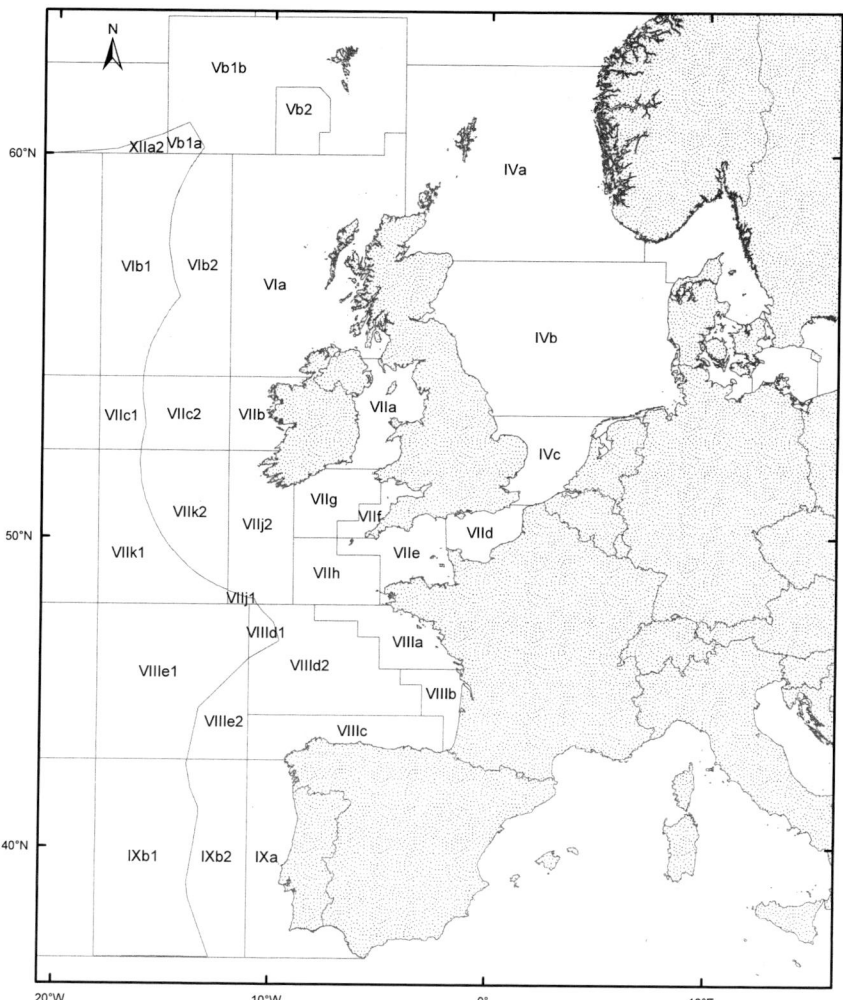

FIG 4. ICES subdivisions of the Northeast Atlantic ecoregions. IV: North Sea; V: Faroes; VI, VII: Celtic Seas; VIII, IX: South European Atlantic Shelf and part of oceanic Northeast Atlantic. (Reproduced courtesy of the General Secretary of the International Council for the Exploration of the Sea)

Trawl samples may be crudely quantified, expressing numbers or biomass of organisms caught in terms of the duration of the trawl, or of the distance, and thus area, covered. Such estimates are still useful in describing broad-scale

FIG 5. The 2 m beam trawl employed in scientific sampling. (J. R. Ellis (CEFAS))

FIG 6. The Agassiz trawl. (J. R. Ellis (CEFAS))

distributions and densities of the larger benthic animals, but benthic ecology did not really begin until marine biologists commenced quantitative sampling of the benthic invertebrate communities of soft sediments. This developed in the early twentieth century, and the pioneer was the Danish fisheries scientist C. G. J. Petersen, who employed the sampling device that still bears his name, the Petersen grab. This consists of two scoops, with toothed edges, hinged together but held apart by a taut chain; on striking bottom the chain is released, and the scoops dig into the sediment, and close together as the grab is hauled in. Petersen's grab collects a sample of sea floor with a defined surface area; this is most often 0.1 m^2, but 0.2 m^2 and 0.25 m^2 grabs are also used. Surveying an area of sea floor by grab sampling produces data on the distribution of sediment types, and the diversity and density of benthic organisms at each sampling point, which can be quantified as numbers, or biomass, per unit area; the sample is essentially a hemicylinder of sediment, deepest along the hinge line of the scoops, but scarcely sampling below the surface along their edges, so faunal densities per unit volume of sediment cannot be measured.

Grab design evolved through the twentieth century as benthic ecologists sought to improve sampling techniques, addressing such problems as premature closure of the scoops, depth of penetration of the sediment, and loss of samples during retrieval of the grab. Many grab designs had only a short existence, being devised by a particular group of biologists and employed for particular surveys, with perhaps specific objectives, and then not used subsequently. Some were later discarded as cumbersome, expensive or, with spring-loaded scoops, potentially dangerous. The Petersen grab is still reliable for sampling soft sediments in relatively sheltered environments, but the most widely used benthic samplers are probably the Day grab and the van Veen grab. In the former the two scoops are supported within a four-sided pyramidal frame, and closure is triggered as the frame settles on the sea floor. The two scoops of the van Veen grab (Fig. 7) are attached to long arms that provide considerable leverage on closure, and maximum penetration of the sediment. The sample retained by the grab is emptied onto a sieve screen and washed through to flush away the sediment and reveal its animal content. The mesh size of the sieve will affect the composition and density of the sample of animals retained. The choice is usually between a 1 mm or 2 mm mesh, although 0.5 mm sieves are also frequently used; one comparison of the influence of mesh size on sample composition (Rees, 1984) showed that a 1 mm mesh sieve retained only 69% of the species, and 70% of the total individuals, recovered using a sieve of 0.5 mm mesh.

Functional efficiency, the efficiency of operation of the grab, and sampling efficiency, the degree to which the sample reflects the actual density of organisms

FIG 7. (a) The long-armed van Veen grab, with buckets open (J. E. Lancaster); (b) ready to be deployed (A. P. Woolmer); (c) hauling aboard the closed grab (J. Turner & K. Mortimer-Jones).

at the sample point, have been calculated for different types of grab under different conditions, and the probability that a particular sample accurately represents the habitat it is drawn from can be estimated, for a particular portion of the fauna. The van Veen grab can be used effectively to sample most mixed sand and mud habitats, but as the coarsest sands grade into gravel and pebbles its efficacy declines sharply. Penetration of the sediment may be inconsistent, and the larger fragments may obstruct closure of the scoops, with the result that a proportion of the sample will be washed out as the grab is hauled in. Yet gravelly habitats may support a substantial fauna of thick-shelled molluscs, as well as numerous small free-living polychaete worms and crustaceans. The Hamon grab was designed especially to sample coarse aggregate-grade sediments. It consists of a robust steel bucket, at the end of an arm pivoting against a heavily weighted rectangular frame; on hauling, the bucket is drawn in a 90° arc through the gravel, closing against a rubber-coated steel plate that prevents loss of any portion of the sample. Two sizes of Hamon grab are in use, sampling 0.1 m² and 0.25 m² of sea floor; the smaller size is usually favoured, for its easier handling, and the lesser time and effort required to process replicate samples.

Numbers of epifaunal animals, those living on the surface of the sediment, are underestimated by all types of grab: a proportion will simply move in response to the impact of the grab, or is washed away by the disturbance it causes. Grab samples provide no insight into the structural characteristics of a benthic habitat, or of the spatial disposition of the fauna inhabiting it. The grade of a sediment – mud, fine sand, coarse sand or gravel – and its organic content are easily established, but it may be of significance to know whether the sediment is layered, and the extent to which it has been reworked by its inhabitants. The vertical distribution of the fauna, and the depth and extent of burrows, may also be of interest. Such data may be collected by means of box corers, which allow vertical cores of sediment to be retrieved undisturbed, preserving the three-dimensional structure of the habitat. However, box corers tend to be large and expensive pieces of equipment, difficult to deploy from small research vessels, especially under any but the calmest weather conditions. They are not routinely used by benthic ecologists, but are important for deep-sea sampling, and in shelf waters for the quantitative sampling of the smallest size categories of invertebrate animals.

The greatest boost to practical marine biology must have been that provided by the easy availability of reliable scuba equipment from the late 1950s. The enthusiast was finally able to observe marine life, at length, in its natural setting, while the professional field biologist gained a completely new methodology for marine ecological research. Scuba allows for precise sampling procedures, free from the uncertainties of dredge and grab, and not only in soft sediment habitats, but for habitats such as rocky reefs, vertical walls and caves, that could not otherwise be sampled. Diver-assisted sampling is also far less damaging to the wider environment than remote sampling from research vessels. Long-term monitoring of benthic communities also becomes possible through use of scuba, with film, television and video technologies for recording temporal change, and field experiments become routine for the diving marine ecologist. Scuba diving has become a conventional technique for marine fieldwork that has enabled scientific exploration of coastal waters worldwide

Depth, temperature and light ultimately limit the effectiveness of the diving biologist, and sampling the benthos offshore is still dependent upon remote techniques. Grabs and dredges continue to be used, and cameras may be mounted on towed sleds to survey habitat types (Fig. 8). However, independently powered, remotely operated vehicles, or ROVs, deployed by cables from research vessels, are also widely used in marine benthic research. Commercial demand for

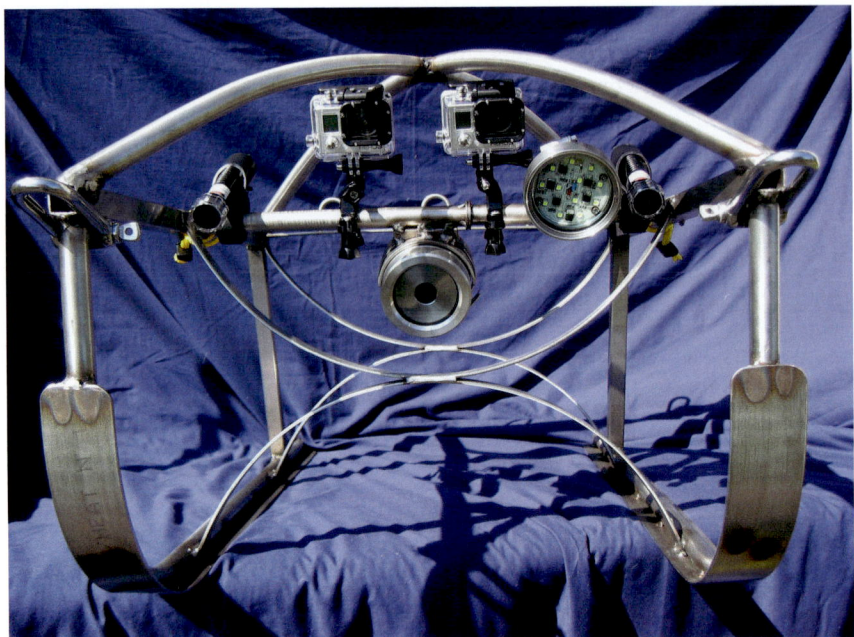

FIG 8. A benthic sled mounted with paired digital cameras, recording video images to a storage card, above a larger camera transmitting real-time images to a shipboard monitor, an LED lamp, and a laser on each side to supply scale. (J. E. Lancaster)

ROVs in oil and mineral exploration, and for submarine engineering, has driven the development of numerous sophisticated submersibles; fitted with cameras, with equipment for recording physical data, and for sampling both water and sediment, some of these may be adapted for marine research. However, ROVs are expensive investments, and costly to operate, and routine surveying of shelf sea floors is more cheaply achieved using acoustic techniques.

There are two contrasting methodologies employed in acoustic surveying of the sea floor. Ship-mounted echo sounders, or acoustic ground discriminating systems (AGDS), generate depth contours, show sea-floor relief and provide insight into the structure and thickness of bottom sediments. Sidescan sonars are towed close to the bottom and gather sonic data from a belt up to 100 m wide on either side of the tow path. While AGDS collect data on topographical and sedimentary features at scales of tens of metres, the low-angle sidescan sonar beam detects sea-floor structures at much smaller, centimetre, scales.

Acoustic data may be difficult to interpret. Variations in tide and current flow, temperature and salinity affect the sonic data, to the extent that the same survey track, sampled on successive occasions, may generate different data each time. However, as sonar equipment and techniques continue to be improved, it is certain that software will also be refined, allowing precise interpretation and integration of sonic data. AGDS and sidescan sonar enable large-scale mapping of the benthic environment. Grab sampling may be used in conjunction with acoustic surveying to verify interpretation of sonic data, a process termed ground truthing, and to correlate acoustic signatures with recognised benthic communities.

It is now much more difficult for the amateur to make original contributions to benthic marine biology. There is no longer a need for extensive surveys, or for comprehensive collecting, and taxonomic research is largely limited to less familiar groups of often small organisms, or to less accessible communities, and as with much benthic ecology frequently requires special techniques and expensive equipment. However, in coastal waters, scuba and underwater photography provide opportunities for the amateur fieldworker to census and monitor populations and communities, collecting data that record changes in time and space, all important at what may prove to be the beginning of a period of sharp and profound change in marine benthic environments.

Marine biology is well served by books. The expanding popularity of scuba diving among amateur naturalists, and as a recreational pursuit, created a demand for marine identification guides, and as the age of the dredge stimulated the flood of illustrated monographs in the nineteenth century, so the late-twentieth-century boom in diving led to the development of popular marine field guides, and to a plethora of inexpensive colour photographic guides. Regional marine field guides can be found for practically every stretch of inhabited coastline, and island group, worldwide. They range from the comprehensive, enabling identification of the most common marine plants and animals, of every taxonomic group, to the specialist, focused on the more familiar, conspicuous or colourful components of the fauna, such as fishes, molluscs and coral. Both categories include practical identification guides, with plates of line drawings or thumbnail photos emphasising diagnostic features, and a brief text of confirmatory characters. Then there are the frankly artistic offerings, displaying selected species, illustrated by high-quality colour images, in their natural habitat.

There are many good comprehensive modern textbooks, such as the students' standard by J. S. Levinton, *Marine Biology* (2001, and subsequent editions), more advanced reviews, such as the multi-authored *Marine Ecology* (Kaiser et al., 2005), and others with specialist focus, for example, *Ecology of Marine Sediments* (Gray & Elliott, 2009). There are presently no popular accounts of the natural history of the northwest European continental shelf environment, but the Internet now allows both amateur and professional marine biologists access to a great deal of primary information, and facilitates easy access to the prime marine research journals, such as the *Journal of the Marine Biological Association of the United Kingdom*, *Marine Biology* and the *Marine Ecology Progress Series*, the *Journal of Experimental Marine Biology and Ecology*, and the invaluable *ICES Journal of Marine Science*, formerly accessible only to members of professional organisations, or through academic libraries. Sampling techniques, experimental protocols, analytical procedures, and the impressive technology now employed in marine benthic ecology are all reviewed at length in *Methods for the Study of Marine Benthos* (Eleftheriou, 2013). The *Handbook of the Marine Fauna of North-West Europe* (Hayward & Ryland, 1995) provides keys for the identification of common benthic invertebrates and fish, together with brief descriptions and illustrations of most species included. More comprehensive guides to selected invertebrate groups are published in the series *Linnean Society Synopses of the British Fauna*.

Local records of marine plants and animals were formerly published by many marine stations, and provided useful guides to the regional distribution of species and community types. The most well known of these is the *Plymouth Marine Fauna*, and others, as single volumes or series, have been produced for the Isle of Man, the Clyde Sea, St Andrew's Bay, and the Northumberland coast. These 'marine faunas' include concise, dated records of individual species, often with notes on habitat, abundance and observed reproductive periods, but each was essentially outdated immediately on publication, because new data continued to accrue. However, they are now important sources of historical distributions, particularly so as marine habitats are steadily degraded by human activity, and as the composition of faunal and floral communities responds to environmental change. Updating a published inventory is an expensive prospect; the *PMF* achieved a third edition in 1957, and the most recently revised fauna to be published in book form is *The Marine Fauna and Flora of the Cullercoats District* (Foster-Smith, 2000). Edition 3 of the *PMF* is now accessible online, and future local faunas and floras will probably be published mostly in electronic format; while the passing of bound volumes of regional faunas and floras may be mourned, electronic records are easily and swiftly updated. A complete checklist of the marine fauna and flora of the British Isles sea area was provided by the *Species Directory of the Marine Fauna and Flora of the British Isles and Surrounding Seas* (Howson & Picton, 1997), while the *European Register of Marine Species* (Costello et al., 2001) covers the whole of the northeast Atlantic region, including the Mediterranean.

CHAPTER 2

The Benthic Environment

The continental shelf of northwest Europe is covered by very shallow seas, less than 100 m deep over two-thirds of their extent (Fig. 9). Along much of the coastline of Britain and Ireland the sea is shallower than 50 m, in some places to 150 km or more offshore. It is especially shallow to the southeast; the 50 m depth contour, or isobath, undulates northeastwards from off Flamborough Head, on the Yorkshire coast, extending into the Skagerrak, the narrow sound separating Denmark and Norway. The southern North Sea is almost entirely shallower than 50 m, and so shallow off parts of East Anglia and northeast Kent that the sea floor has been cultivable land in historical times. The shelf break has been taken conventionally to be marked by the 200 m isobath, although along much of the northwest shelf edge it is shallower than this, on average at around 180 m. The 200 m isobath defines the boundary of a narrow band of deep shelf extending from west of the Shetland Isles and the Hebrides, and around the western coasts of Ireland. To the south of Ireland, in the region generally referred to as the Celtic Sea, the deep shelf broadens, to about twice the width of the Irish, Hebridean and West Shetland shelf. The broadest area of deep shelf lies east of the Shetland Isles; here, the shelf break runs north-northwest, along the edge of a narrow trough of much deeper water, the Rinne, isolating it from the narrow shelf of western Norway. Much of the inner shelf of western Norway lies deeper than 50 m. Three detached fragments of shelf lie northwest of the Shetland Isles. The two smaller constitute the submarine Faroe Bank and Bill Bailey's Bank, while the larger supports the archipelago of the Faroe Islands. A deepwater ridge, named for the Victorian marine biologist Wyville Thomson, runs southeast, below 200 m depth, towards the West Shetland Shelf, and is an important influence on

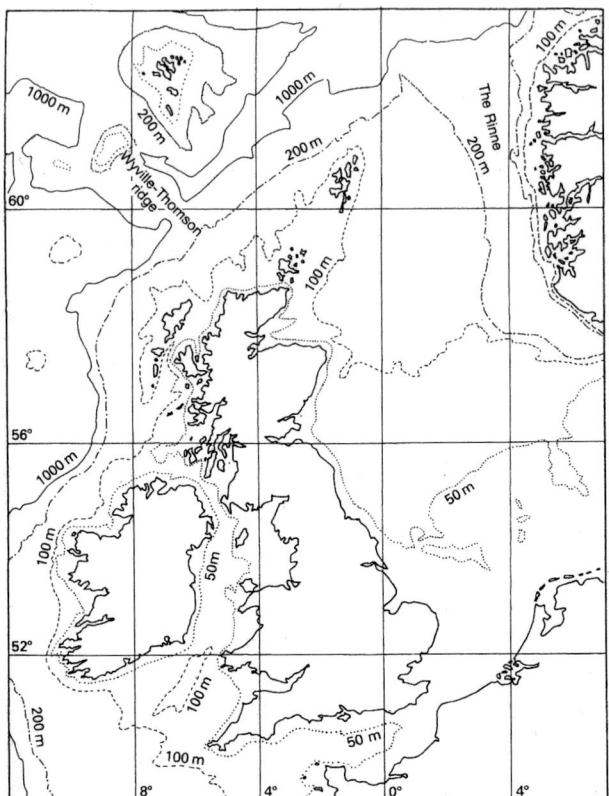

FIG 9. Bathymetry of the northwest European shelf. (From Hayward & Ryland, 1995)

deep-water circulation. The physical, hydrographical and biological features of the northwest European shelf, and its economic resources, were mapped in the elegant *Atlas of the Seas around the British Isles* (Lee & Ramster, 1981).

THE PHYSICAL ENVIRONMENT

The principal oceanic influence on the marine environment of northwest Europe is the **N**orth **A**tlantic **C**urrent (NAC), a warm flow that originates off the eastern coasts of the USA as the Gulf Stream, crosses the Atlantic to reach the European shelf at around 50° N, and then flows northeastwards into the Norwegian Sea. The track of the NAC passes between Iceland and Scotland, flowing at depth along the 100 m isobaths, with a branch passing to the southeast along the edge of the Rinne. Surface flow enters the northern North Sea between Orkney and Shetland, and through the Pentland Firth down the eastern coasts of Britain. Surface currents enter the southern North Sea through the Dover Strait, flowing northeast to the Danish coasts. From the Irish Sea to the Minch – the channel separating the Western Isles and mainland Scotland – surface circulation is wind-driven and variable, and a significant factor affecting both shallow circulation and the flow of the NAC is a climatic phenomenon referred to as the **N**orth **A**tlantic **O**scillation, or NAO (Pingree, 2005).

The NAO relates to irregular, and unpredictable, fluctuations in air pressure differentials between boreal low-pressure and subtropical high-pressure systems, which have their most important environmental effects during the winter period, December to the end of March (Ottersen et al., 2004). It is expressed as an index, derived from differences between normalised values for sea-level air pressure measured at Stykkisholmur, Reykjavik, Iceland, and either Lisbon, Portugal, or Ponta Delgado, Azores; the index oscillates between positive and negative values on either side of the long-term mean (Fig. 10). During winters with a strong subtropical high-pressure system over the Azores, and a deeper than normal Icelandic low, the large pressure differential between the two systems results in a high positive value of the index (NAO+), while weak

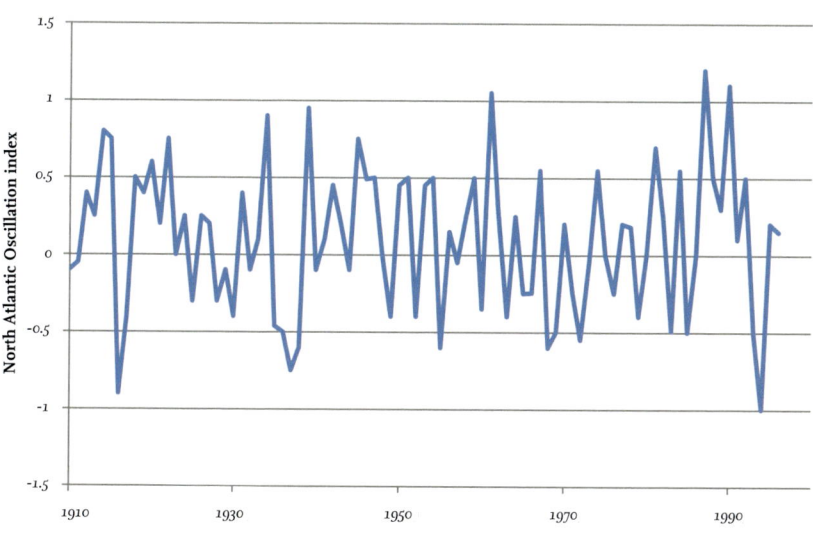

FIG 10. Annual North Atlantic Oscillation (NAO) index values, 1910 to 2000. (Data plotted from Garcia-Soto & Pingree, 2012)

systems lead to a lower pressure gradient and a negative index value (NAO–). The North Atlantic Oscillation is responsible for most variability in weather conditions during the winter months; wind speed and direction, temperature range, rainfall, and rate of change in all those parameters, are related to the NAO index. In NAO+ years strong westerly winds predominate over the mean southwesterly regime, the winter is mild and wet, and the summer is cool. In NAO– years westerlies are reduced and storms track further south, with increased rain over southern Europe, and the winter is cold. Winter inflow of Atlantic water into the northern North Sea increases with positive NAO index conditions, but decreases in negative years, and in extreme NAO-negative years North Atlantic Current inflow north of Shetland almost ceases. In winter months, inflowing NAC water is up to 3 °C warmer than mean winter seawater temperature for the North Sea, and there is thus a positive correlation between sea surface temperature and the NAO index. Winter sea surface temperatures are lower during NAO-negative years, and as they rise through the spring there may be a lag of one month before they reach levels usual for NAO-positive years (Fig. 11).

Fluctuations in the North Atlantic Oscillation Index do not follow regular patterns through time, although variations over large time scales have been recorded. For the first three decades of the twentieth century the winter NAO index was most frequently positive, and winters were mild. Harsh winters experienced in the 1960s marked a predominance of negative winter NAO values, while since 1980 the index has been mostly positive. Sea surface temperatures (SSTs) are also indicators of a long-term, low-frequency climatic phenomenon referred to as the Atlantic Multidecadal Oscillation (AMO), which has a

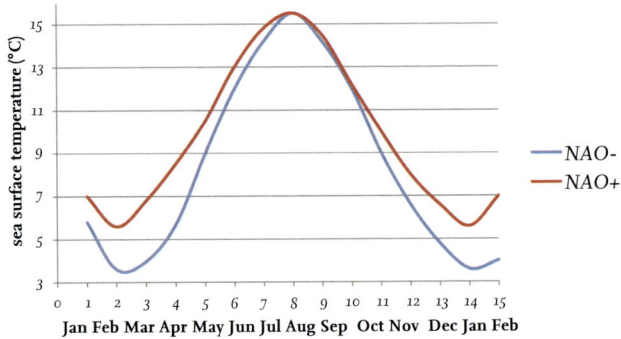

FIG 11. Annual cycle of sea surface temperature (SST) for the North Sea, in NAO-negative and NAO-positive years. (After Pingree, 2005)

periodicity of around 50–70 years (Garcia-Soto & Pingree, 2012). The AMO is also expressed as an index, derived from annual average SST anomalies, that is, the negative or positive deviation, in degrees Celsius, from the long-term mean SST for the region (Fig. 12). The factors driving the AMO are not fully understood, but it is a result of interactions between ocean and atmosphere, perhaps through shifts in large-scale circulation patterns of both, and perhaps relating to drift in the earth's orbit. The AMO seems to have persisted through much of the past 8,000 years, though with varying intensity (Knudsen *et al.*, 2011). SST variability also correlates with sunspot activity. The mean annual number of sunspots shows an 11-year cycle between maxima, and an overlying multidecadal periodicity of around 100 years, and has shown a long-term linear increase through the past three centuries (Garcia-Soto & Pingree, 2012). When the c.70-year AMO cycle is at its minimum, its short-term variability more or less corresponds to the 11-year sunspot cycle (Fig. 13).

Temperature and salinity are important physical factors affecting the benthic communities of shelf habitats. The salinity of seawater is defined as the proportion of the ten most abundant inorganic salts dissolved in a standard unit of seawater, expressed as grams per kilogram, and it has a mean value of around 35 for full-strength seawater. In practice, salinity values are a measure of the electrical conductivity of seawater, increasing as salt content increases, denoted as practical salinity units (psu). Surface-water salinity varies seasonally in relation to summer evaporation and winter rainfall maxima, and also as a consequence of wind-driven surface circulation (Fig. 14). Bottom salinity shows little seasonal variation, and above the deep shelf regions to the north and west

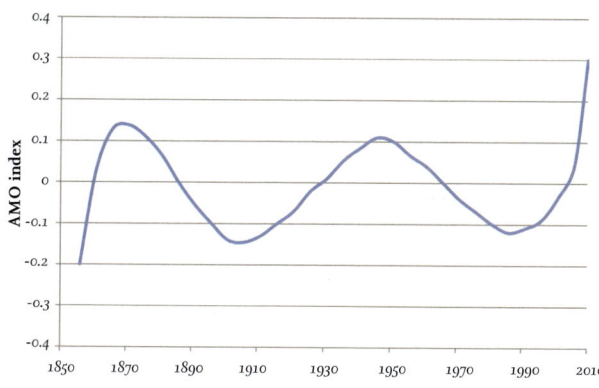

FIG 12. The Atlantic Multidecadal Oscillation (AMO) index, plotted as a trend line for the period 1856 to 2012. (After Garcia-Soto & Pingree, 2012)

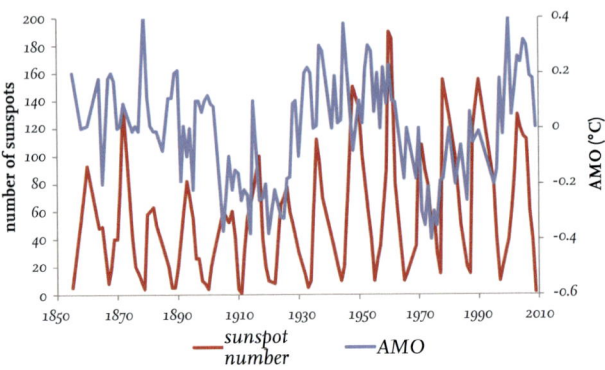

FIG 13. Annual number of sunspots and AMO index values, 1855 to 2009. AMO values reflect the 11-year sunspot cycle during periods of AMO minima (1910 and 1975/80). (Data plotted from Garcia-Soto & Pingree, 2012)

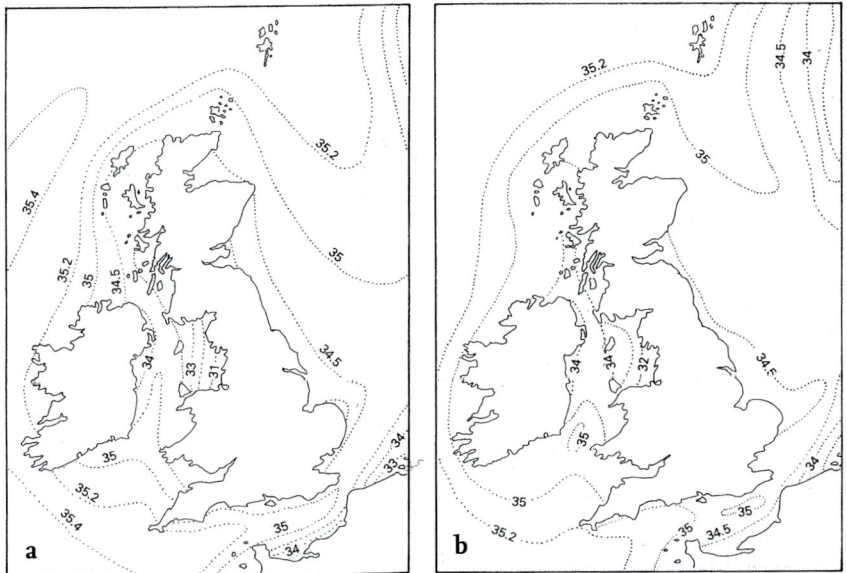

FIG 14. Mean surface water salinity (psu): (a) winter; (b) summer. (From Hayward & Ryland, 1995)

surface salinity values show least seasonal fluctuation, although winter storms may drive full-salinity oceanic water southwards into the central North Sea. The greatest variation in seasonal salinity occurs in the south and southeast of the region, and is most marked in the Skagerrak and along the west coast of

Denmark, where the 33 psu contour, or isohaline, moves well offshore during the summer.

Seasonal sea temperature variation is more marked than salinity variation, particularly in the North Sea and through the English Channel (Fig. 15). To the west of the British Isles and Ireland the annual range in SST may be as much as 7 or 8 °C, and exceptionally 11 °C off northwest England, where the winter mean may be as low as 5 °C and the summer mean as high as 16 °C. From the western approaches of the Channel to the shelf edge, mean winter SST is in the range of 8–10 °C, with summer means of 13–16 °C. Across the deep shelf of the northern North Sea the surface temperature regime is similar to that of the western shelf, with winter means in the range of 7–8 °C and maximum summer means of 16 °C, but the shallow waters of the southern North Sea are subject to often extreme seasonal variation. Along the southwestern coast of Denmark, and in the inshore waters of Germany and the Low Countries, sea surface temperatures may exceed 20 °C in summer, and drop below 2 °C in winter, an average annual range of 18 °C, while the coasts of southeast England experience winter means as low as 5 °C and maximum summer means of 17 °C. Bottom water temperatures, and the sea floor to surface temperature gradient,

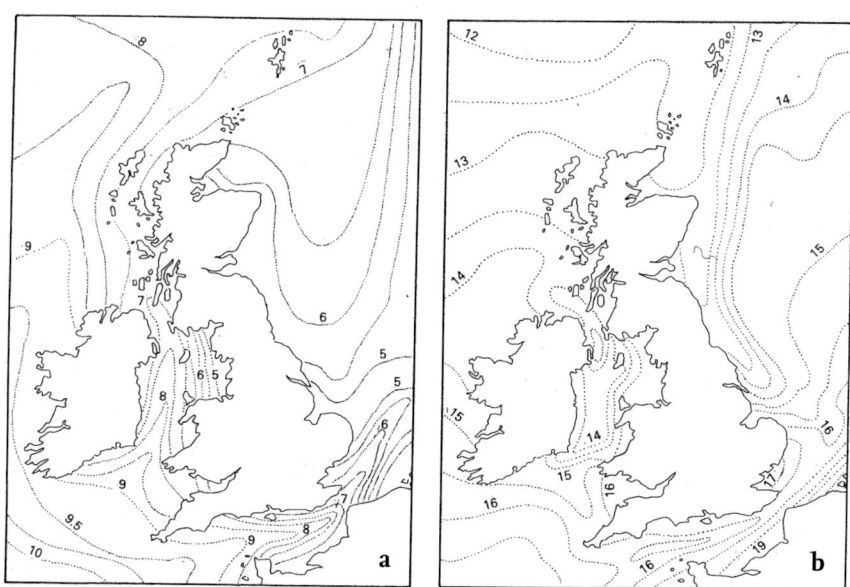

FIG 15. Mean annual sea surface temperatures (°C): (a) winter; (b) summer. (From Hayward & Ryland, 1995)

are physical factors as important as surface temperatures and, again, may show most extreme variation in the shallow southeast. Over much of the western shelf and the northern North Sea there is little difference between mean water temperature at the surface and at the sea floor during winter, while in summer bottom temperatures may be 5 °C lower than surface temperatures over the western shelf, and as much as 7 or 8 °C lower over the deep northern North Sea. In contrast, the difference between mean surface and bottom temperatures is least in the southern North Sea, just 1 or 2 °C in both winter and summer. One consequence of this is that during the summer months, to the north and west, the water column is stratified, with a sharp boundary, the thermocline, between warmer upper layers and deeper, colder waters. In the North Sea, within the 50 m isobath, through the Channel, most of the Irish Sea, and inshore around the whole of the British Isles, the water column is well mixed through the entire year, and no thermocline is established.

The surface layers of the temperate northeast Atlantic constitute a discrete body of water, or water mass, termed the East North Atlantic Central Water, defined by its temperature and salinity characteristics. These two physical parameters determine the density of the water mass: density increases as salinity increases, and as temperature decreases. Very cold, salty water to the north of the region is especially dense and constitutes another water mass, the Atlantic Subarctic Upper Water, and the boundary between the two is referred to as a front. However, while the pelagic environment of the East North Atlantic Central Water is theoretically continuous, it is not homogeneous, and in the neritic zone, over the continental shelf, heterogeneity is created by further frontal zones. A front may be marked by sharp discontinuities in temperature and salinity regimes, but also by changes in current flow, by rapid shallowing or by riverine outflow. Sharp changes may result in upwelling on one side of the boundary and downward water movement on the other, mixing well-oxygenated and warmed surface waters with cold bottom waters. Fronts form along the shelf edge, as a consequence of sudden shallowing, and the slowing and deflection of currents, often related to interaction with tidal currents; estuarine outflows have a similar effect, stimulating upwelling of saline bottom waters, creating a mixing zone referred to as a plume front.

Most of the northwest European shelf is layered with loose, unconsolidated sediments. There are comparatively small areas of fine-grained muds, and more extensive stretches of clean sand, but for most of its extent bottom deposits of the shallow sea consist of mixed muddy sands and gravels. Bare bedrock is rare, except along the Atlantic margin, where rocky ridges are interspersed with coarse sands and gravels. Inshore, storm-beaten coasts may retain cobbles and boulders, with patches of coarse, pebbly sands and gravels, while finer material is carried offshore. Much of the sediment of the shelf is glacial in origin, dumped by retreating glaciers at the end of the Pleistocene, winnowed by melt waters, and sorted and transported by the encroaching sea. Glacial material and the products of later erosion are washed into the sea by river outflow, augmented through the past few centuries by agricultural and industrial by-products, and by every kind of human waste. Some of the sediment has marine biogenic origins: the tests and shells of molluscs and crustaceans, and of foraminiferans, tubeworms and bryozoans, contribute to the formation of carbonate gravels. Calcium carbonate is easily eroded, perforated by shell-boring organisms, fracturing and dissolving readily, but the thick, heavy shells of large bivalves such as the Dog Cockle, *Glycymeris glycymeris* (Fig. 16), are slow to break down and form the greater bulk of the coarsest shell gravels. The plates of barnacle tests, bryozoan colonies and tubes of serpulid worms are light enough

FIG 16. Robust shells of the Dog Cockle, *Glycymeris glycymeris*. Scale bar: 5 cm.

to be transported by moderate bottom currents, and may form patches of finer calcareous sediment offshore.

Hydrodynamic forces – wind-driven wave action and tidal currents – are primary influences on the benthic environment. Fine material is suspended and transported from areas of strong water flow, falling out of suspension where flow slackens. High tidal ranges generate strong tidal currents, as in the Bristol Channel and the Gulf of St Malo. In the central regions of the Bristol Channel the sea floor is formed of rocky ridges and coarse sands, with increasingly finer, muddy sands shorewards, and localised mud deposits in sheltered embayments such as Bridgwater Bay. Tidal flow is also enhanced where water passes through narrow straits, such as the eastern end of the English Channel, and past promontories such as the Cotentin Peninsula: fine sands and muds are carried away and the bottom sediment consists of coarse sands, gravels and pebbles, with patches of exposed bedrock. In general, sediments along the northwest edge of the shelf are coarse-grained, between ridges of bare rock; beyond the shelf break benthic deposits are increasingly fine-grained and muddy. The benthic environment of the North Sea, the English Channel and the western approaches is a patchwork of muddy sands and gravels, interspersed with areas of cleaner, better-sorted sands and gravels. Soft mud is limited to sheltered inshore areas, or to deeper water offshore, and rocky areas with gravel and pebbles occur along the coasts of northeast England and eastern Scotland, and to the north of the Heligoland Bight.

THE BIOLOGICAL ENVIRONMENT

Practically all marine food webs are founded upon the photosynthetic production of the euphotic zone. Exceptions are hydrothermal vent and cold seep habitats, where often rich benthic communities are supported entirely by sulphide-oxidising bacterial populations, but the proportion of marine energy flow represented by these very special communities is minimal in global terms. "Marine primary production" denotes the synthesis of organic compounds by seaweeds, the three phyla of which are commonly termed macroalgae, and by single-celled phytoplankton, the most important of which is a single phylum, the Bacillariophyta, or diatoms. Photosynthesis is a sequence of chemical reactions through which sunlight, radiant energy, is transformed into chemical energy, through the mediation of photosynthetic pigments, most significantly chlorophyll a, and utilised in the formation of carbohydrates, lipids, amino acids and proteins. Water, carbon dioxide (CO_2), and inorganic nitrates and phosphates

$$CO_2 + H_2O \leftrightarrow H_2CO_3 \leftrightarrow HCO_3^- + H^+$$

$$HCO_3^- \leftrightarrow CO_3^- + H^+$$

FIG 17. The cycling of dissolved carbon dioxide in surface waters.

are the materials from which these organic compounds are formed, and carbon is the most important of the elements composing them. Atmospheric CO_2 dissolves readily in seawater, where it exists at higher concentrations than in air, because it also originates from chemical and biological processes within the water column. Free, dissolved CO_2 combines with water (H_2O) to form carbonic acid (H_2CO_3) part of which dissociates to form charged bicarbonate (HCO_3^-) and hydrogen (H^+) ions; a proportion of bicarbonate ions also dissociates, as carbonate (CO_3^-) and more hydrogen ions. These reversible reactions (Fig. 17) are fundamental to the most important biological processes in the sea. Atmospheric CO_2 is augmented by CO_2 generated by respiration, by the dissolution of calcium carbonate ($CaCO_3$), and through remineralisation, a process whereby organic compounds are broken down into their inorganic components by benthic decomposers and detritivores. Most of this dissolved CO_2 exists as bound carbonate and bicarbonate ions, but as photosynthesis and calcification remove free CO_2 the chemical reactions reverse to restore its levels. Seawater is slightly alkaline, with an average pH range of 8 ± 0.5; increasing concentrations of carbonate and bicarbonate ions also increase the number of hydrogen ions, resulting in a lowered pH, indicating rising acidity, but photosynthesis and calcification lead to a reduction in H^+ ions and maintain a buffered solution with a narrow pH range.

The nitrates, phosphates and minerals, particularly iron and silica, essential to growth, are mostly in short supply in seawater, and pelagic primary production depends on constant recycling. Silica is mostly utilised in coastal waters by diatoms, and recycling simply requires seasonal overturn of bottom sediments, and the siliceous diatom frustules accumulated through the summer, to restore dissolved silica levels in the water column. Organic phosphates are also swiftly hydrolysed in the sediments to inorganic phosphates, but nitrogen cycling is a much more complex process. Nitrogen sources in seawater include a small proportion of free dissolved N_2, but the principal store consists of inorganic nitrate (NO_3^-), nitrite (NO_2^-) and ammonia (NH_4^-) ions, as well as organic compounds such as urea ($CO(NH_2)_2$). Nitrate ions comprise the greatest proportion of these chemical forms, and are utilised as such by phytoplankton, but the others generally need to be converted to other states before they can be recycled (see page 88), and the rate at which these other sources become available to primary producers may limit productivity.

Rainfall, riverine outflows and wind-blown dust all contribute to the mineral and nutrient budgets of marine ecosystems, but as these materials are incorporated into biological entities, and diffuse through marine food webs, there is a constant depletion of concentrations in the water column. Over the deep oceans this represents a permanent loss: nutrients and minerals are carried to the sea floor in a rain of dead plant and animal material, which will support communities of abyssal consumers, but will not be returned to the surface waters to be re-incorporated into pelagic production cycles. Shallow shelf seas are of profound importance to all stages of marine production, and their benthic communities play critical roles in the rapid processing and recycling of pelagic production. In temperate coastal seas, at depths of less than 50 m, the water column is almost always well mixed, benthic growth and production is closely coupled to that of the pelagic primary producers, and pelagic production passes through diverse pathways – encompassing decomposers, detritivores, herbivores and secondary consumers – to re-mineralisation and recycling, and is only temporarily consigned to a sink below the 50 m isobath by the establishment of the seasonal thermocline.

Primary production is expressed as units of carbon fixed, per unit of area or volume, per unit of time. For example, one estimate of primary production for the North Sea was 90 grams of carbon per square metre per year, the figure representing the mean of a succession of values through the water column. Primary production varies widely across the world ocean; it may be as low as 50 g C per square metre per year in the Arctic Ocean and above 500 for the highly productive eastern boundary currents on the coasts of Peru and Namibia. For temperate shelf seas primary production may exceed 300 g C per square metre per year along the shelf-break frontal zone.

In the temperate northeast Atlantic region phytoplankton production displays two annual maxima, in spring and in late summer. Growth begins to accelerate in March, stimulated by increasing daylight and rising solar radiation. Dissolved CO_2 is always at a level sufficient for photosynthesis; the concentrations of inorganic nutrients, nitrates and phosphates in the euphotic zone have been enriched by turbulent winter mixing, stirring up bottom deposits, and provided that a few important trace elements, such as iron and silica, are present in sufficient quantity, phytoplankton growth increases rapidly. As production reaches its first peak herbivorous zooplankton, the primary consumers, also begin to show rapid increase and achieve a peak of biomass just as the peak in primary production has begun to decline (Fig. 18a). Grazers, especially copepods and cladocerans, are partly responsible for the ensuing slump in diatom biomass, but the most significant factor in its decline is the depletion of nutrients in the

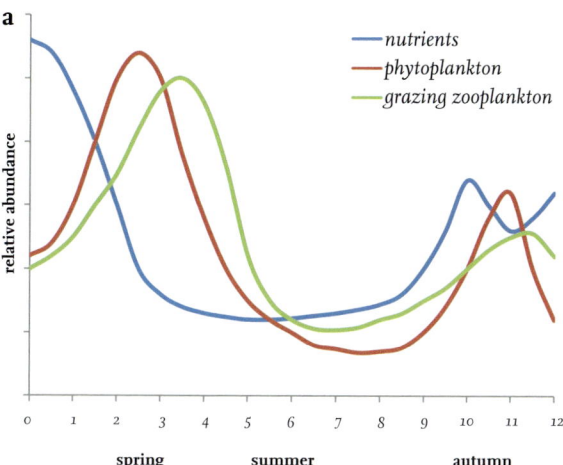

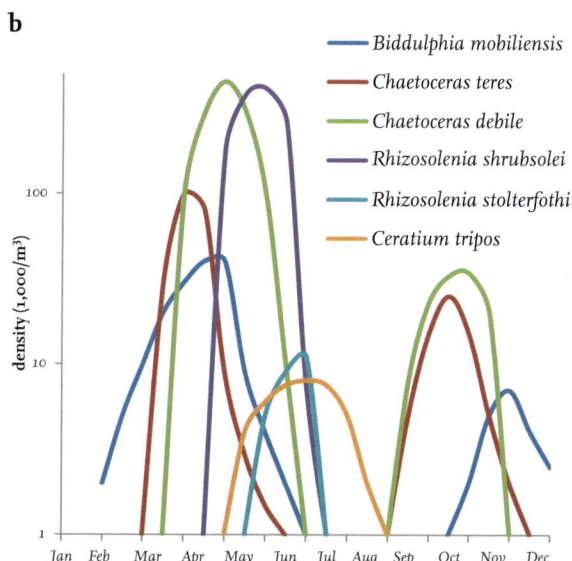

FIG 18. (a) Seasonal cycles of nutrients, phytoplankton and grazing zooplankton abundance in the temperate shelf seas of northwest Europe. (b) Seasonal succession of the most abundant phytoplankton species in the Irish Sea, averaged over 14 years. (After Barnes & Hughes, 1982)

upper layers of the water column. The composition of the bloom changes as nutrients and essential elements are depleted at differing rates (Fig. 18b); silica is exhausted fairly swiftly and the large-celled diatoms that predominate early in the season are succeeded by smaller-celled species, and as these decline, non-siliceous plankton, especially small flagellates, are the dominant fraction. This is reversed in late summer when autumn gales overturn the water column once more, restoring nutrient levels and stimulating a second bloom in phytoplankton populations, which finally slump as the sun moves south, daylength decreases and solar radiation falls to winter levels. Along frontal zones phytoplankton production may be more protracted, especially along plume fronts, where continual inflow of nutrients into continuously mixed water sustains persistent diatom blooms.

Primary production in the euphotic zone, principally by phytoplankton but augmented by coastal macroalgae, supports at least three trophic levels within the neritic zone of temperate shelf seas. The herbivorous zooplankton is consumed by small planktivorous fish, such as sardines and herring, which are then eaten by larger fish; small carnivorous zooplankton, such as arrow worms, could be considered as a further trophic level, while toothed whales, seals and mankind, as top predators, would constitute a fifth. Pathways between these trophic levels are numerous, as most pelagic consumers feed on a variety of prey species, and resulting food webs are dense. Primary and secondary production in the upper layers of the water column are the ultimate sources of energy for benthic communities. Decayed phytoplankton cells, casts, corpses and faeces of primary, secondary and tertiary consumers provide for the nutrition of the benthos. Much of this will enter benthic food webs as finely comminuted detritus, termed particulate organic matter (POM), while mucus and other exudates form a pool of dissolved organic carbon (DOC), which is absorbed directly from the seawater by some components of both pelagic and benthic communities. Until comparatively recently it was assumed that phytoplankton production entered benthic food webs only through these pathways, but it is now well established that not only do many benthic species, especially molluscs, polychaetes and brittlestars, feed directly on fresh phytoplankton, but that the cycles of growth and reproduction of those species are closely attuned to growth cycles of phytoplankton communities.

In subarctic waters high-summer day length may be 24 hours but the total period of spring and summer illumination is so short that only a single peak of phytoplankton production occurs; nutrient salts are never depleted but primary production is limited by day length. In warm temperate and tropical seas sunlight is never limited but dissolved nutrients are at permanently low concentrations and phytoplankton production is at continually low levels, with only small peaks perhaps reflecting localised seasonal nutrient enhancement, for example by monsoonal river outflow. In the summer-stratified areas of the northwest European shelf seas, the thermocline prevents any vertical mixing of water layers and the primary production peaks may be sharply defined, with a deeper minimum between them than in mixed coastal waters. The finest POM may then accumulate at the thermocline as flocculent material often described as 'marine snow'. This supports an additional, seasonal trophic step, referred to as the microbial loop. The marine snow supports bacterial populations which are consumed by the smallest planktonic species, especially flagellates and non-photosynthesising dinoflagellates, which then pass back into the pelagic food web via larger carnivorous zooplankton. There are further consequences for the benthos here. The thermocline slows or even stops the fall of organic material to the benthos, and benthic production consequently declines. However, when the thermocline breaks down in late summer, the accumulated mass of marine snow and its bacterial communities may be so great that, falling to the sea floor, it blankets the benthos, and the oxygen demand of the decaying material is so great that the surface sediments are completely depleted and the benthos dies through hypoxia.

BENTHIC COMMUNITIES

Benthic hydrodynamic regimes determine the physical nature of the benthic environment. High-energy regimes, of strong tidal current or wave action, are characterised by coarse sands and gravels, often well sorted and size-graded, well oxygenated but with a low content of organic matter. In lower-energy environments sediment grades are finer and often mixed, with increased proportions of silt; they are less readily flushed through, and interstitial water is low in oxygen, but organic content is often high. Rocky substrata are swept clear of sediment by constant turbulent water flow, and the density and diversity of any encrusting plants and animals they support may depend upon the degree of sand scour they are subject to. Where this is significant, scour-resistant organisms such as honeycomb worms, *Sabellaria spinulosa*, and Horse Mussels, *Modiolus modiolus*, may be the framework of the community, providing shelter for smaller, more delicate species.

Rocky coastal margins of the northwest European shelf are fringed with a band of macroalgae, with a vertical and horizontal extent determined by water clarity and the gradient of the sea floor. On precipitous coasts the macroalgal band may be narrow, but extending perhaps to as deep as 20 m; on more sheltered, gently shelving coasts the seaweed fringe may extend offshore for some kilometres, but limited to less than 10 m depth by turbid waters. The bulk of the macroalgal biomass consists of kelps, principally species of *Laminaria*. Kelps are large seaweeds with a potential life span exceeding ten years, and their perennial stipes and holdfasts support a shrubby community of red seaweeds, while between them there is typically an understorey of smaller seaweeds. Where kelp populations thin out, towards their depth limit, or into higher-energy environments beyond their hydrodynamic tolerances, hard substrata are colonised by encrusting coralline algae, which persist as long as light is sufficient for photosynthesis. Some of these coralline algae develop knobbly branching forms that accumulate as coralline gravels, referred to as maerl. Kelp forests and maerl beds occupy only a small proportion of the shallow benthic environment of northwest Europe, but they have a considerable significance through the complex three-dimensional habitats they provide, and are the foci of high species diversity. They are explored at length in Chapters 5 and 6, as key benthic habitats.

Benthic faunas may be categorised by habitat, and by lifestyle and feeding modes of the constituent species. The broadest division is between a surface-dwelling epifauna, on both hard and soft substrata, and a burrowing infauna in soft, unconsolidated sediments. It is also possible to recognise a suprabenthic fauna that lives on or just above the sea floor, but moves up into the water column to feed. The soft-sediment epifauna includes many echinoderms, especially starfish and brittlestars, numerous species of decapod crustaceans, and a great variety of molluscs, especially large species of bivalve and many gastropods, and most are motile. Some of the epifauna are suspension feeders, such as the bivalves and some brittlestar species, others are predatory, including all starfish and numerous species of gastropod and decapod, but probably the largest proportion consists of omnivores and scavengers. On rocky substrata the epifauna consists of sessile suspension feeders (Fig. 19), such as sponges, cnidarians, bryozoans, tube-dwelling polychaetes and often dense communities of ascidians, together with associated carnivores and omnivores, most of which will be motile.

On soft substrata, the infaunal community is taxonomically more diverse than the epifauna, embracing all invertebrate animal types, and some species of fish; it includes active burrowers and sedentary tube dwellers, and all feeding modes, from suspension feeding and deposit feeding to active carnivory. The infauna is further categorised by size into a macrofauna, of animals larger than 1 mm, and a meiofauna of smaller animals that will pass through a 1 mm sieve mesh but are retained by a mesh of 0.063 mm. The smallest organisms constitute the microfauna. While this category does include the smallest animals it also

FIG 19. A small Edible Crab, *Cancer pagurus*, amongst an epifauna of sessile suspension feeders, including sponges, cnidarians, bryozoans and sea squirts. (J. S. Porter)

embraces single-celled protists and bacterial populations, and is thus often referred to as the 'microbiota'. These categories have functional significance. The microfauna encrusts the surfaces of sand grains, meiofauna live in the interstices between sand grains without disturbing sedimentary structures, while the tunnelling and burrowing macrofauna dislodge and disrupt them. Ecologists also recognise a megafauna, characterised by function rather than size: these are animals that construct deep and permanent burrows, or excavate large pits whilst feeding, and are responsible not simply for disrupting sedimentary structures but for large-scale translocation of sediment, a process termed bioturbation.

Infaunal community types are partially determined by sediment grade. Coarse sands support many species of suspension-feeding bivalves, each long-lived and with relatively large individual body size; densities of each species may be moderate but the total community biomass may be high. Fine sands with low organic content will be inhabited by a different array of smaller bivalve species, and by numerous species of small polychaetes and crustaceans; small individual body sizes will give only a low total biomass but densities may be extremely high. Mixed muddy sands and gravels tend to accumulate high levels of organic material; the burrowing activity of the infauna continually re-suspends a surface layer of fine silt that may discourage some suspension feeders, while the proportion of deposit-feeding animals increases. Species diversity and density are both high, and the total number of individuals may be of the order of tens of thousands per square metre of substratum. In fine muds organic content may be extremely high, but oxygen levels may be so low that within just a few centimetres from the surface the sediment is entirely anoxic. Diversity of both the epifauna and the infauna is lowest in these kinds of habitats, especially where they are affected by additional environmental stresses, such as the low and fluctuating salinities typical of estuarine habitats. However, density of those species able to tolerate the conditions associated with muddy habitats, principally small polychaete and oligochaete worms, may be exceptionally high.

It is a general rule that diverse benthic communities are stable through time and resistant to perturbation, whilst the least diverse communities tend to be unstable, with population densities showing significant fluctuation over short time scales. These characteristics illustrate the two extremes of a life-strategy continuum: diverse communities are composed predominantly of long-lived species, with large individual body size and low reproductive output, termed K-selected, while low-diversity communities consist of small, short-lived species with high reproductive output, referred to as r-selected. The r–K spectrum is explored further on pages 44–45.

Petersen's sampling of benthic sedimentary habitats of Danish coasts (page 11) showed that distinctive suites of species recurred consistently, and these he classified into a series of assemblages, each indicative of a particular benthic habitat. Petersen's scheme was adopted and refined by successive generations of benthic ecologists, some of whom defined benthic communities more precisely in terms of the dominant constituent species, while others defined each in terms of the sedimentary environment (Table 1). A suggested correlate of the former classification was that the assemblages of species characteristic of each sediment type were 'ecological units', that is, communities of ecologically interdependent species linked by biological interactions. In the latter case it was argued that simply by knowing the grain size composition of a sediment the community of species inhabiting it could be predicted. Much of the shelf environment of northwest Europe has now been sampled repeatedly, and for some regions bottom assemblages have been mapped (Hiscock, 1998). Depth and sediment type are the principal factors defining these assemblages, while temperature regimes

TABLE 1. A summary of soft-sediment marine invertebrate assemblages defined for northwest European shelf habitats, with some common representative species.

Boreal shallow mud (*Macoma* communities)	Boreal offshore mud (*Brissopsis* communities)
Macoma balthica	*Brissopsis lyrifera*
Arenicola marina	*Amphiura chiajei*
Nephtys hombergii	*Calocaris macandreae*
	Turritella communis
	Nephrops norvegicus

Boreal offshore muddy sand (*E. cordatum/ A. filiformis* and *Abra* communities)	
Echinocardium cordatum	
Abra alba	
Ophiura ophiura	
Callianassa subterranea	
Amphiura filiformis	
Ensis ensis	
Nucula nitidosa	

Boreal shallow sand	Boreal offshore sand (*Venus* communities)
Tellina tenuis	*Chamelea striatula*
Donax vittatus	*Echinocardium cordatum*
Arenicola marina	*Gari fervensis*
Mactra stultorum	*Ensis siliqua*
Spisula solida	*Dosinia lupinus*
	Acanthocardia echinata

Boreal offshore gravel	Boreal offshore muddy gravel
Glycymeris glycymeris	*Upogebia deltaura*
Venus casina	*Nucula nucleus*
Polititapes rhomboides	*Venus verrucosa*
Spisula elliptica	*Turritella communis*
Echinocardium flavescens	*Astarte sulcata*
Spatangus purpureus	*Modiolus modiolus*
	Ophiothrix fragilis

are significant at wider scales and salinity may be important at narrow, local scales. Shallow-water mud deposits, found especially in sheltered environments in or adjacent to estuaries, are characterised by the small bivalve *Macoma balthica* (Fig. 20a), together with polychaete worms and amphipods, while in shallow sand habitats the predominant bivalve is *Tellina tenuis* (Fig. 20b), and the lugworm *Arenicola marina* is typically common. In deeper water, offshore, clean sands are populated by the heart urchin *Echinocardium cordatum* (Fig. 21a), and an array of thick-shelled bivalve species, but with an admixture of mud the latter are

FIG 20. Shells of (a) the Baltic Tellin (*Macoma balthica*), and (b) the Thin Tellin (*Tellina tenuis*). Scale bar: 2 cm.

FIG 21. Two burrowing heart urchins: (a) *Echinocardium cordatum*; (b) *Brissopsis lyrifera*. (J. R. Ellis (CEFAS))

replaced by nut shells, *Nucula*, the Pelican's Foot Shell, *Aporrhais pespelecani*, and large burrowing shrimps, and as the mud content increases the dominant heart urchin is *Brissopsis lyrifera* (Fig. 21b). Gravel deposits tend to be inhabited by further stout clam species, particularly the Dog Cockle (Fig. 16), and the shallow-burrowing Purple Heart Urchin, *Spatangus purpureus*, but with a significant content of mud burrowing shrimps, worm-like sipunculans and the Auger Shell, *Turritella communis*, become the dominant indicator species. However, sharp boundaries on maps do not reflect reality, and both sediment types and benthic assemblages tend to intergrade. Distributions of benthic organisms are discontinuous in both space and time: they show patchiness resulting from constant interaction of physical and biological processes. Species diversity and individual species abundance may be subject to regular seasonal and annual fluctuation, and may also change in response to random events, such as habitat disturbance, or as a result of reproductive failure, and sampling programmes demand careful design in order to yield data suitable for comparison between sampling sites and through time.

Functional group analysis is another approach towards the categorisation of benthic communities, especially of soft-sediment habitats (Pearson, 2001). Its central tenet is that species may be classified, grouped, according to functional similarities. Trophic group analysis is the most familiar aspect of this approach: species are grouped according to feeding mode (Table 2). Feeding modes of infaunal communities might be divided initially into suspension

TABLE 2. Some functional group attributes applied to soft-sediment benthos, with examples drawn from the text.

Trophic group	*Mobility*	*Feeding mode*	*Examples*
Suspension feeder	Sedentary	Ciliated tentacles	*Owenia fusiformis*
		Captorial tentacles	*Cerianthus lloydii*
	Low motility	Ciliary currents	*Kurtiella bidentata*
	Actively mobile	Ciliary currents	*Pecten maximus*
		Mucus trapping	*Amphiura filiformis*
Surface deposit feeder	Sedentary	Ciliary currents	*Maxmuelleria lankesteri*
	Low motility	Ciliary currents	*Abra alba*
	Actively mobile	Particle selection	*Monoporeia affinis*
Subsurface deposit feeder	Sedentary	Sediment sorting	*Nucula nitidosa*
	Low motility	Bacterial stripping	*Callianassa subterranea*
	Actively mobile	Sediment ingestion	*Brissopsis lyrifera*
Carnivore	Low motility	Non-selective	*Aphrodita aculeata*
	Low motility	Prey selection	*Astropecten irregularis*
	Actively mobile	Prey selection	*Asterias rubens*

feeding, surface deposit feeding, subsurface deposit feeding and carnivory. A further division might relate to feeding structures: suspension feeders filtering fine material by means of ciliary currents could be distinguished from those employing captorial tentacles or mucous nets, and ciliary feeding could include separate categories for bivalves, exploiting siphonal respiratory currents, and bryozoans, employing cones of ciliated tentacles. Some ecologists make the first distinction between microphagous feeders, which might be suspension feeders or deposit feeders, and macrophagous feeders, herbivorous or carnivorous but selecting large particles, while others accept just three categories – suspension feeders, detritus feeders and carnivores. Other functional groupings are based on manner of feeding, whether the animal employs ciliary currents, an extensible proboscis, venomous stinging cells (cnidocytes), toothed radular ribbon or jaws; on degree of motility, whether free living, occupying impermanent burrows or permanent tubes; and even on the type of bioturbation its activities create. It is not clear that such finely defined functional groupings increase insight into the structure of benthic communities. Certainly, along depth and hydrodynamic gradients one observes changes in sediment and faunal characteristics: at one point functional groups are principally fast-moving jawed predators, siphonal suspension feeders and small detritivores, while at another point surface detritivores, employing ciliated palps or proboscis, and subsurface deposit feeders, selecting large particles from inhalant currents, may predominate. However, many benthic invertebrate species are simply categorised as omnivorous, for the lack of accurate information on their feeding mode, while others are known to adapt their feeding mode in response to change in hydrodynamic regime, often at quite a short timescale, and perhaps also to food supply, switching readily from suspension feeding to deposit feeding, either on or below the surface.

Classification of benthic habitats remains useful in establishing baselines for recording change, and determining conservation priorities, and a scheme currently favoured categorises benthic habitats into a series of biotopes. A biotope is defined by the combination of all physical and biological characteristics of the habitat: the substratum, sediment type and grade, and all physical environmental parameters affecting them; species diversity and richness, and dominant biota. Several hundred benthic biotopes have been catalogued for European coastal waters (Connor et al., 1997); while some are sharply defined, limited in extent and occurring at just a few localities, many perhaps reflect points on a continuum of more broadly defined habitat types, with community composition changing in relation to depth and substratum, and to other, often unpredictable, factors.

THE BENTHIC ENVIRONMENT · 39

LARVAE AND LIFE CYCLES

The life cycles of almost all marine benthic invertebrates include a sexually produced larval stage. In a minority of species the fertilised egg undergoes development and passes through a larval phase enclosed within a capsule, or in a dedicated maternal brood chamber, and hatches, or is released, as a fully formed juvenile. Whelks are a good example: all species of the whelk family, Buccinidae, parcel their eggs in protective capsules, loosely glued together in a spongy mass (Fig. 22), and miniature whelks emerge after a lengthy larval period passed within the capsules. In the majority, around 80%, of benthic invertebrates the larva is a free-swimming stage which may either have a prolonged pelagic existence or spend just a brief period above the bottom before adopting a sedentary benthic existence. A second major distinction is seen between pelagic larvae which feed within the plankton as they grow, planktotrophs, and those in which growth and

FIG 22. Familiar strandline objects: dried egg masses of the Common Whelk, *Buccinum undatum*. Scale bar: 5 cm.

metamorphosis to the juvenile form is fuelled by yolk reserves provided by the parent, lecithotrophs.

The larvae of benthic invertebrates form a significant proportion of the coastal zooplankton during the spring and summer peaks. They are collectively termed the meroplankton, distinguishing them from the holoplanktonic invertebrates, the life cycles of which are entirely pelagic. Early stages in the life cycles of many benthic species are thus influenced by physical and biological factors particular to the pelagic environment. Each of the major taxonomic groups of benthic invertebrates is characterised by a particular larval form, which may pass through a series of morphologically distinct developmental stages. In some groups the larval series is compressed, with the early stages passed within an egg membrane, or capsule, while in others the entire developmental succession occurs in the plankton. Polychaetes, for example, have a distinctive larval type, the trochophore (Fig. 23a), with a girdle and apical tuft of cilia, but free-swimming trochophores are rather rare. Instead, most polychaete species having a larval phase release a more advanced larval form, the metatrochophore, which already has its first series of body segments, with recognisable polychaete setae, and the late larvae of tube-building species are

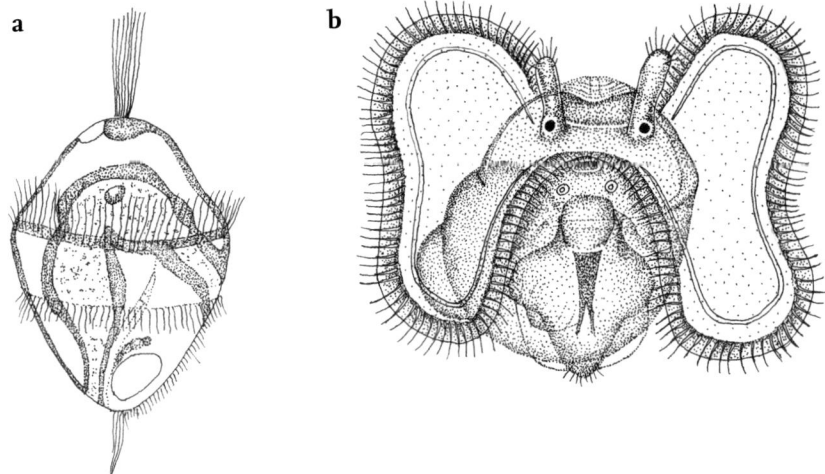

FIG 23. (a) Trochophore larva of a serpulid polychaete, with its characteristic girdle of swimming cilia (after Newell & Newell, 1963). (b) Veliger larva of the prosobranch gastropod *Nassarius reticulatus.*, with cilia distributed along the edges of the velum (after Thorson, 1946). Not to scale.

often nestled within their first transparent tubes. The trochophore form is also seen in the very earliest larval stages of some bivalve species, but it soon develops into a larval type termed a veliger, in which the ciliary girdle extends onto prominent lateral lobes, the velum. The veliger is enclosed in the first delicate, transparent bivalved shell, the prodissoconch. In those prosobranch gastropods that have pelagic larvae the trochophore is suppressed and the first larval stage is a veliger (Fig. 23b), with an expanded, lobed and ciliated velum, and a coiled larval shell, the protoconch.

Larval morphology may be rather complex among the Crustacea, in which growth is episodic, punctuated by regular moulting, or ecdysis, and each intermoult stage may be morphologically different from its predecessor. There are three main larval stages, common to almost all crustacean groups. The first, the nauplius (Fig. 24a), is a simple, unsegmented form, with three pairs of appendages and a single median eye. Stage 2 larvae have a partially segmented thorax and a developing abdomen, while third-stage larvae are recognisable as a type termed the zooea (Fig. 24b), in which all three body regions, head, thorax and abdomen, are clearly segmented. Each of these three stages may comprise two or more intermoult instars, and each ecdysis is marked by increasing morphological complexity. Among the Decapoda some species may pass through as many as nine successive larval instars, the last of which displays the taxonomically definitive morphology of the adult, and is generally termed the megalop (Fig. 24c).

Echinoderm species with a pelagic larval phase shed their embryos at an early developmental stage, as a ciliated, ovoid blastula, but this quickly differentiates into a bilaterally symmetrical larval type termed a dipleurula, with a characteristic morphology for each class. Brittlestars and echinoids have rather similar larvae, the echinopluteus (Fig. 25a) and ophiopluteus (Fig. 25b) respectively, while crinoids and holothurians have a barrel-shaped larval type. In starfish the first-stage larva is a stumpy form termed a bipinnaria, which develops elongate arms in the second stage, the brachiolaria (Fig. 25c).

Polychaetes, crustaceans, molluscs and echinoderms provide the bulk of the invertebrate meroplankton, are readily distinguished and separated in quantitative plankton studies, and in most cases can be identified at least to the level of genus; many later-stage larvae may even be identified to species level. A few other distinctive invertebrate larval forms may be found commonly in coastal zooplankton, though perhaps for only brief periods. Most distinctive are the tornaria larvae of hemichordates (Fig. 26a), the tiny triangular cyphonautes produced by a few species of membraniporine bryozoans (Fig. 26b), and the curious, club-shaped larvae of the horseshoe worms, or phoronids (Fig. 26c).

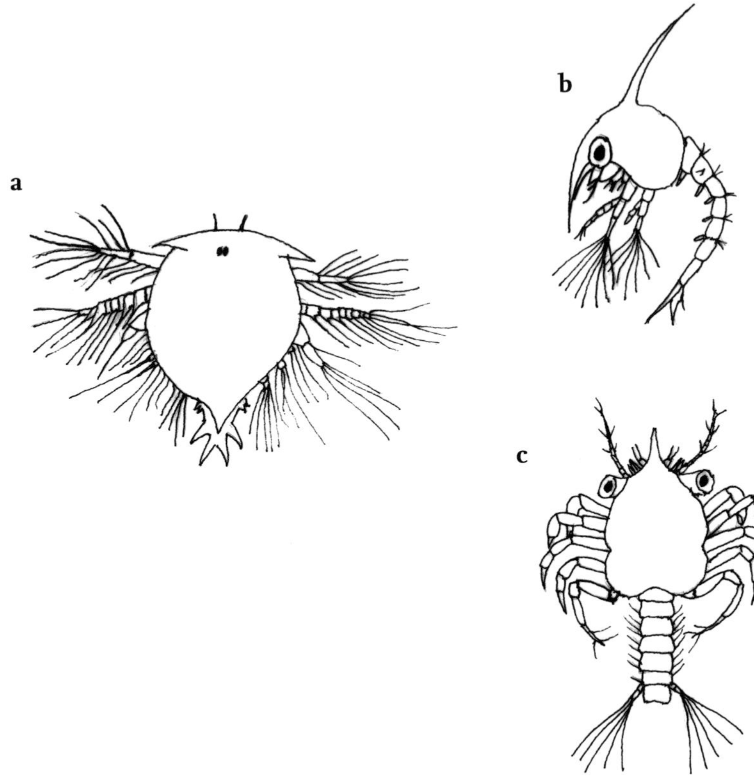

FIG 24. Crustacean larval stages. (a) Nauplius of the acorn barnacle *Chthamalus montagui*. (b) Stage 3 zooea, and (c) megalop of the Green Crab (*Carcinus maenas*). Not to scale.

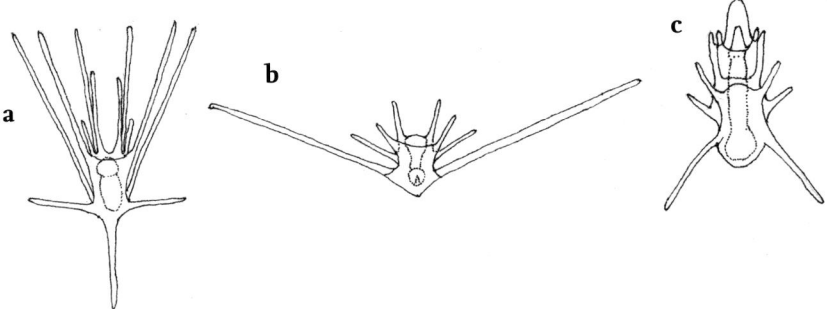

FIG 25. Some echinoderm larval types. (a) Echinopluteus of the heart urchin *Echinocardium cordatum*. (b) Ophiopluteus of the brittlestar *Ophiothrix fragilis*. (c) Brachiolaria of the starfish *Asterias rubens*. Not to scale.

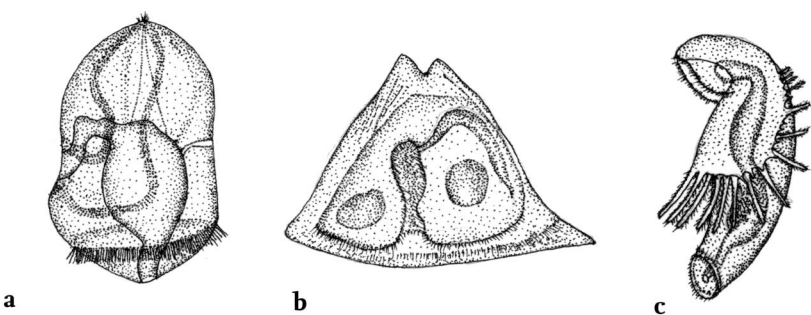

FIG 26. (a) The tornaria larva of the hemichordate *Balanoglossus*. (b) The cyphonautes larva characteristic of the bryozoan genus *Membranipora*. (c) A phoronid actinotroch larva. Not to scale.

Systematic study of marine invertebrate larvae began with the work of another influential Danish scientist, Gunnar Thorson, who was the first to classify larval types, and to relate them to life-cycle patterns. Thorson distinguished between brooded larvae, which are released as fully developed juveniles, and three categories of pelagic larvae. Firstly, he recognised planktotrophic larvae with a long free-swimming period, up to two weeks in summer but as long as three months in winter-spawned larvae. Next, a minority of planktotrophic larvae could be characterised by a short pelagic period of just a few days, and the final category, lecithotrophic larvae, have a variable free-swimming period, but are independent of planktonic food sources. Two observations arise from this classification: first, planktotrophic larvae typically hatch from small eggs that are spawned in large numbers, while lecithotrophic and non-pelagic larvae result from smaller batches of large, yolk-rich eggs; and second, the proportion of each larval type produced by benthic communities varies with latitude.

With respect to the former, Thorson contended that high larval mortality through predation resulted in species with pelagic development producing many more eggs than species that lacked a pelagic larval phase. Further, as mortality rate increased in proportion to the duration of the pelagic phase, species with planktotrophic larval modes must produce many more larvae than species with lecithotrophic larvae. This hypothesis was published in 1950 and accepted as intuitively reasonable for the next half-century. Mortality rates of larval populations are difficult to measure, but plankton populations vary between years, and reproductive success of benthic species must also fluctuate annually. Peaks in larval densities in the water column tend to be reflected in peaks in the numbers of newly settled benthic juveniles, but there is no further correspondence, and it has become apparent that adult population densities of benthic invertebrates are determined by benthic ecological processes rather than by larval population densities.

The second observation, that the proportion of different larval modes produced by benthic invertebrates varies with latitude, proves to be more an approximation than a rule. Thorson (1950) published data on the reproductive modes of prosobranch gastropod species along the East Atlantic margin, from east Greenland to the Canary Isles, and demonstrated that the proportion of species with non-planktonic larval development was highest, at 90%, off east Greenland and declined to the south. Around Iceland less than 10% of prosobranch species produced planktonic larvae, while at the southern end of the range, 60% of Canary Isles prosobranchs were characterised by pelagic larval development. In cold polar waters, low temperatures lead to long development times; food resources are sparse, and most prosobranch species invest their reproductive output in few, large, yolk-rich eggs that are brooded, or encapsulated, and hatch as advanced juveniles. In temperate waters to the south the seasonal production maxima are predictable, while in the warm temperate waters around the Canary Isles primary production, though low, is continuous, and in both cases the appropriate reproductive strategy is to produce large numbers of small eggs that hatch as planktivorous pelagic larvae. Whether this rule has general application, either to prosobranch gastropods or more widely to all benthic invertebrates, has still to be tested adequately. Along the Chilean coast the proportion of prosobranch species showing non-planktonic larval development has been shown to increase towards higher latitudes, and cold polar waters (Gallardo & Penchaszadeh, 2001). Conversely, on the other side of the continent, the broad, muddy Argentinian Shelf has been the site of a substantial adaptive radiation of small carnivorous whelks, and there was no apparent pattern in the latitudinal distribution of reproductive modes among southwest Atlantic prosobranchs. It might also be noted that a substantial proportion of Antarctic echinoderm species have a pelagic larval phase, while practically all species of Bryozoa brood their embryos, irrespective of latitude.

Larval mode tends to reflect the life history of the adult. Species with small individual size and short life spans, that mature early and produce large numbers of eggs, hatching as planktotrophic larvae, are termed r-selected; the parameter r is the population growth rate, while K refers to the maximum population size, or carrying capacity of the habitat. Species with r-selected life histories are viewed as opportunistic: they are characteristic of variable habitats, their populations respond swiftly to favourable environmental change through high reproductive output to colonise newly available habitat, but are easily eliminated by competitive pressure, or unfavourable environmental change. Stable habitats become dominated by populations of K-selected, or equilibrium, species, with densities close to the maximum sustainable by the habitat. These are individually large, long-lived organisms with low reproductive output that produce lecithotrophic, and frequently non-planktonic, larvae. However, where frequency of disturbance and reproductive output are both low, neither opportunistic nor equilibrium species will be predominant; benthic communities will include a wide range of life strategies; this is the dynamic equilibrium model of species diversity (Fig. 27).

Larval type may also be related to other features of the life cycle (Levin & Bridges, 1995), and larvae may be categorised by criteria other than those defining the four basic archetypes recognised by Thorson (Table 3). Nutritional mode may be far more diverse than is suggested by the simple dichotomy between planktotrophy and lecithotrophy. About 80% of all marine invertebrate larvae are planktotrophic, but there is an additional proportion of facultative planktotrophs, exemplified by the pelagosphaera larvae of sipunculans

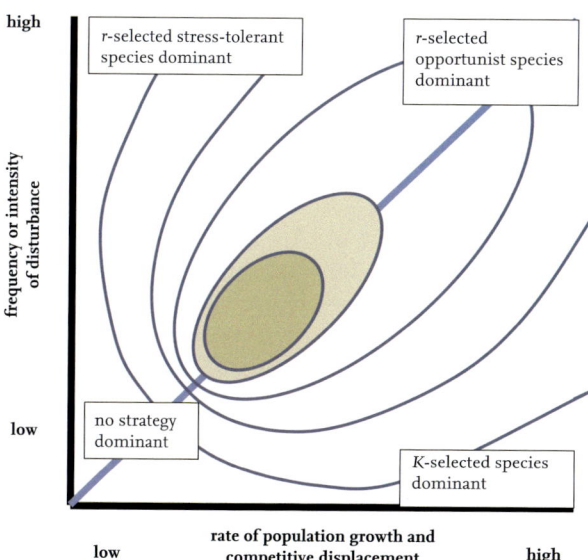

FIG 27. The dynamic equilibrium model of species diversity. Where disturbance and rate of population growth and competitive displacement are low, species diversity is highest. The contours indicate declining diversity optima. (Based on Huston, 1994, and Gray & Elliott, 2009)

TABLE 3. Classification schemes for marine invertebrate developmental modes. (From Levin & Bridges, 1995)

Nutritional mode			
	Planktotrophy		
	Facultative planktotrophy		
	Maternally derived	Lecithotrophy	
		Adelophagy	
		Translocation	
	Osmotrophy		
	Autotrophy	Photoautotrophy	
		Chemoautotrophy	
		Somatoautotrophy	
Site of development			
	Planktonic		
	Demersal		
	Benthic	Aparental	Solitary
			Encapsulated
		Parental	Internal brooding
			External brooding
Dispersal potential			
	Teleplanic		
	Actaeplanic		
	Anchiplanic		
	Aplanic		
Morphogenesis			
	Indirect	Free-living	
		Contained	
	Direct		

(Fig. 28a), which complete their development as lecithotrophs, but then prolong their pelagic existence, once their yolk reserves have been exhausted, by a switch to plankton feeding. Lecithotrophic larvae develop from yolk-rich eggs, and especially typical of this larval mode are the amphiblastula larvae of sponges and the bryozoan coronate larva (Fig. 28b), but in some polychaetes and echinoderms, and particularly among species of whelk, in the molluscan order Gastropoda, nutrition for the developing embryo is provided in the form of nurse eggs. As few as 1% of the eggs produced by each female in some whelk species undergoes embryogenesis, the other 99% simply supplying a food source for the developing minority. In some sponge species brooded embryos obtain nourishment through the ingestion of nurse cells, and in some bryozoans embryos developing in a maternal brood chamber are supplied with nutriment through cellular

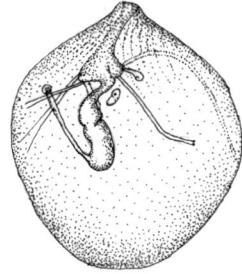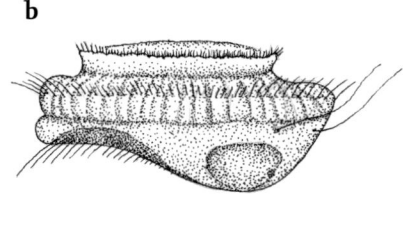

FIG 28. (a) The pelagosphaera larva of a sipunculan, a facultative planktotroph. (b) A lecithotrophic coronate larva characteristic of most bryozoans. Not to scale.

transport via a placental analogue. These two nutritional modes are referred to as translocation.

A nutritional mode only recognised comparatively recently is osmotrophy, which refers to the uptake of organic nutrients, especially amino acids, from the surrounding water, and is seen in many larval echinoderm species, as well as molluscs. Autotrophs synthesise their own food, through photosynthesis (photoautotrophy), as in the planula larvae of reef corals, which carry the same intracellular symbiotic zooxanthellae as the adult coral, or through the agency of symbiotic chemosynthetic bacteria (chemoautotrophy). Somatoautotrophy was coined to describe the veligers of some gastropods, in which the swimming velum is ingested by the larva prior to settlement, probably providing an energy boost needed for metamorphosis.

An interesting alternative criterion for categorising larval types is dispersal potential, which may be useful in describing distributional patterns of adult benthic invertebrate populations. Aplanic larvae undergo brooded or encapsulated embyogenesis, and have no potential for dispersal during the developmental stage. Anchiplanic larvae are planktotrophs with a short free swimming period and limited dispersal capability, while actaeplanic describes Thorson's planktotrophic larvae with a long pelagic phase. Actaeplanic larvae are the most common larval type in temperate seas, produced by about 70% of benthic invertebrate species, and, with a free swimming period of up to two months, may be significant in dispersal, and in the ultimate distribution of adult populations. Teleplanic, or 'far wandering', larvae are encountered in oceanic as well as shelf waters, and were unknown to Thorson, who considered that larvae carried off the shelf edge represented 'larval wastage'. They have pelagic periods measured in months, commonly up to a year or more, and are seen among the otherwise very sedentary sipunculans, tropical gastropods and some polychaetes, including the familiar North Atlantic *Chaetopterus variopedatus*. With such an extended pelagic existence, teleplanic larvae are of significance in establishing widespread, often transoceanic, distributions, and perhaps in maintaining gene flow between distant populations.

The assumption that pelagic larvae are subject to very high rates of mortality is difficult to substantiate, because it is equally difficult to quantify larval mortality. It is possible to estimate the fecundity of a given population of a particular species, by reference to the density and demography of the population, and the average individual reproductive output, but larval mortality rates can only be established by tracking the development of single larval generations, or cohorts. This has been achieved for some species by maintaining experimental populations in large tanks – mesocosms – but extrapolation from experimental results to wild populations poses numerous problems. In a majority of benthic invertebrates fertilisation occurs externally, following the spawning of eggs and sperm into the water column, and this point may mark the first reproductive loss of the population. Synchronous spawning within each population might increase fertilisation success, but both gametes and fertilised eggs may be consumed shortly after spawning by benthic suspension feeders, or dispersed beyond the limits of suitable habitat by bottom currents, and this loss is unquantifiable for almost all benthic invertebrates. Mortality of developing larvae might then result from physiological stress, such as suboptimum temperatures and salinities, perhaps following over-dispersal, or from lack of food, or from predation. The relative importance of these factors has been tested experimentally for the larvae of many fish species, and for commercially important species of decapod and mollusc. For some species the mortality of specific cohorts may be estimated through frequent sampling of wild populations, but for most benthic invertebrates estimates of larval mortality are few, and probably unreliable. For the most abundant fraction of the meroplankton, the larvae of coastal polychaetes, molluscs, decapods and echinoderms, larval survival from hatching to metamorphosis has been estimated at from 10% to 2%, and for most species the latter value will probably be an overestimate.

Mortality is also assumed to be high at two other critical junctures in benthic invertebrate life cycles: settlement and recruitment. Settlement is that stage at which the larva has completed development and is competent to metamorphose, and switches from a pelagic to a benthic habit. In seashore habitats, and among the encrusting epifaunal communities of sublittoral rocky environments, larval choice is often important in determining the settlement site. Competent larvae explore surfaces and are induced to settle, attach and metamorphose by very specific chemical cues. Settlement behaviour has been demonstrated experimentally for many sessile species, but the larvae of benthic soft-sediment species are less easily manipulated, and processes determining their eventual settlement sites are less well known. 'Settlement' thus describes a sequence of biological events attending a critical stage in an organism's life cycle; 'recruitment' describes the point at which an individual joins the adult population, rather than a biological process. Defining a recruit is not straightforward: attainment of a specified size has been suggested as the most important criterion, although, for any species, that size may not necessarily correlate with age, as growth rates vary between species, and habitats, and the time lapse between settlement and recruitment may be very variable. Recruitment has also been suggested as that point at which the juvenile attains a size at which it is safe from further post-settlement mortality. In practice a recruit is defined by the size at which the individual becomes apparent in samples, and contributes to assessments of population density. For sessile species that size depends entirely on sampling procedures: it is usually impossible to census newly settled larvae among established sessile communities, and rates of settlement are derived from numbers of settlers recorded on newly cleared surfaces or on experimental plates. For mobile epifaunal and infaunal species the definition of a recruit depends on the mesh size of the sieve employed in the sampling, and so the smallest individuals in a sample are termed 'sieve recruits'. The new settlers may only become apparent in samples through the use of meiofaunal sieves, and then only when the frequency of sampling is appropriate to their growth rates; sampling at intervals of six weeks or more, for example, may indicate low or no recruitment, simply because new settlers have grown to a perceived juvenile size.

Larval plankton varies in density and composition both seasonally and annually, in response to environmental factors. Physical environmental factors, particularly temperature, have direct effects on plankton populations, or modulate the timing and intensity of phytoplankton blooms, and thus, indirectly, the reproductive cycles of benthic invertebrates. At three fixed stations in the southern North Sea sampled at monthly intervals over a two-year period (Bosselmann, 1989), total plankton densities showed clear seasonal fluctuations, with sharp differences also between sites, and between years (Fig. 29). In 1985 peak larval densities were recorded by June at all three sites, and there was a second peak in early winter at two of them. In the following year, while plankton populations began to show high densities as early as March, larval plankton densities did not peak until August and September, and were still high in early winter at the same two sites. High-diversity plankton

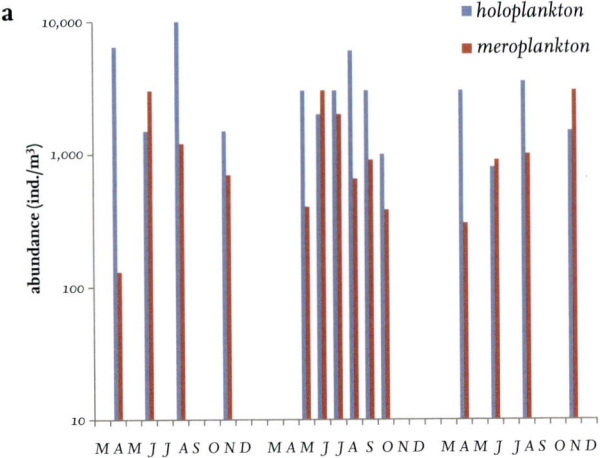

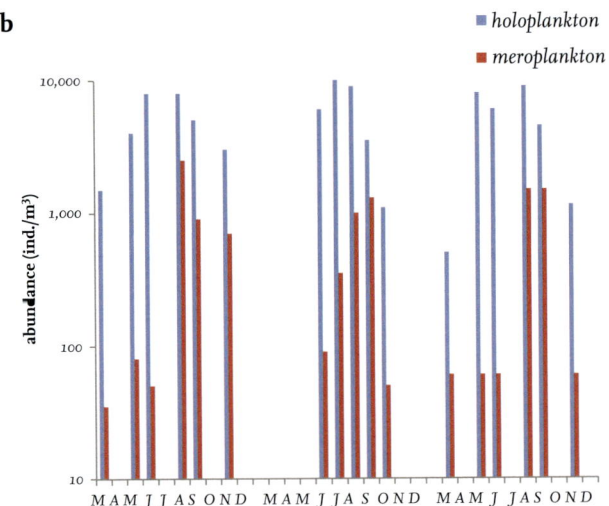

FIG 29. Seasonal and annual variation in plankton density at three stations in the southern North Sea. (a) 1985; (b) 1986. (After Bosselmann, 1989)

populations were evident early in the summer, at all three stations, in 1985, but species diversity was greatest in September in 1986. The composition of the meroplankton also varied seasonally and annually: larvae of the heart urchin *Echinocardium cordatum* were predominant in June 1985, but in August in the following year.

Recruitment also varies with season and between years, but this variability is not always reflected in the densities of adult populations. Further, it is not always evident that recruitment patterns can be related to larval populations. Recruit densities represent the sum of all settlement and post-settlement events, incorporating all sources of mortality, and may be further modified by emigration of settlers, or by immigration of allochthonous (i.e. foreign) settlers into the local breeding population. A survey of benthic polychaete and bivalve populations in a South Carolina bay examined the relationship between pelagic larval densities and the densities of new settlers and recruits in the benthos (Feller et al., 1992). Plankton and benthic infaunal populations were sampled at two-weekly intervals over a five-year period, sieving the sediment samples through macrofaunal (0.5 mm) and meiofaunal (0.063 mm) meshes. The short sampling interval ensured that all new settlers would be censused before they grew sufficiently to be retained by the macrofaunal sieve. There was a distinct match between peak densities of pelagic larvae and new settlers, for both taxonomic groups, but no correspondence between either larval or meiofaunal peaks and macrofaunal densities, suggesting high and variable post-settlement mortality. Peak settlements were not reflected by high recruitment.

Whether recruitment may be related to adult densities of infaunal benthic communities was investigated in another study in the southern North Sea, in which two fixed sites were sampled at intervals from March to December (Künitzer, 1992). One station was at 54 m depth, northeast of the Dogger Bank, with a fauna of 99 species, and the other was a coastal station, 38 m deep, northwest of Heligoland, with a fauna of 71 species. All juvenile individuals, 0.125–1.00 mm long, were sieved from the samples, and categorised in four taxonomic groups: polychaetes, molluscs, crustaceans and echinoderms. The physical environment of the deep station was relatively stable, with narrow annual temperature ranges, and only minimal sediment disturbance, while the coastal station experienced extreme summer and winter temperatures, and during the winter months often profound sediment disturbance. Three peaks in the density of juvenile infauna were recorded at the deep station (Fig. 30a), corresponding with a spring peak for all four taxonomic groups, and then a high density of crustaceans in summer, and of molluscs in early winter. Each peak showed a predictable decline, but total densities remained high through the winter. At the coastal station there was a small peak in spring, but then a single huge peak in late summer (Fig. 30b), when total density exceeded that at the deep station. This single peak was attributable to fast growth of post-settlement juveniles in the warm coastal waters, and then an abrupt and high mortality induced by low temperatures and winter storms. At the deeper site,

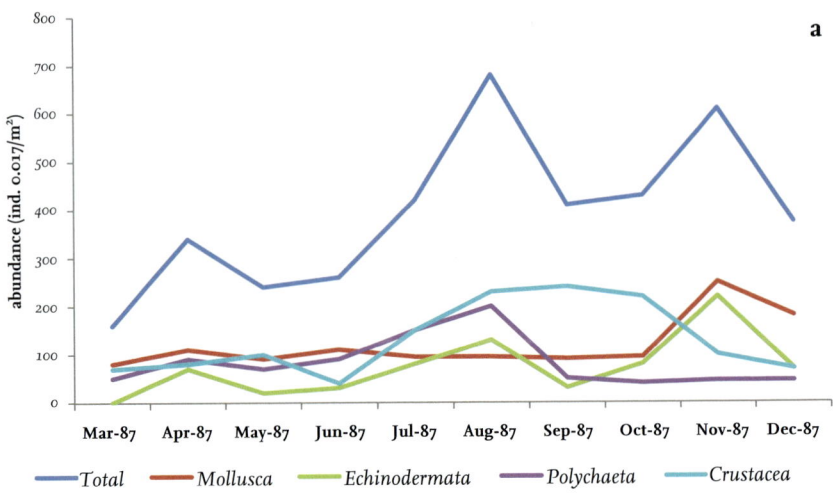

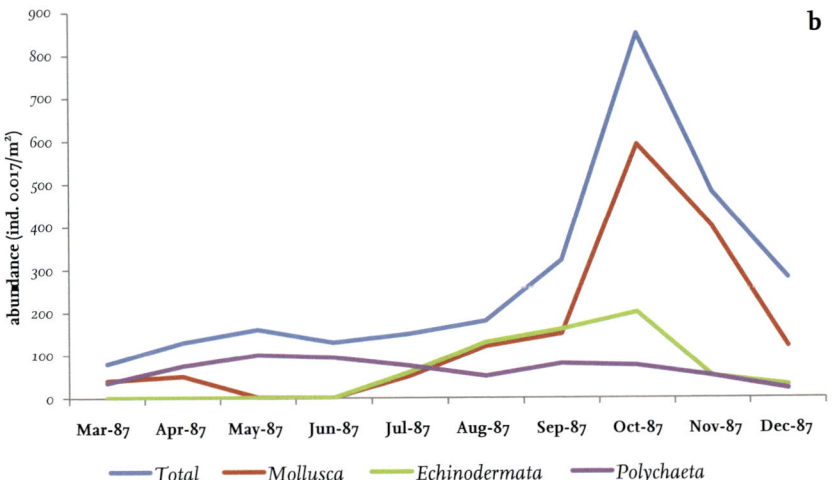

FIG 30. Seasonal variation in the density of juvenile macrofauna (< 1 mm) at two stations in the southern North Sea. (a) Offshore, 54 m; (b) coastal, 38 m. (After Künitzer, 1992)

post-settlement juveniles grew more slowly and recruited at a slower rate, with individual species showing peak recruitment at different times. Adding these data for sieve recruits to adult population densities showed that at the coastal station high numbers of settlers and rapid growth led to an immediate increase in population density and biomass: seasonal changes in benthic population densities reflected seasonal changes in recruitment. Conversely, at the deep, offshore station, density, biomass and diversity of the benthic infauna varied only minimally through the year, the community was stable and recruitment had little discernible effect on community dynamics.

Post-settlement, pre-recruitment mortality has been variously attributed to physical disturbance, to ingestion by, or disablement by currents created by, suspension feeders, including those of conspecific adults, and to the effects of bioturbation. It is also apparent that high mortality of post-settlement juveniles results from predation, particularly by small infaunal predators. A comparison between larval abundances and infaunal densities of juveniles of the Sand Gaper clam, *Mya arenaria*, in the Dutch Wadden Sea showed sequential peaks, in May and June (Günther, 1992). New settlers, with shell length < 0.250 mm, were recorded at densities as high as 13,000 per square metre; at shell length 0.25–0.5 mm these were considered to have recruited to the population, and at 0.5 mm they were defined as juveniles. Densities of these two latter size classes dipped sharply through the three months following settlement, but stabilised at a shell length of 4 mm (Fig. 31), at which size the juvenile Sand Gapers achieved a refuge from predation. Mortality was greatest among the smallest size classes, and possibly a result of meiofaunal predation, an effect referred to as the

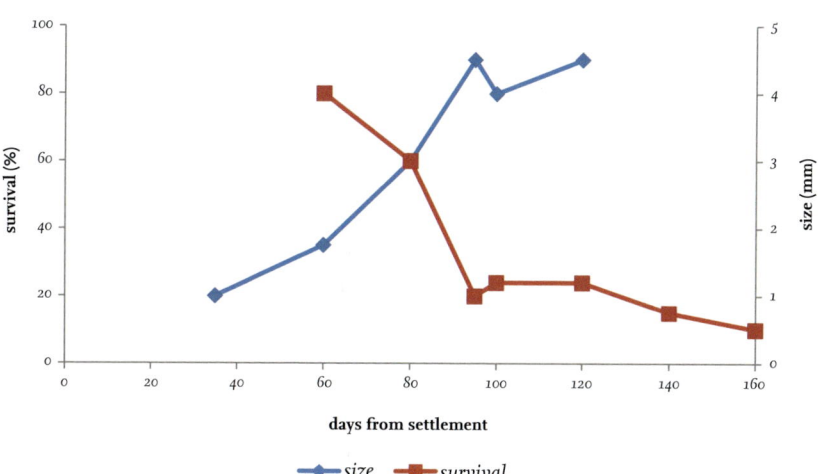

FIG 31. Shell size and survival in Sand Gaper clam, *Mya arenaria*, recruits > 0.5 mm. (After Günther, 1992)

'meiofaunal bottleneck', although the identity of the predator, or predators, was not established. However, in another example, post-recruitment mortality of a prey species could be related to a specific predator (Ejdung & Elmgren, 1998). The small infaunal bivalve *Macoma balthica* (Fig. 20a) is abundant in low-salinity inshore muddy sands along much of the coastline of northwest Europe. In some Baltic habitats it recruits at densities as high as 80,000 individuals per square metre during the midsummer period and suffers huge mortality through late summer to October. An amphipod, *Monoporeia affinis*, is also a common member of the deposit-feeding infauna, and its juveniles recruit at densities of 10,000 per square metre in late winter, growing to a length of 3.5 mm by the time the bivalve spat recruit, at just 0.2 mm long, providing for a brief period an important food source for the young amphipods. Predation of *Macoma balthica* by juvenile *Monoporeia affinis* is density-dependent, i.e. the greater the density of bivalves the greater the rate of predation, which rises steeply with just a small increase in prey. *Macoma balthica* is safe from predation once it achieves a length of 1 mm, its size refuge, but the intensity of predation in the first few months following settlement is such that over much of the Baltic its populations are probably largely limited by predatory amphipods.

DISTRIBUTIONS AND DIVERSITY

Shelf sea environments are under relentless pressure in an overcrowded, resource-hungry world. More than a third of the world's human population is concentrated on coastal lands and along river catchments feeding into coastal waters. Shelf seas are both a source of food and a repository for rubbish, and are exploited for oil, gas, minerals and aggregates, the engineering and extraction processes associated with which generate even more rubbish. Coastal marine environments are critically important to biogeochemical cycling, which may be seriously perturbed by human activities that result in increasing eutrophication (inorganic nutrient enrichment). Protection and conservation of the marine environment is now seen as an urgent priority by all European maritime states, although it has taken decades to reach a consensus. In northwest Europe the grim potential of oil pollution was realised in 1967 with the wreck of the tanker *Torrey Canyon,* and led to the 1969 Bonn agreement between west European nations to cooperate in confronting the problem of oil pollution in the North Sea. A further move towards a cleaner marine environment was marked by the Oslo Convention of 1972, the signatories to which agreed to ban the 'over-the-side' dumping of rubbish by all ships, a practice that had by then created thick strandlines of non-degradable, largely plastic, material on long stretches of the sandy coastlines of western Europe. In 1974 the even more radical Paris Convention banned the offshore dumping of industrial and domestic waste generated by the industries and cities of western Europe. By 1992 the two agreements had been united as the Oslo/Paris – OSPAR – Convention for the Protection of the Marine Environment of the North East Atlantic, which began effective operation in 1998, 31 years after the *Torrey Canyon* disaster. The OSPAR Convention is observed by the 12 west European maritime nations, together with Finland, and Luxembourg and Switzerland, which encompass significant proportions of the Rhine catchment and thus contribute to the riverine input to the southern North Sea. Its mandate is to identify threats to the marine ecosystems of the northeast Atlantic, to implement measures to counteract perceived threats, to monitor and assess marine environmental quality in relation to internationally agreed goals, and, finally, to ensure that the members of the convention implement the measures necessary to achieve these objectives (Heslenfeld & Enserink, 2008; Johnson, 2008). The work of the convention is focused upon four strategic areas: the protection and conservation of ecosystem and biological diversity, eutrophication, hazardous waste and radioactivity.

All parts of the marine environment have changed continuously as a consequence of human exploitation and utilisation, at increasing rates as human populations expanded, and over much of the past century marine environmental change can only be described as a process of profound degradation. The nature and pace of change also reflect more indirect stresses resulting from anthropogenic pressures on the earth's climate, and while direct factors such as overfishing, mineral extraction, waste dumping and eutrophication can be measured, and their effects predicted, the consequences of climate change are still only poorly understood. There is a need to monitor and chart the distributions and diversity of marine benthos and benthic assemblages, and to track changes as precisely as possible, in order to try and distinguish changes related to natural environmental cycles from those induced by direct anthropogenic pressures, and those that might be attributable to sustained climate change.

The OSPAR Commission is committed to the development of policies for the management of human activities in the marine environment that ensure sustainable use of marine resources in a healthy marine environment, and these policies are necessarily based on the best available scientific data on the structure and function of all marine ecosystems. From 2000 the International Council for the Exploration of the Sea (ICES) began to advise OSPAR on the development of ecological quality objectives (EcoQO) for defined ecological quality issues, such as commercial fish stocks, fish communities, seabirds, marine mammals and benthic communities, as well as threatened habitats and species, and eutrophication (Table 4). Ecological quality (EcoQ) is the overall measure of the structure and function of a marine ecosystem, incorporating physical, chemical, biological and climate aspects, and human activities, and may comprise one or

TABLE 4. Examples of OSPAR EcoQ issues and elements, with accepted objectives. (After Heslenfeld & Enserink, 2008)

EcoQ Issue	EcoQ Elements	Objectives
1. Commercial fish species	1.1. Biomass of spawning stock of North Sea species	Above agreed precautionary reference points.
2. Marine mammals	2.1. Seal population trends in the North Sea	(A) No decline of ≥ 10% in 5-year running mean for Harbour Seals in any of 11 subareas of the North Sea. (B) No reduction of ≥ 10% in Grey Seal Pup production in any of 9 subareas of the North Sea.
	2.2. Bycatch of Harbour Porpoises	Objective: annual catch should be reduced to below 1.7% of best population estimate.
3. Seabirds	3.1. Proportion of oiled birds among Common Guillemots found dead/dying on shore	Proportion should be ≤ 10%.
	3.2. Mercury concentrations in seabird eggs	
	3.3. Organochlorine concentrations in seabird eggs	
	3.4. Plastic particles in seabird stomachs	
	3.5. Local availability of sand eel prey to Kittiwakes	
4. Fish communities	4.1. Changes in proportion of large individuals (average weight and average maximum length)	

THE BENTHIC ENVIRONMENT · 57

5. Benthic communities	5.1. Average level of imposex (sterility resulting from development of penis in females) in *Nucella lapillus* or other indicator species of whelk	Level should be consistent with exposure to tributyltin contamination at concentrations below the agreed environmental assessment criterion.
	5.2. Density of sensitive species	
	5.3. Supporting EcoQ 9.1.5	
6. Plankton communities	6.1. Supporting EcoQs 9.1.2, 9.1.3	
7. Threatened/declining species	7.1. Status of threatened/declining species in the North Sea relative to initial OSPAR list	
8. Threatened/declining habitats	8.1. Restore/maintain quality and extent, as shown on initial OSPAR list	
9. Eutrophication	9.1. Eutrophication status of North Sea	All parts should have status of non-problem areas as assessed under the OSPAR common procedure.
	9.1.1. Winter nutrient concentrations	Concentrations of dissolved inorganic nitrogen and phosphates below a justified salinity-related/area-specific % deviation from background, < 50%.
	9.1.2. Phytoplankton chlorophyll *a* concentrations	Maximum and mean in growing season below area-specific % deviation from background, < 50%.
	9.1.3. Phytoplankton eutrophication indicators	Area-specific indicator species below nuisance/toxic levels; no increase in average duration of bloom.
	9.1.4. Oxygen	Concentrations decreased by nutrient enrichment above area-specific assessment levels, within range 4–6 mg/l.
	9.1.5. Zoobenthos kills resulting from eutrophication	No kills resulting from oxygen deficiency and/or toxic phytoplankton species.

more elements. For example, the proportion of oiled Common Guillemots, *Uria aalge*, among dead or dying individuals, or sand eel, *Ammodytes* spp., availability to breeding Kittiwakes, *Rissa tridactyla*, the proportion of plastic particles in the stomach contents of Fulmars, *Fulmarus glacialis*, or, more generally, trends in local seabird populations (Heslenfeld & Enserink, 2008) are elements of the EcoQ issue relating to seabirds. Ecological quality objectives may then be defined for each element: for example, fewer than 10% of dead Common Guillemots should be oiled; no more than 2% of stomach contents of samples of 50–100 stranded Fulmars should consist of plastic granules.

In 2008 the European Union established a framework (Directive 2008/56/EC of the European Parliament and of the Council of 17 June 2008) for community action in the field of marine environmental policy, the Marine Strategy Framework Directive – naturally, the MSFD – which requires member states to implement target-driven goals to achieve and maintain 'good environmental status' (GES) for all marine habitats by 2012, and to 'support the strong position taken by the Community, in the context of the Convention on Biological Diversity, on halting biodiversity loss'. OSPAR EcoQs, quantified by programmes of monitoring and assessment, provide evidence of habitat resilience, stability, productivity, diversity and trophic structure, and contribute to 'common indicators' of the status of each criterion defining GES (OSPAR Commission, 2013).

The plethora of acronyms serves to illustrate the extraordinary complexity of the problems attending the development of marine environmental management policies, even in such a relatively small proportion of the eastern North Atlantic region as the North Sea. The taxonomic diversity – or species richness – of benthic habitats is greatest under conditions of long-term environmental stability, and decreases along gradients of increasing physical disturbance and rising inorganic nutrient enrichment. Stable benthic communities tend to be resilient, recovering swiftly from natural perturbation; taxonomically diverse communities usually possess a degree of ecological redundancy, with perhaps numerous species fulfilling similar ecological roles, so that the loss of any is balanced by the expansion of others. However, as disturbance increases, in both frequency and intensity, communities change in terms of diversity, composition and abundance, as stress-intolerant species decline. Stable habitats are populated by long-lived, *K*-selected species, while highly disturbed habitats are characterised by short-lived, *r*-selected, opportunistic species, adapted for swift colonisation of depopulated habitats. These contrasting community types may be simply expressed as *K*-dominance curves: species are ranked according to abundance in the sample, and the proportional representation of each species is plotted on a cumulative curve (Fig. 32). A shallow, concave curve indicates an assemblage

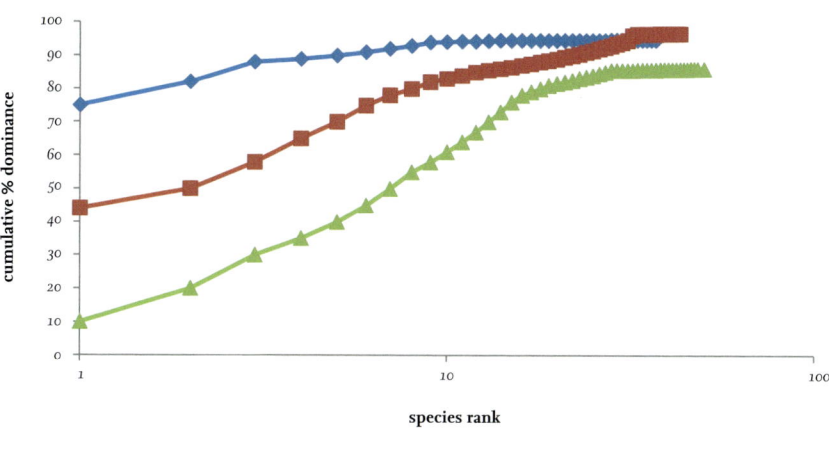

FIG 32. *K*-dominance curves for infaunal assemblages characteristic of three contrasting sediment grades. The flat curve for the mud habitat represents a community of *r*-selected species, while the shelly sand community consists predominantly of *K*-strategists.

of predominantly *K*-selected species, with no numerically dominant species; a flat curve commencing high on the ordinate axis is typical of a community of *r*-selected opportunists, with just two or three species comprising the majority of the individuals. This relationship can be further emphasised by plotting both number of individuals and total biomass against species. For an undisturbed, stable community the curve for cumulative biomass lies above that for abundance (Fig. 33a), with many large-bodied species contributing similar proportions to the community biomass, while the reverse is the case for disturbed habitats (Fig. 33b), in which the greater part of the biomass is just two or three abundant species. Biological diversity is a critical attribute of 'good environmental status'; numbers of species and total number of individuals, species composition and relative abundance, and biomass, are all determinants of biological diversity, and all are likely to fluctuate in response to environmental change, with consequent effects on community structure and function.

Each species displays annual and seasonal population cycles. Extreme weather events are significant natural causes of disturbance and change in shallow coastal habitats, but may occur only infrequently. Long-term climatic factors may be driving low-amplitude change on a large geographical scale, while anthropogenic pressures arising from dredging and aggregate extraction

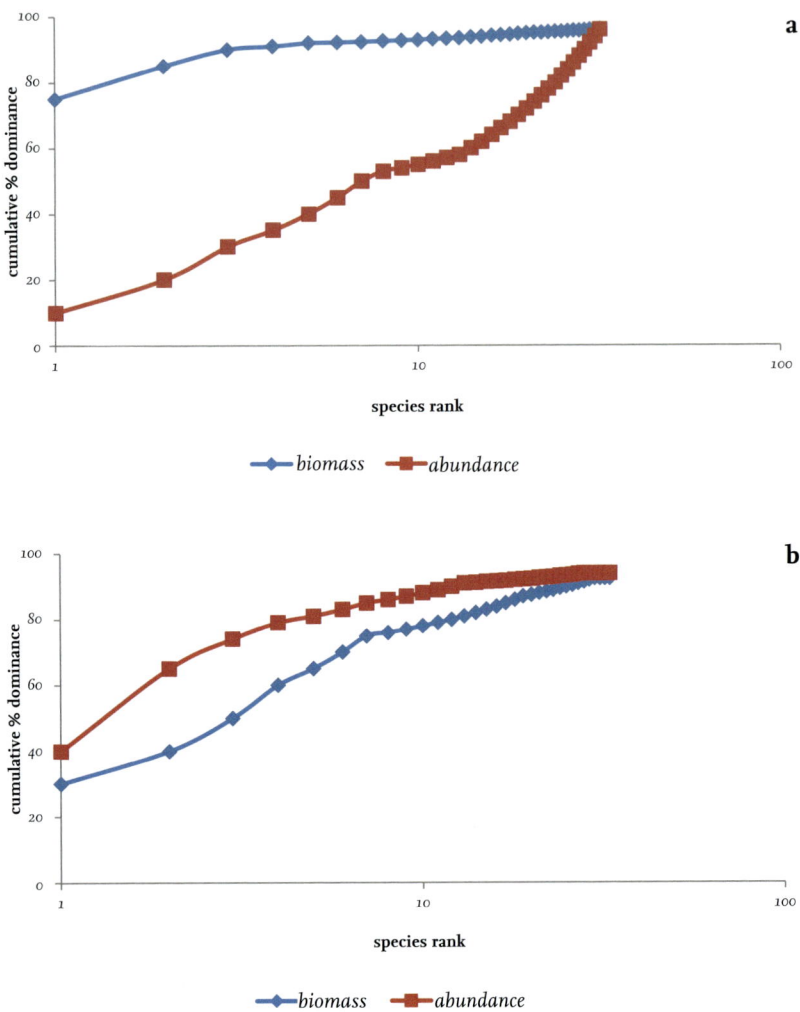

FIG 33. Abundance/biomass plots. (a) An undisturbed, stable community, with the curve for biomass lying above that for abundance: large-bodied K-strategists predominate. (b) For a perturbed community, the curve for abundance lies above that for biomass, indicating numerous small individuals.

may lead to severe high-amplitude change at local scales. In northwest Europe the winter of 1995/96 was a prolonged period of extreme weather: it was exceptionally cold, and in the southeast North Sea, within the 50 m isobaths, the mean and minimum winter bottom temperatures were 6.8 and 1.9 °C respectively (Neumann et al., 2009), and were matched by negative sea surface temperature anomalies that persisted through 1997. The epibenthos suffered unusually high winter mortality, and a consequent lower reproductive output through the spring and summer of 1996, but there was a greatly increased recruitment of a few opportunistic species, including the small polychaetes *Spiophanes bombyx* and *Scoloplos armiger*, and the brittlestar *Ophiura albida*, which became especially abundant by 1998, with population densities close to 2000/m². *Ophiura albida* tolerates low winter temperatures; it spawns in May, juveniles settle in October, and high recruitment would have been achieved following the great mortality of predators during the cold winter months. However, *O. albida* populations along the Frisian coast then dropped sharply in 1999, and continued to decrease through 2002, as did populations of its congener *O. ophiura*, the hermit crab *Pagurus bernhardus*, and the sea squirt *Ascidiella scabra*, their decline coinciding with rising sea surface temperatures from 1998 to 2008. The area-averaged SST for the North Sea in September 2002 was the highest recorded since 1971, and the winter and spring SSTs for that year were also above normal following the lack of cooling in February and March. By 2003 the area-averaged SST, at 18 °C, was the highest since 1968, and from 2000 to 2007 winter bottom temperatures in the German Bight were 3 °C higher than normal. Brittlestar populations continued to shrink, and increases in the number and abundance of other species led to changes in the composition and structure of benthic communities in the coastal waters of the southeastern North Sea. Rising SSTs and a longer period of primary production resulted in enhanced chlorophyll *a* concentrations, and hence increased food inputs, which correlated with rising abundance and biomass of the epibenthos, which would also have been increased by low winter mortality and high spring reproductive output through the mild winters of 2003 to 2008. Benthic predation might also have been a significant agent in the changing composition and structure of benthic communities during this period of warming. Populations of small predatory benthic fish increased considerably, especially those of Solenette, *Buglossidium luteum*, Scaldfish, *Arnoglossus laterna*, and the sand goby *Pomatoschistus minutus*, all of which are generalists, with diets likely to include brittlestar arms and small polychaetes. Another generalist, the swimming crab *Liocarcinus holsatus*, displays accelerated development, and enhanced recruitment success, with rising sea temperatures, and is a further potential source of increased predation pressure. Warmer bottom temperatures in winter may also drive change in benthic community structure as larvae dispersed beyond the range of adult populations are enabled to settle, recruit and survive, eventually establishing new, self-sustaining populations. The Angular Crab, *Goneplax rhomboides*, is commonly distributed along the western coasts of Britain and Europe, but extended its range eastwards along the English Channel, and by 2003 had become established in the southeastern North Sea. *Goneplax rhomboides* is an ecologically important habitat engineer, constructing permanent branching burrows in sandy muds, each individual burrow ramifying through perhaps a quarter of a cubic metre of sediment. As its population densities increase the Angular Crab is likely to have a considerable impact on the composition, and structure, of local benthic assemblages in its new habitats.

Many naturalists will have a confident conception of biological diversity: it is self-evident that if one habitat, or sample site, supports a greater number of species than another, it must be biologically the more diverse. Many ecologists, and perhaps most biostatisticians, in contrast, will continue to debate the definition of biodiversity, how to measure it, and indeed what one is actually measuring. Biodiversity encompasses genetic diversity, within and between species, and taxonomic diversity, within and between communities, as well as ecological or functional diversity, within and between habitats. The two metric elements of biodiversity are the number of species and the number of individuals, or biomass, that a given habitat supports. K-dominance curves are one technique for presenting such data, and demonstrate sharply the significance of the proportional representation of each species (Figs 32 & 33). The number of species, or species richness, S, and the number of individuals, N, may be very similar in each of two or more habitats, but the relative proportion of each may be quite dissimilar. Such plots may also mask additional, ecologically important, information: two similar curves may indicate similar habitat stability, or vulnerability, but may have been generated by two different suites of species, and thus perhaps two functionally different communities. Species richness data, together with the numbers of individuals recorded for each species, or evenness, can be employed to derive indices of diversity.

There are many types of diversity index (see Magurran, 1996, for example), and the two indices frequently encountered in marine ecological studies are the Simpson index, D, and the Shannon–Wiener index, H', worked examples of both of which have been provided by Magurran (1996) and Gray & Elliott (2009) (Box 1). The Simpson index increases as diversity decreases, and is usually expressed as its reciprocal, which will increase with rising diversity. However, the statistic D is biased by the abundance of the most common species in the sample, and is thus more sensitive to dominance than evenness. The Shannon–Wiener index increases as the number of species in a sample, or habitat, increases, but it also increases as the proportion of individuals per species becomes more uniform,

Box 1

Two frequently employed diversity indices applied to a quantitative sample of, for example, six infaunal invertebrates:

	Species	Number of individuals	Proportion
	A	240	0.432
	B	180	0.324
	C	60	0.108
	D	30	0.054
	E	25	0.045
	F	20	0.036
Total	S: 6	N: 555	0.999

The Simpson index of diversity, D, is expressed as:

$$\frac{\Sigma (n_i (n_i - 1))}{(N (N-1))}$$

and for the example above would be calculated as:

$(240 \times 239)/(555 \times 554) + (180 \times 179/(555 \times 554)) + (60 \times 59)/(555 \times 554) \ldots$

$= 0.186 + 0.105 + 0.011 + 0.003 + 0.002 + 0.001 = 0.308$

D is the reciprocal: $1/0.308 =$ **3.55**

The Shannon–Wiener Index, H', is expressed as:

$-\Sigma p_i \ln p_i$

where the proportional abundance of each species is multiplied by the natural log of its value, and for the above example would be calculated as:

$(0.432 \times \ln 0.432) + (0.324 \times \ln 0.324) + (0.108 \times \ln 0.108)$

$= 0.362 + 0.363 + 0.240 + 0.158 + 0.139 + 0.120 =$ **1.384**

(note: the negative sign beginning the formula cancels the negatives arising from the calculations).

Evenness of the sample is calculated as

$E = H'/\ln S = 1.384/1.792 =$ **0.772**

and is thus sensitive to species evenness. Each diversity index tends to measure a particular aspect of biodiversity, but while species richness and evenness provide more informative measurements of diversity than simple counts of species, there are also potential factors that further confound the difficulty in assessing biodiversity. Stable habitats are characterised by large numbers of persistent species, by definition K-selected, with none numerically dominant. Disturbed habitats are populated by short-lived, small, r-selected species; just a very few will comprise the numerical bulk of the community, which nonetheless may include many other small opportunists. Such communities may be taxonomically diverse, but they are likely to consist of very similar, closely related species, with similar trophic modes and ecological functions.

Diversity indices based on measurements of S and N are useful tools in comparative ecological studies at small spatial and temporal scales, with controlled sampling protocols, but are less reliable when applied over broad spatial and temporal ranges. Firstly, they tend to be sample-size dependent, with numbers of species and individuals both increasing as the sample size increases. Also, the number of species tends to rise in relation to the number of samples; this is the familiar species accumulation curve (Fig. 34). The number of species recorded rises steeply at first as sampling effort increases but eventually levels off to an asymptote, a point beyond which continued sampling results in no significant increase in the number of species collected. Diversity measures based on S and N may thus be unreliable simply because the sampling effort is almost always insufficient to have reached the asymptote for the habitat, and there will always be a proportion of rare species that are not sampled. Another difficulty is that it cannot be predicted where on the disturbance gradient a

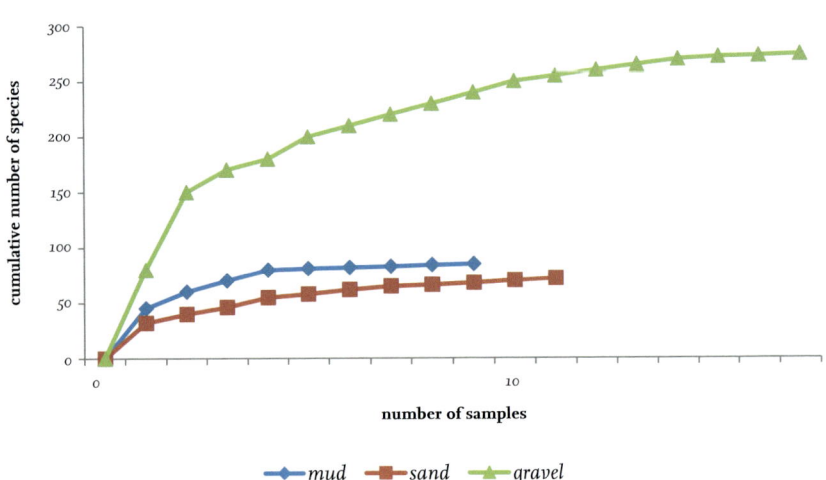

FIG 34. Species accumulation curves for three sediment grades. (Data from Newell *et al.*, 2001)

habitat lies when sampling commences. Long-term stability may result in a loss of some species from a habitat as a consequence of sustained competitive pressure, while an intermediate level of disturbance may actually enhance diversity by providing space for colonisation by additional species. Continual, chronic perturbation is likely to lead to ecological succession stalling, and to habitats occupied by a low diversity of pioneer colonising species. The functional diversity of a benthic assemblage – trophic mode and degree of mobility, for example – will vary in relation to habitat type. Fine sand habitats tend to support more macrofaunal and meiofaunal species than both finer and coarser sediments, but changes in diversity may not be apparent; a habitat dominated by a single trophic group may support a greater diversity of species than another inhabited by fewer species, but with a range of trophic modes, and thus greater functional diversity. Further, in the central and southern North Sea natural environmental variability, climatic fluctuation and anthropogenic pressure are presently driving change in benthic assemblages leading to the decline of a number of large-bodied animals, while the diversity and density of amphipods and small polychaetes increase; overall species richness does not change, despite a significant loss of broader taxonomic diversity. However, in a long-term study at sites off the Frisian coast species richness, evenness and the Shannon–Wiener index showed a slight increase through periods of rising North Atlantic Oscillation index from the 1980s, although decreasing through the cold winters of 1982/83, 1984/85, 1985/86, and 1995/96, 1996/97 (Kröncke & Reiss, 2010).

A realistic measure of biodiversity needs to incorporate data on the taxonomic relatedness of individuals within a sample, to express the total systematic range of the community sampled. Recently designed indices of taxonomic diversity, D, and taxonomic distinctness, D^*, employ the classic Linnaean hierarchical classification to this end (Warwick & Clarke, 2001). At each hierarchical level, from species to phylum, related taxa, from species to class, are linked at a branching point, or node, and the pathways between nodes are accepted as proxies for the branches of a phylogenetic tree. Taxonomic diversity, D, represents the average path length between every pair of individuals within a sample, and thus individuals identified as belonging to one or more species within a single genus have a shorter average path length between them, than between individuals assigned to different genera, families, orders or classes (Fig. 35). There is still some potential for bias in the calculations, because D will be partly dependent on the individual abundance of species. To take account of this, D^* is defined as the average path length between two individuals of different species, and it is therefore the most realistic

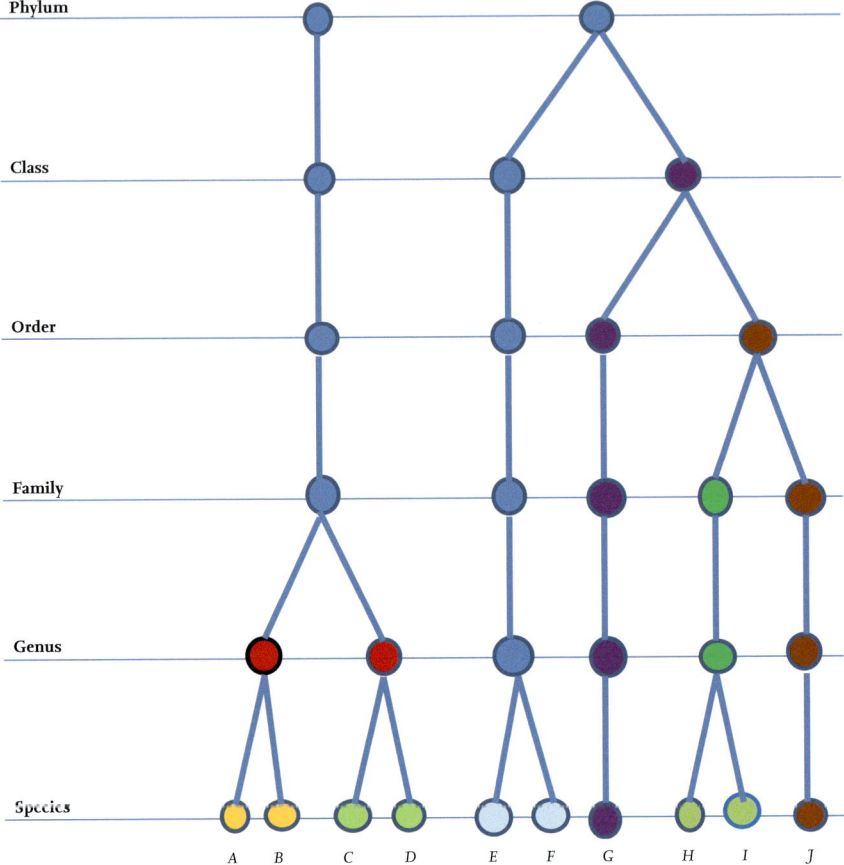

FIG 35. Hierarchical taxonomic trees for two contrasting assemblages. The taxonomic distinctiveness of each sample is a measure of the number of hierarchical levels between each pair of species. See Box 2 for further explanation.

expression of the overall taxonomic relatedness – or systematic spread – of the sample (Box 2).

While an index of taxonomic distinctness may be independent of both S and N, it demands accuracy in the recognition of, and discrimination between, all of the species in a sample. This requires not only experience on the part of the researcher but also a reliable modern taxonomy, and appropriate identification guides; for some marine invertebrate groups it can be extremely difficult even for

Box 2. Calculating taxonomic distinctness, D*, for the two samples shown in Figure 35

	B	C	D	P
A	1	2	2	5
B		2	2	4
C			1	1
D				

	F	G	H	I	J	P
E	1	5	5	5	5	21
F		5	5	5	5	20
G			4	4	4	12
H				1	3	4
I					3	3
J						

For each sample the matrix shows the number of hierarchical levels separating each pair of species.

L is the number of hierarchical levels, or linkages, between all species. Thus, in the small sample all species are encompassed by two levels, genus and family, while the larger sample embraces five, from phylum to species.

P is the total number of hierarchical levels, or pathway, between each pair of species. Thus, for species E, 21 hierarchical steps are required to link it with all other species.

$D^* = P/L$ $= 10/3$ $= 3.33$ for the small sample
             $= 60/5$ $= 12$ for the large sample

---

the specialist to distinguish between superficially similar species. The index is also likely to be sensitive to the systematic hierarchy currently used by specialists of each group of organisms, and for some groups this may not reflect phylogenetic reality. Ideally, systematic hierarchies should be based upon phylogenies derived from molecular genetic data, and representing scales of genetic relatedness, but for some groups such data are presently unavailable and for others it is unlikely that they ever will be. The classifications of well-studied taxonomic groups, such as fish, bivalves and gastropods, tend to be rich in genera, with perhaps five hierarchical levels between species and class, while for other groups species may be lumped into fewer, more broadly defined genera, and taxonomic categories such as subgenus and superfamily are not consistently employed.

Diversity indices based on functional characteristics of soft-sediment benthos have been developed for the purpose of monitoring the ecological quality (EcoQ) of estuarine and coastal environments. The first to be designed, and now the most widely applied, is the AZTI Marine Biotic Index (Borja *et al.*, 2000); AZTI-Tecnalia is the organisation that created the software for calculating the index, now simply denoted AMBI. Macrobenthic species are assigned individually to one of five ecological groups (EG), each defined according to the sensitivity of the assemblage to environmental disturbance, from EGI, most sensitive, to EGV, least sensitive (Table 5). The AMBI is then calculated from the proportional representation of species assigned to each ecological group within a sample. A further elaboration of the index incorporates species richness and Shannon–Wiener diversity measures in a multivariate index, M-AMBI (Muxika *et al.*, 2007).

Marine biotic indices were devised to monitor change in soft sediment benthic communities in response to anthropogenic disturbance and environmental pollution, and thereby provide assessment of EcoQ status, as required by the European Water Framework Directive (Muxika *et al.*, 2007). While their primary purpose was to monitor environmental quality from an anthropocentric perspective, MBIs also record changes in the composition and diversity of benthic assemblages driven by natural environmental

TABLE 5. Ecological groupings of soft-sediment benthos. (After Borja *et al.*, 2000)

EGI		Species very sensitive to organic enrichment and disturbance, usually present only in the most stable, unpolluted habitats. Typically *K*-selected species, persistent and long-lived. They include specialist carnivores, large burrow-dwelling deposit feeders, and other species responsible for structuring the assemblage.
EGII		Species indifferent to enrichment or disturbance, always present in low densities that show little temporal variation. They include many suspension feeders, non-specialist carnivores and scavengers.
EGIII		Species tolerant of organic enrichment. They may occur under unpolluted conditions, but populations are stimulated by organic enrichment. They include many surface deposit feeders.
EGIV		Second-order opportunistic species. Typified by small, subsurface-feeding polychaetes.
EGV		First-order opportunistic species, resistant to high disturbance and pollution. Deposit-feeding species that may be abundant in reduced sediments.

fluctuation and climatic change (Kröncke & Reiss, 2010). MBIs are generated by sophisticated software, far distant from the pencil and paper techniques employed in the calculation of simpler diversity indices, but they are still reliant on the accurate recognition and identification of each taxon, and a sound understanding of its ecological functions. Obstacles arising from the former requirement are obvious: there are too few experienced taxonomic specialists and, as yet, no cheap and effective technique for the discrimination of taxa within large samples. The second caveat is more interesting. In a pilot study applying AMBI methods in surveys of soft-sediment communities in bays of southern California, local taxonomic specialists disputed the ecological classification of 308 out of 630 taxa identified, and for 42 species the difference was by more than one ecological group (Teixeira et al., 2012).

The classification of benthic taxa in five ecological groups, each characterising a stage along a gradient of disturbance, was developed by marine biologists for European soft sediment habitats, and has subsequently been adapted for use in many other regions, from Greenland to North Africa, the Mediterranean, the western North Atlantic from Canada to Florida, and for one location in the southwest Indian Ocean and another at Hong Kong. Taxa with wide geographic ranges are generally classified in the same ecological group to which they were assigned in the European classification. When a taxon does not appear in the AMBI classification, local specialist opinion is consulted to seek a consensus as to which EG it should be assigned to; lack of consensus may arise from doubts over the identity of the taxon, or from a paucity of information on its ecological functions.

Species with apparently wide geographical distributions, re-examined using molecular genetic techniques, are often found to comprise two or more distinct biological species, with discrete or overlapping geographical ranges, and perhaps differing life strategies. A good example would be the two morphologically similar bivalves, *Chamelea striatula*, with a north European distribution, and *C. gallina*, with a Mediterranean distribution, overlapping along the Iberian coasts, and which appear to have different life histories (Backeljau et al., 1994) and perhaps as a consequence different ecological tolerances (see page 285). Some invasive species, such as the Manila Clam, *Ruditapes philippinarum* (see page 314), seem to have established practically cosmopolitan geographical distributions, in temperate to tropical environments, but each population must have adapted to local environmental conditions, and even when separate populations show genetic congruence it must be considered possible that the ecological functions and tolerances might vary in relation to local environmental conditions, habitat type and population structure.

Analysing differences, both spatial and temporal, between samples and habitats is now generally based upon multivariate data that include habitat characteristics such as sediment grade, depth, temperature ranges, as well as species richness and evenness, abundances and biomass, and diversity indices. Multivariate analyses can reveal degrees of similarity, or dissimilarity, between pairs of samples, or sites, and show which species contribute most to the similarity or differences in the composition of each assemblage sampled. Calculation of similarity coefficients is complex, but software packages such as the Plymouth Routines in Multivariate Ecological Research (PRIMER) are available to amateur as well as professional marine ecologists.

CHAPTER 3

Mud, Sand and Gravel

The primary sources of variation in the composition of benthic assemblages of marine sediments are hydrodynamic regimes and the nature of the sediment, and the two factors are closely interrelated. Water flowing over the sea floor is subject to a frictional drag at the interface, or boundary, between water and sediment. This lower, boundary, layer of the water body is slowed relative to the upper layers, and if the degree of slowing is pronounced the smooth, laminar flow of the water is disturbed, and the boundary layer breaks up into eddies of turbulent flow. Above a particular critical velocity, dependent on sediment grade, turbulent water flow results in displacement and transport of the sediment. Feeding modes of the benthos are also related to flow regimes: a low-velocity boundary layer allows suspension feeders to extend tentacles or siphons into the water column, and actively collect food particles. With increasing flow velocity more robust organisms function as passive suspension feeders, straining food from the water, while others resort to surface deposit feeding, employing siphons, palps or tentacles to collect food from the sediment surface. Below the surface, animals occupying permanent burrows or tubes may function as suspension feeders, collecting food particles from water flow created by the occupants, while those constructing only temporary burrows are subsurface deposit feeders, although the depth below the surface at which they feed may be quite narrowly restricted.

The sedimentary environment includes both inorganic and organic components; the former consists largely of silica, as sand, gravel and pebbles, while the latter comprises not only carbonate gravels (page oo) but also finer material, such as decayed phytoplankton and macroalgal detritus, and all of the by-products of biological processes within the sediment. This organic material contributes significantly to the nature of the infaunal habitat, but in defining its character it is useful, as a first step, to analyse the particle sizes of the inorganic component, measuring the proportions of fine, medium and coarse sand, and the degree to which the sediment has been sorted by wave and current action. Sand particle size varies continuously and infinitely, so a stacked series of sieves is employed to divide sand samples into a series of size classes or grades, referred to as the Wentworth scale (Table 6). Thus, coarse sand passes through a sieve mesh of 1 mm but is retained by a mesh of 0.5 mm; medium sand is retained by a 0.25 mm mesh and fine sand by a 0.125 mm mesh. A sieve mesh of 0.0625 mm marks the boundary between very fine sand and silts, while the finest material, passing through a mesh of 0.005 mm, consists of clay minerals. The largest material in a sediment sample is classed successively as very coarse sand, granules, pebbles and cobbles.

Conventionally, the sieve mesh sizes have been transformed from mm to log base 2 values, to give single numbers termed phi (the Greek letter, Φ) units. This has no analytical purpose, and ecologists now often use the millimetre values in plotting graphs, although the use of phi units does simplify graphical

TABLE 6. The Wentworth scale for the classification of sediments.

Type of sediment	Diameter (mm)	Φ units
Boulder	> 256	> −8.0
Cobble	64–256	−6.0 to −8.0
Pebble	4–64	−2.0 to −6.0
Granule	2–4	−1.0 to −2.0
Very coarse sand	1–2	0 to −1.0
Coarse sand	0.5–1	1.0 to 0
Medium sand	0.25–0.5	2.0 to 1.0
Fine sand	0.125–0.25	3.0 to 2.0
Very fine sand	0.0625–0.125	4.0 to 3.0
Coarse silt	0.0312–0.0625	5.0 to 4.0
Medium silt	0.0156–0.0312	6.0 to 5.0
Fine silt	0.0078–0.0156	7.0 to 6.0
Very fine silt	0.0039–0.0078	8.0 to 6.0
Coarse clay	0.00195–0.0039	9.0 to 8.0
Medium clay	0.00098–0.00195	10.0 to 9.0

presentation of sediment data. The sediment retained by each sieve is weighed and the weight expressed as a percentage of the total weight of the sample, and the values for each particle size class may be plotted as a cumulative frequency curve (Fig. 36), from which it is possible to infer important characteristics of the sedimentary habitat sampled. The sediment type may be described by reference to its median Φ value, the value at which 50% of the sample is finer and 50% coarser, or by its proportion of coarse (0.5–1.0 Φ), medium (0.25–0.5 Φ) or fine (0.125–0.25 Φ) sand, and the degree to which it is sorted. This is estimated by the range of particle sizes in the sediment, and may be read from the graph as the range of Φ values between the 25% (first quartile) and 75% (third quartile) values. A flat curve with a large quartile deviation (QD) indicates a poorly sorted sand with a wide range of particle sizes, while a steep curve and a narrow QD is characteristic of a well-sorted sand. Phi values have usually been plotted along the axis of the graph from negative to positive, the coarsest categories thus to the left and the finest to the right, so a steep curve to the left indicates coarse, well-sorted sand, and to the right well sorted and fine. A sorting coefficient may be derived by plotting the cumulative percentage of sediment on a probability scale and reading the Φ percentile values for 16% and 84%. The graphic standard deviation coefficient is then calculated as:

$$(\Phi_{84} - \Phi_{16})/2$$

The inclusive graphic standard deviation (IGSD) is given by the equation:
$$(\Phi_{84} - \Phi_{16})/4 + (\Phi_{95} - \Phi_{15})/6.6$$

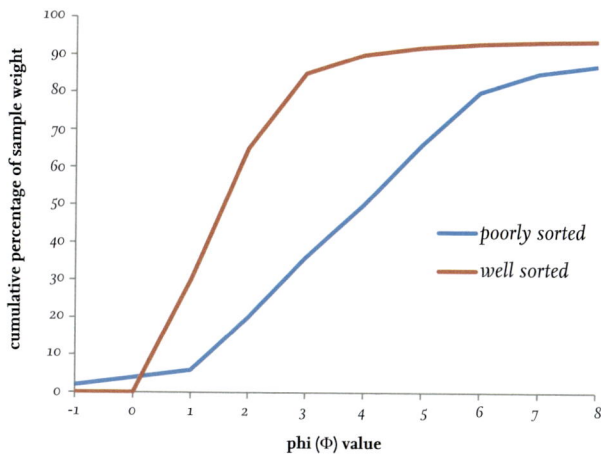

FIG 36. Grain size frequency distribution, plotted as cumulative percentages of total sample weight, for a well-sorted and a poorly sorted sediment.

which covers more than 90% of the particle size distribution of the sample. The sorting classes recognised using the IGSD (Table 7) are more precisely defined than those calculated from the interquartile deviation.

Well-sorted sands are characteristic of high-energy hydrodynamic environments, with the sediment progressively coarser as current speed and wave energy increases. They have a high proportion of pore space and, except in the finest sands, are very permeable, so that they are regularly flushed with clean water, and are well oxygenated. High porosity, high permeability and constant flushing result in a very low content of silt and organic matter, with consequences for the types of infaunal assemblages such clean, well-sorted sands support. Species diversity is often low or moderate, densities of individual species are similarly modest, and most are suspension feeders, relying on food particles filtered from the overlying water column; deposit-feeding species are usually small and few, or absent, but small predators may be frequent.

The invertebrate macrofauna of marine sediments includes representative species of most phyla. In northwest European shelf habitats there are no strictly infaunal sponges or brachiopods, while ctenophores (comb jellies) and chaetognaths (arrow worms) are everywhere pelagic (with one exception – the benthic chaetognath *Spadella cephaloptera*), but there is an enormous variety of infaunal polychaetes, crustaceans, molluscs and echinoderms. For some phyla, such as echiurans (spoon worms), hemichordates (acorn worms) and sipunculans (unbelievably, 'peanut worms') burrow dwelling is the usual life habit, while hydrozoans and tunicates (sea squirts) are mostly epifaunal organisms with a few interesting infaunal species. The epifauna is perhaps less diverse than the infauna, but its biomass is usually very much greater. For much of the shallow shelf the epifauna is dominated by echinoderms, especially large starfish, and

TABLE 7. Sediment sorting categories defined by the IGSD. (After Gray & Elliott, 2009)

Φ values	Categories
< 0.35	Very well sorted
0.35–0.50	Well sorted
0.51–0.70	Moderately well sorted
0.71–1.00	Moderately sorted
1.01–2.00	Poorly sorted
2.01–4.00	Very poorly sorted
> 4.00	Extremely poorly sorted

large species of gastropod, bivalve and crustacean, and almost all of them are free-moving scavengers and carnivores. The meiofauna consists largely of nematode worms and harpacticoid copepods, but also includes a number of obscure phyla, in particular gastrotrichs, kinorhynchs and loriciferans, which consist entirely of interstitial, infaunal species, and is augmented during the reproductive season by the smallest juvenile stages of many macrofaunal species.

THE INFAUNA

The structure of a benthic community in a habitat of well-sorted sand was revealed by a survey off Roscoff, on the coast of Brittany, where sediments consisted of 30% by weight of very coarse sand, 0 to 1.0 Φ, and 35.6% of granules, –1.0 to –2.0 Φ, with an organic content of just 0.14%. The macrofaunal community consisted of 191 species, among which polychaetes and bivalves were predominant. Polychaetes were overwhelmingly most abundant, comprising more than 50% of the total number of individuals recorded, and the ten most abundant species were all polychaetes, with densities ranging from 5.5 to 24.6 individuals per square metre. However, these were mostly small, free-burrowing species, and they together contributed only 2.95% to the total biomass of the community. In contrast, the bivalves comprised only 25% of the total number of individuals, but 83.9% of the total biomass: one species, the Dog Cockle (Fig. 16), with a mean density of only 14 per square metre, accounted for 71% of the total recorded biomass. Seven species of bivalve comprised 90% of the mean biomass per square metre of substratum, and all were suspension feeders; carnivores were much more abundant than suspension feeders, but they were all small species of polychaete. This coarse-sediment community thus consisted of a modest diversity of large, suspension-feeding bivalves, dependent upon energy inputs from outside of the habitat, practically no deposit-feeding animals, but numerous species of small polychaete carnivores. What they might have been feeding on was suggested by a study of the infauna of linear sand banks off the coast of Belgium. These lie in parallel series, trending southwest to northeast, and may be up to 25 km long, with a width of 3–6 km and an average height of 25 m. This is a high-energy environment, the banks rippled by strong tidal currents and subject to periodic overturn by winter storms, and the upper few centimetres of sediment are probably in almost constant motion. Prevailing tidal streams create a gradient of well-sorted sediment from fine sand, 2 Φ, in the southwest to coarse sand, 1 Φ, in the northeast. The fauna of one of these banks, the Kwinte Bank, was investigated by a team of Belgian marine biologists (Willems *et al.*, 1982). Ten stations were sampled along the crest of the bank on a single September day, using a van Veen grab of 0.1 m², the samples were washed through a 0.25 mm mesh sieve, unusually fine for macroinfaunal surveys, and a total of 73 species was recorded. Twelve bivalve species comprised 4.9% of the total number of individuals recorded for the ten samples, and the most frequently recorded of these, *Spisula solida*, achieved maximum densities of 97 and 63 per square metre at two of the sampling sites. Total densities of macrofauna ranged from 50 to 1,533 per 0.1 m² grab sample, which scaled up to 500–15,330/m² with a mean value of 4,910. The most abundant polychaete species, *Hesionura augeneri*, occurred at densities of up to 1,980/m² in fine sand, and to an extraordinary 7,880/m² in the coarsest sand. Polychaetes were the predominant component of the fauna, in both number of species and total number of individuals. The 41 species identified comprised from 27% to 93% of the total number of individuals recorded in the ten samples, with a mean of 82%. Published data suggested that densities of shallow benthic infauna might range from 800 to more than 1,000 individuals per square metre for fine sand habitats, and from around 600 to beyond 1,500 for coarse sands, with numbers of species ranging from 23 to 34. Mean density of individuals for the Kwinte Bank samples, at 4,910, thus seemed rather high.

Each of the ten stations was also sampled with a small box corer, which yielded sediment cores 10 cm deep, with a surface area of 10.17 cm², which were washed through a very fine sieve, with mesh size 0.038 mm. These sampled the meiofauna, which consisted principally of nematode worms and harpacticoid copepods, with a minimal representation of hydrozoans, polychaetes, ostracods and halacarid mites. Nematodes were the majority, with 136 species identified, an average density of 384 individuals per core and constituting an average 60% of the total individuals per sample. The 66 species of harpacticoid copepod occurred at an average density of 366 individuals and represented a mean 33.4% of individuals per sample. Each core contained 20–50 species of nematode and from 5–37 copepod species. These are apparently enormous densities, but both nematodes and harpacticoid copepods had been recorded in much higher numbers in other studies, the former at up to 8,750, mean 1,650, and the latter at up to 244, for 10 × 10 cm² cores. The density of the Kwinte Bank meiofaunal community was thus judged to be on the low side for well-sorted, and thus well-aerated, sand. As in the case of the Roscoff infaunal community, the Kwinte Bank macrofauna consisted of small numbers of individually large, filter-feeding clams, and extremely large numbers of small carnivorous and omnivorous polychaetes. Although figures were not provided, it is reasonable to infer that the low numbers of suspension feeders comprised the bulk of the macrofaunal biomass, with the bivalve *Spisula solida* probably representing the greatest proportion. The meiofaunal component, largely nematode and copepod detritivores, provided food for the polychaete carnivores, and the infaunal food web was supported by low inputs of phytoplankton detritus.

Formerly, both the Roscoff and Kwinte Bank benthic communities would have been described as a 'boreal offshore sand association', or a '*Spatangus/Venus fasciatus* community'. Both inhabited current-swept, well-sorted sands, with very low organic content. The Roscoff sediment seems to have been rather coarser than any of the Kwinte Bank sediment samples. No further comparisons should be made between the two benthic communities, as different methodologies were employed in the two studies, but some interesting contrasts might be explored. At Roscoff 191 macrofaunal species were recorded, while the Kwinte Bank macrofauna totalled just 71. However, while all of the Kwinte Bank samples were collected on a single day, the Roscoff site was sampled at quarterly intervals over four years, and recorded species diversity tends to rise with repeated sampling (Fig. 34). Mean density of individuals was higher in the Kwinte Bank samples, but these were sieved through a 0.25 mm mesh, while the smallest mesh employed in the Roscoff study was 1 mm. Sieving the Kwinte Bank samples at 1 mm would probably have given the impression of a very much poorer, less diverse and less dense, community than at Roscoff, while use of a meiofaunal sieve at the latter site would have revealed a much richer infaunal community. However, while numbers of species and individuals increase substantially with the use of finer sieves, total biomass increases only minimally, as the additional species are all individually small.

In high-energy environments the nature of the benthic community is largely determined by physical factors. The upper few centimetres of sand may be constantly disturbed; the infauna consists essentially of free-living animals, able to rebury swiftly if dislodged, and permanent burrows and tubes cannot be maintained. Low organic content supports only the smallest detritivores, and only suspension feeding can support a significant biomass. As the effects of tidal scour and wave energy lessen offshore, and with increasing water depth, physical factors have less influence on benthic community structures, which are instead modulated by biological interactions. Sediments are less mobile, fine material accumulates, and the benthic environment becomes relatively stable. The infauna includes many large burrowing organisms constructing labyrinths of permanent tunnels, and tube builders occupying and maintaining deep, substantial tubes. Organic material settling from the water column is augmented by faecal material, mucus and metabolic by-products of the fauna, and a wider range of feeding modes is apparent. Deposit feeding becomes a common trophic mode, both within the sediment and on the surface; some large filter-feeding species are disadvantaged by the loose suspension of finer material created by deposit feeders, but other suspension-feeding modes appear as some tube dwellers utilise respiratory currents to collect food. Particle size analysis of the sand fraction is still useful in characterising the infaunal habitat, in particular the sorting coefficient, but parameters such as the percentage of silt and clay minerals, and the organic content of the sediment, are equally important indicators of both the habitat and its infaunal community.

Predominant among the burrowing infauna are decapod crustaceans, which are often individually large, and generally comprise the bulk of the megafauna, and some species of which occur in dense populations. The Norway Lobster, or Dublin Bay Prawn, *Nephrops norvegicus*, ignominiously 'scampi' (Fig. 37), may achieve a length of around 24 cm from claw tip to tail, and excavates substantial burrows, a metre or more in length with a diameter of about 10 cm, in the top 30 cm of sediment. It is most common in fine muds, in deep offshore waters, or in sheltered, low-energy environments, such as areas of the Clyde Sea and the Minch, western Scotland. *Nephrops norvegicus* populations commonly occur at densities of one individual per square metre over large areas of sea floor. The

FIG 37. The Norway Lobster, *Nephrops norvegicus*. (J. R. Ellis (CEFAS))

Angular Crab (Fig. 38) is another large decapod miner, with a rectangular carapace up to 3 cm broad, and the males with slim chelipeds – claws – up to 10 cm long. It inhabits similar habitats to those of *Nephrops*, but occurs also in sandy sediments in much shallower water, and is also a shallow burrower, constructing horizontal, straight or branching tunnels up to 40 cm long.

The greatest diversity of burrowing megafaunal crustaceans is seen in the thalassinidean decapods, a group of animals distributed in shelf environments worldwide, and generally referred to as 'mud shrimps' (Atkinson & Taylor, 2005). There are nine species of mud shrimp recorded from the northwest European shelf, four of which are common and widespread, and one of which, *Jaxea nocturnea*, is perhaps more common than is apparent from published records. It is a slender animal, with a maximum total length of 6 cm, which burrows deeply into sandy muds. It has been recorded from the Clyde Sea, from the Irish Sea and western Ireland, and from localities southwards to the Mediterranean, but it is possibly under-recorded as it burrows to at least 90 cm into the sediment,

FIG 38. The flamboyant male Angular Crab, *Goneplax rhomboides*. (J. R. Ellis (CEFAS))

beyond the depth sampled by conventional grabs. The four common mud shrimps, *Calocaris macandreae*, *Callianassa subterranea*, *Upogebia deltaura* and *U. stellatus*, are found throughout the British sea area. They are the most abundant of the burrowing crustacean megafauna, and show interesting contrasts in their distribution and lifestyles. *Calocaris macandreae* (Fig. 39), which has a maximum total length of 5 cm, is the most widely distributed of them (Pinn & Atkinson, 2010). It burrows in fine sandy muds, with a minimum 20% by weight silt/clay content, at depths of 15 to more than 1000 m. Its burrows consist of branching galleries at two distinct horizons in the sediment, at 10 cm and 20 cm beneath the surface. At the lower level the shrimp excavates semicircular passages, while the branching upper galleries typically have triple junctions marked by surface openings in groups of three; a single burrow system, sheltering just a single shrimp, has been found to have as many as 17 openings to the surface. *Calocaris macandreae* has several different feeding modes, and it has been suggested that it is this adaptability that enables it to occupy such a broad range of depths. It is primarily a deposit feeder, ingesting organic matter from within the sediment, but it is also known to feed on polychaetes and small bivalves, though whether as an active predator or as an opportunistic scavenger is not clear. In laboratory experiments *C. macandreae* has been observed to consume dead polychaetes and bivalve tissue offered to it, but also to cache such food items, burying them within

FIG 39. The mud shrimp *Calocaris macandreae*. (J. E. Lancaster)

the walls of its burrow. This odd behaviour is difficult to explain, but the shrimp has also been noted to mine anoxic, sulphide-rich sediment and redistribute it through its burrow system. These two curious activities perhaps serve to stimulate bacterial population growth, and suggest that *C. macandreae* may be a microbial gardener, the bacterial production providing a further food source for the shrimp.

Callianassa subterranea is similar in size to *Calocaris macandreae*, and is perhaps the most abundant of the common mud shrimp species (Rowden & Jones, 1994). It has been recorded at densities as high as 65 individuals per square metre, and over much of the shallow southern North Sea population densities have been found to range from 38 to 59/m². It has a narrower bathymetric range than *C. macandreae*, from the low intertidal to around 100 m, and a more restricted geographical range to the south, reaching the Mediterranean but not the West African shelf. *Callianassa subterranea* favours muddy sands, constructing branching, horizontal passages deep within the sediment, with few vertical shafts to the surface. It is a deposit feeder, stripping the organic material coating sediment particles, and selects sand grains of 12–30 μm (0.012–0.03 mm) diameter, which, being so small, have the maximum surface area to volume ratio, and thus potentially the greatest bacterial coating.

Upogebia deltaura is the largest of the northwest European mud shrimps, with a maximum total length of 15 cm, and its burrow structure and lifestyle contrast strikingly with the two species previously discussed (Howe et al., 2004). This heavyweight mud shrimp is a suspension feeder, filtering food from an inhalant water stream driven by its beating abdominal legs, or pleopods. It builds a U-shaped burrow, generally with single inhalant and exhalant openings. In contrast to *C. subterranea*, *Upogebia deltaura* maintains a permanent burrow, plastering and stabilising its walls with mucus secretions, and drawing through it a constant flow of water, at a rate equivalent to about 150 ml per hour. The burrows of the deposit-feeding *C. subterranea* are only semi-permanent, their extent changing constantly as the shrimp excavates fresh sediment, and the through flow of water is little more than 50 ml per hour. The burrow of *U. deltaura* is constantly flushed with well-oxygenated water, and bacterial populations flourish, but whether these provide an additional food source for the shrimp is unknown. *Upogebia stellata*, the last of the four common mud shrimps, is much smaller than *U. deltaura*, with a maximum total length of 5 cm. The two species may occur together in low intertidal sands; *U. stellata* is limited to coastal habitats, shallower than 40 m, but ranges into estuaries, may be common in seagrass beds, and seems to favour finer sediment than its larger congener.

The two common species of heart urchin in northwest European shelf environments also show contrasting lifestyles (Hollertz, 2002). *Echinocardium cordatum* (Fig. 21a) is the more familiar of the two; it can be found low in the intertidal on sheltered sandy beaches where its dried test, devoid of its golden pelage of fine spines and appropriately described as the 'sea potato', is often found along the strandline. It is a bulky animal, with a maximum length of around 6 cm; it ranges offshore, to the shelf edge, but is most abundant in coastal waters, burrowed into moderately sorted sands, fine to coarse in grade but with a low silt/clay content. *Brissopsis lyrifera* (Fig. 21b) is similarly sized to *E. cordatum*, but with sparser, duller spines; it occurs from the shallow subtidal downwards, usually below 10 m depth, burrowing in mixed muddy sediments, and its population density often exceeds 50 individuals per square metre over much of its range. Both of these heart urchins are deposit feeders that select nutrient-rich particles from the surrounding sediment; the elongate tube feet that maintain the vertical respiratory shaft to the surface are employed secondarily to sweep particles from around the rim of the surface opening into the inhalant current, providing an additional source of food. *Echinocardium cordatum* may burrow to 20 cm depth, and its respiratory current, which has a recorded flow rate as high as 90 ml per hour, is sufficiently strong to supply a significant proportion of its food requirements as suspended material. While *E. cordatum* is thus a facultative suspension feeder, *B. lyrifera* is able to create a flow rate of only around 11 ml per hour, too low to provide a significant additional food source. However, *B. lyrifera* does not burrow as deeply as *E. cordatum*, usually no deeper than 10 cm, and laboratory experiments have suggested that it lives mostly within the top 5 cm of sediment, and may adjust its depth according to the food resources available to it. When phytoplankton and macroalgal detritus accumulate on the surface, *B. lyrifera* will emerge from its burrow to feed, burrowing again once these surface supplies are depleted.

The Cnidaria includes several large-sized infaunal species that may be locally abundant in special habitats. Three species of sea pen, the fleshy *Pennatula phosphorea* (Fig. 40) that grows to a length of 25 cm, the wand-like *Virgularia mirabilis*, to 60 cm, and *Funiculina quadrangularis*, which may reach 2 m in length, inhabit vertical tubes in fine muds, into which they withdraw swiftly on disturbance. They are found in deep waters offshore, but also as disjunct populations in the deep still environments of Scottish sea lochs. *Cerianthus lloydii* (Fig. 41) is a stout-bodied cnidarian up to 15 cm long, with a crown of 70 slender tentacles; it resembles a sea anemone, order Actiniaria, but belongs to a different taxonomic group, the tube anemones, order Ceriantharia, which are characterised especially by the inner cycles of tentacles being very much shorter than those of

FIG 40. Small specimens of the sea pen *Pennatula phosphorea*.

the outer cycles. It creates a permanent tube, in mixed, muddy coastal sediments which it lines with soft felt-like material.

Infaunal polychaete species probably outnumber decapod species in most unconsolidated sediments, and in the low-energy environments of mixed, muddy deposits they display the greatest variety of feeding mode and life habit.

FIG 41. The tube anemone *Cerianthus lloydii*. (J. S. Porter)

Deposit feeders include the bamboo worms, species of the family Maldanidae; these are large worms, some species exceeding 15 cm length, living head-downwards in permanent mucus-bound tubes. The bizarre Parchment Worm, *Chaetopterus variopedatus*, is another sedentary tube dweller, up to 25 cm long, occupying a U-shaped tube of layered, toughened mucus, but it is a suspension feeder, spinning a mucus net within its tube to trap fine particles carried in its respiratory current. *Lagis koreni*, a tubicolous deposit feeder, constructs a rigid, tapered tube of mucus-cemented sand grains, but it is free-living rather than sedentary, digging head-first deep into the sediment to feed. Many polychaete species occupy non-permanent, mucus-lined galleries in the top few centimetres of sediment, which probably serve to maintain a respiratory current while the animal moves through the sediment; some are particulate deposit feeders, while others are scavengers or carnivores. Most of the active carnivores, however, do not maintain tubes or burrows but instead tunnel constantly through the sediment as they forage. Most familiar of these are the fast-moving species of Glyceridae (Fig. 42) and Nephtyidae, all robust worms 10–35 cm in length.

The spoon worm *Maxmuelleria lankesteri* is a large, unsegmented worm belonging to the enigmatic phylum Echiura. It has a soft, sacciform body up to 30 cm in length, and a retractile proboscis almost as long; fully contracted, the animal has a total length of around 20 cm. It burrows deeply, to 80 cm, in fine muddy sediments, with minimum silt content of 25%, and feeds by extending its proboscis from the burrow entrance to collect fine material from the sediment surface. *Maxmuelleria lankesteri* was thought to be a rather rare animal. It was discovered in muddy sands off the Isle of Man in the late nineteenth century, and subsequently collected on a few occasions from elsewhere in the Irish Sea, and in Scottish sea lochs. However, recent research suggests that it might instead be rather common, and it has now been recorded off most coasts of the British Isles (Hughes *et al.*, 1996a). Its previous apparent rarity is ascribable partly to the

FIG 42. A predatory polychaete, *Glycera fallax*. (J. S. Ryland)

depth of its burrow, deeper than the penetration depth of most grabs, and partly to its fragility, the body readily fragmenting in the process of sampling. Further, *M. lankesteri* withdraws swiftly into its burrow on disturbance and its extended proboscis is rarely seen by divers. Populations in Scottish sea-loch habitats have been found to achieve densities of around 3 per square metre, but some Irish Sea populations are now known to exceed $30/m^2$.

The sedimentary environment is continually modified by the activities of the infauna (Pearson, 2001). Material excavated by burrowers is passed to the surface; tube builders select and accumulate particles within an often narrow size range; fine material is drawn into tubes and burrows by respiratory currents, food items are selected, and rejected particles, together with faecal material, are passed back to the surface. Sediment is thus continually reworked, sorted and redistributed by infaunal communities. This is termed bioturbation: it encompasses all of the biological processes that determine the structure of infaunal habitats, and it is of critical importance to the benthic environment, as well as to overlying pelagic systems. Bioturbation has numerous functional consequences. Material settling from suspension – sedimenting – is mixed into the sediment, often to a considerable depth. The sediment is thereby aerated and decomposition of sedimented organic matter is accelerated, providing food for infaunal deposit feeders, and for microbial populations. Respiratory currents generated by burrow dwellers are another source of oxygenated water, while exhalant streams return fine sediment, enriched by passage through the gut of the animal, and faecal pellets, to the surface. These organically rich ejecta are a food source for surface deposit feeders.

Burrowing, tunnelling and tube building, by a variety of differently sized animals, create a more heterogeneous environment than those provided by undisturbed, layered sediments or current-swept sands. Environmental heterogeneity promotes greater species diversity, and greater community stability. However, bioturbation is also a source of perturbation, or disturbance; populations are dislocated by the activities of, especially, large, shallow-burrowing species such as the heart urchin *Brissopsis lyrifera*. A moderate level of perturbation has a positive effect in preventing the dominance of one, or a few, competitively dominant species, and tends to encourage species diversity (Widdicombe & Austen, 1999). *Brissopsis lyrifera* is a bulky animal, up to 7 cm long, feeding on and within the top 5 cm of sediment, and functionally it has been described as a 'bulldozer'. The distinctive bristly polychaete *Aphrodita aculeata* (Fig. 43), with the apt vernacular name, Sea Mouse, is another such, but it is a mobile predator, specialising on other polychaetes, and ploughs its way through the top centimetre or so of sediment. Other common bioturbators include the bivalves *Ennucula*

FIG 43. The Sea Mouse, *Aphrodita aculeata*, in ventral (left) and dorsal view.

tenuis, a subsurface deposit feeder, *Abra alba*, which functions as both a surface deposit feeder and a suspension feeder, and the short-siphoned *Astarte sulcata* (Fig. 44), an obligative suspension feeder. These are all very small animals compared with *B. lyrifera* and *A. aculeata*, but at high densities their sediment shifting capabilities are significant. A feature of all of the most important bioturbators is that their population densities may show considerable spatial and temporal variation. In Loch Etive, western Scotland, *E. tenuis* populations ranged from 417 to 2,231 individuals per square metre between successive years, and in the Dutch Wadden Sea *A. alba* populations, with mean densities of 1,220–3,200

FIG 44. Shells of the infaunal bivalve *Astarte sulcata*. Scale bar: 2 cm.

individuals per square metre, had a range of zero to 8,907 over a 14-year period. Such fluctuation leads to patchy distributions, and another scale of heterogeneity, and further promotes and sustains infaunal diversity.

Infaunal communities, directly or indirectly, also modify the chemistry of their habitat, and this is the least obvious but most important functional role of bioturbating organisms. Pelagic primary production, by photosynthesising phytoplankton, is critically dependent on resources of inorganic nutrients in the form of nitrate (NO_3) and phosphate (PO_4) ions (page 26). In temperate seas these are quickly depleted during plankton blooms, utilised in the formation of organic compounds. Organic waste must then be broken down through decomposition to replenish levels of nutrient required to fuel the next bloom. This latter process is termed mineralisation or remineralisation, and refers to the decomposition of organic compounds to form inorganic ions. The constant recycling of nutrients is essential to the functioning of pelagic ecosystems, and also to the benthic ecosystems that depend upon them. Mineralisation occurs in the water column through the oxidation of organic material by bacterial metabolism, but much primary production is sedimented, as dead phytoplankton cells, zooplankton corpses, exuviae and faeces, to decompose in the benthic ecosystem. In aerated sediments nitrifying bacteria oxidise ammonia, first to nitrite and then to nitrate ions; in anoxic muds and fine sands this nitrogen cycle is reversed, as a consequence of anaerobic bacterial metabolism, and nitrate is reduced to nitrite and ammonia, which are then incorporated into chemosynthetic processes within the sediment. Burrowing and burrow irrigation supply oxygen for nitrifying microbial populations, and burrow walls provide substrates for growth and expansion of bacterial populations; inorganic nitrates are ultimately returned to pelagic production cycles, and the bacterial populations provide a regenerating food source for the infauna.

The importance of bioturbation to benthic ecological processes should be stressed. In fine-sand habitats, at 10–30 m depth in the Irish Sea, *Maxmuelleria lankesteri* occurs at densities as high as 35 per square metre (Hughes et al., 1996a). The volcano-shaped mounds of material spreading from the exhalant aperture of each individual burrow create a distinctive sea-floor topography: mounds may be up to 30 cm high, with a basal diameter of 40 cm, and a central opening through which flow pulses of fine-grained material, mixed with faecal pellets 2–4 mm long. This flow has been estimated to average 24–35 cm^3 per day, for each spoon worm, equivalent to 13 g of dry matter (Hughes et al., 1996b). Compute these figures for a population density of $35/m^2$, and the volume of sediment overturn achieved annually by just one bioturbating species must be considerable.

A comparative study of infaunal communities at three sites in the shallow southern North Sea and a deep-water locality in the Skagerrak demonstrated the influence of bioturbation on infaunal community structure (Dauwe et al., 1998). The most westerly site, Broad Fourteens, at 28 m off the Netherlands coast, was a high-energy environment, with a surface current velocity of up to 40 cm per second and a sediment of coarse sand. Pelagic production was high, and the high ratio of chlorophyll *a* to total organic carbon in the water column indicated a predominance of fresh phytoplankton, but practically no organic material was sedimented. The second site, further to the northeast and deeper, at 39 m, was situated at the Frisian Front zone between permanently mixed and seasonally stratified waters; surface current velocity was much lower and the sediment was poorly sorted and fine-grained. High spring and summer primary production was matched by a high rate of sedimentation of organic material, but with a lower proportion of fresh phytoplankton. The final shallow-water site, at 20 m, was located close to the mouths of the Elbe and the Weser, in the corner of the sea area known as the German Bight. Surface current velocity here is variable, sometimes approaching that recorded at the first site, but a clockwise gyre ensures retention of suspended material within the region. Riverine outflow contributes both sediment and nutrients; there is no seasonal thermocline, and pelagic production is high and characterised by dense algal blooms. Organic material had the highest ratio of chlorophyll *a* to total organic carbon, sedimented at such a rate that the top layer of sediment was finegrained, black and partly anoxic, with a bacterial biomass twice that of the well-mixed sediment at site 2. At the deep-water site in the Skagerrak, at 270 m, organic material settled at a rate equivalent to the mean value for the German Bight site, but it consisted of decayed material transported by lateral current flow rather than deriving from pelagic production in the overlying water column.

Faunal structure and the vertical distribution of the fauna within the sediment differed strikingly between the four sites. At the high-energy site (Fig. 45a) the numerically dominant organisms were the heart urchin *Echinocardium cordatum* and species of amphipod, both restricted to the upper

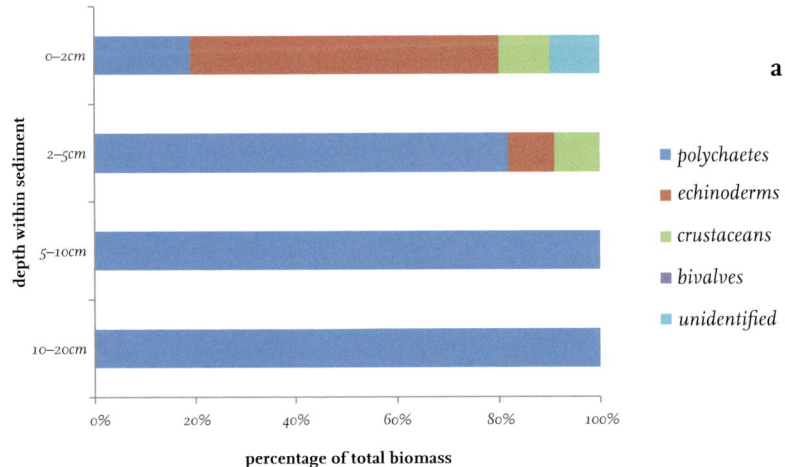

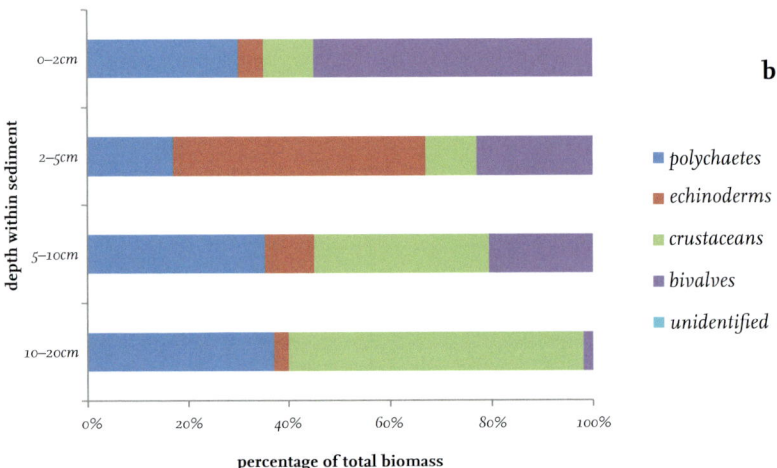

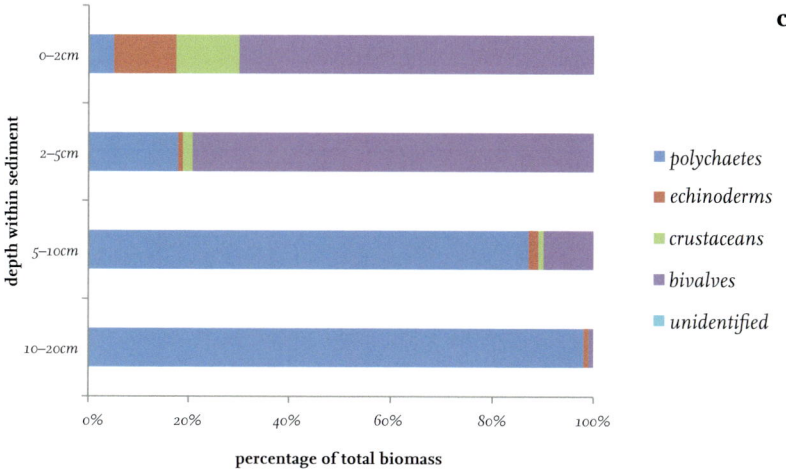

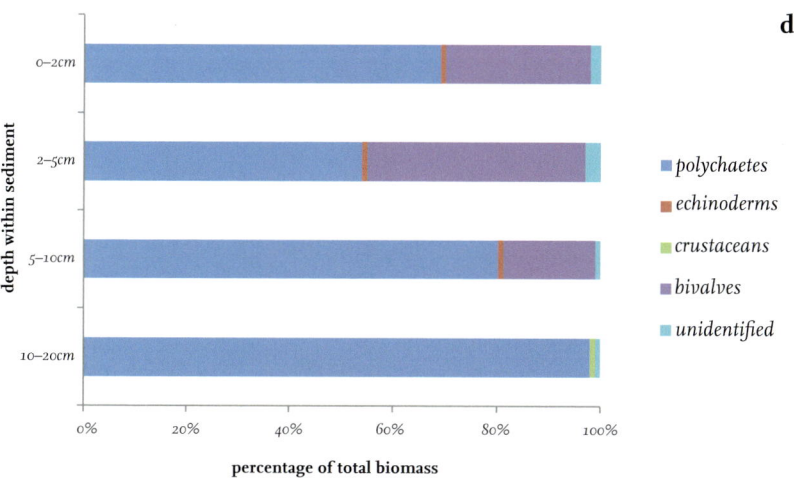

OPPOSITE AND ABOVE: **FIG 45**. Percentage of total biomass, of the four main infaunal groups, at successive levels in the sediment, at four contrasting North Sea locations. (a) Broad Fourteens; (b) Frisian Front. (c) German Bight; (d) Skagerrak. (After Dauwe et al., 1998)

layers of sediment, while small polychaetes were present down to 20 cm depth. Mean total density for the dominant fraction of this community was 2,361 individuals per square metre, and the mean biomass was 1.8 g ash-free dry weight (AFDW) per square metre; the greater part of the latter was probably attributable to *E. cordatum*, which might have been only a transitory population. The upper layers of sediment would have been maintained in almost permanent suspension by strong current flow; suspension-feeding organisms were absent simply because larval recruits could not settle, and surface deposit feeding was the most viable trophic mode. Here, bioturbation had little effect on depth distribution of the fauna, and organic content of the sediment was low.

The Frisian Front habitat (Fig. 45b) supported a diverse and stable community, with a mean total density of 3,435 individuals per square metre and a mean biomass of 13.5 g AFDW/m^2 for the ten most abundant taxa. Two bivalves, *Kurtiella* (formerly *Mysella*) *bidentata* and *Abra alba*, and the brittlestar *Amphiura filiformis*, comprised the bulk of the community, both numerically and as biomass, in the top 5 cm of sediment. Polychaetes were distributed through the entire sediment core in a more or less constant proportion; they were taxonomically the most diverse of the infauna, but the largest proportion of polychaete biomass consisted of just one species, the Parchment Worm, *Chaetopterus variopedatus*, at depths below 10 cm. The biomass of crustaceans increased with increasing depth within the sediment, and this was attributable to the deep-burrowing mud shrimps, *Callianassa subterranea* and *Upogebia stellata*; amphipods comprised 40% of the total number of crustacean species, but, as tiny animals, only 2% of the biomass. The largest burrowing worms and crustaceans mixed organic material deep into the sediment, and their respiratory currents accelerated bacterial decomposition. All trophic modes were represented in the community; surface and subsurface deposit feeding were most frequent, but suspension feeders were also present.

In the German Bight habitat (Fig. 45c), mean density of individuals for the ten most abundant species exceeded 20,000/m^2, with a mean biomass of 38.1 g AFDW/m^2. Four bivalve species, *Abra alba*, *Spisula subtruncata*, *Ensis directus* and *Nucula nitidosa*, together comprised more than 70% of the total biomass in the top 5 cm of sediment. The first three, all suspension feeders, showed a decline in biomass with increasing depth within the sediment, while the deposit-feeding *N. nitidosa* achieved maximum biomass between 2 and 5 cm depth. The polychaete tube builder *Owenia fusiformis* was the fifth most abundant species in the community, but contributed almost 25% to the total biomass. This was recorded for the 5–10 cm sediment depth horizon, a potentially confusing result as the worm moves upwards to feed from the top of the tube, extending ciliated tentacles onto the sediment surface. The bulk of the fauna fed on the freshly sedimented material, as suspension or surface deposit feeders, and the most significant bioturbation effect was the deposition of faeces and burrow ejecta by these shallowly burrowing species.

The infaunal assemblage at the deep-water site in the Skagerrak (Fig. 45d) was equally dense, with a mean 21,000 individuals per square metre for the 14 most abundant taxa, but mean biomass was only 11.7 g AFDW/m^2. Two small bivalves, *Abra nitida* and *Thyasira flexuosa*, together comprised 20% of the density and around 16% biomass. Both were limited to the top 5 cm of sediment, but while the former achieved peak biomass between 2 and 5 cm, *T. flexuosa* was almost entirely limited to the top 2 cm. Polychaetes were the dominant taxonomic group, with a much higher diversity than at the shallow-water sites. They were distributed throughout the sediment column, and biomass generally increased with depth as a result of the abundance of larger, tube-building species, especially Maldanidae and Ampharetidae. The ratio of chlorophyll *a* to total organic carbon in the water column was only 3% of that recorded for the German Bight, and the organic input to the community must have had very low nutritional value, but it was incorporated deep into the sediment, and surface and subsurface deposit feeding were the predominant trophic modes.

THE EPIFAUNA

The soft-sediment epifauna includes the most familiar benthic animals, the commercially important fish, crabs and shellfish, starfish and sea urchins, and a number of large gastropod species, as well as some rather enigmatic organisms, the affinities of which are not immediately obvious. While bottom fishing has a long history in northwest European waters, the ecology of epifaunal communities is less well understood than that of the infauna, in part because scientific sampling of the epifauna is quite difficult. Grabs collect too small a surface sample to provide meaningful estimates of the density and distribution of large epibenthic animals: some species appear to aggregate while populations of others may be widely dispersed, and in either case successive grab samples may simply fail to sample them at all. Bottom trawling is the alternative option, but trawled samples are always likely to be biased by the design of the equipment used, and its deployment, and data collected at different times and locations are not readily comparable. Commercial fishing promoted the development of large, faster trawls, to cover maximum area and thus to maximise the catch, and these are necessarily selective: small individuals of the target species, and individually small species of the epifauna, are under-represented in commercial trawls. Quantitative sampling of the epifauna requires standard-sized trawls, and standardised tow speeds and trawl duration, in order to yield data that may be comparable between sampling dates and sites, and susceptible to analysis for environmental effects on the benthic communities. Fishery science still requires large otter trawls and long haul duration to sample fish populations adequately, and fishery ecologists have developed analytical techniques appropriate to their sampling techniques. Fishery science is often viewed as a separate discipline within marine ecology, and fish will not be considered at length here, except in relation to the interactions between fish species and the rest of the benthic community. This dichotomy between fisheries science and benthic ecology is not entirely unnatural, as most bottom-dwelling or bottom-feeding fish species are migratory to a greater or lesser extent, with life cycles that often incorporate extensive movements, while the invertebrate species that form the bulk of the epibenthos, though mostly motile, are largely non-migratory.

Trawl surveys of northwest European shelf habitats show that the bulk of the soft-sediment epifauna consists of relatively few, abundant species. For example, in 1998 marine biologists from the Centre for Environment, Fisheries and Aquaculture Science (CEFAS) conducted a benthic survey of the eastern English Channel and adjacent areas of the southern North Sea, and the Bristol Channel and Irish Sea (Ellis & Rogers, 2000). They employed a 4 m beam trawl, with a 40 mm mesh net, towed for 30 minutes at each of 200 sampling stations. Each trawl tow covered an area of 15,000 m^2, and the catch of large epifaunal animals was recorded as numbers and weights per unit of time trawled. Of the mean total biomass per station, fish comprised 32.6% and echinoderms 28.7%, while the third largest proportion, 10.4%, consisted of two species of bryozoan, *Alcyonidium diaphanum* (Fig. 46) and *Flustra foliacea* (Fig. 47). Crustaceans, principally large crabs, were the next largest component, with an average 9.4% of the total biomass per station, but the next 7.0% was composed of two bulky cnidarians, the Plumose Anemone, *Metridium senile* (Fig. 80), and the soft coral *Alcyonium digitatum* (Fig. 82).

The substantial biomass of echinoderms recorded in this survey is particularly interesting (Table 8). A total of 24 species was identified, but in the western block of sampling stations the ubiquitous common starfish *Asterias rubens* (Fig. 48) accounted for 86.4% of echinoderm biomass, while at the eastern stations *A. rubens* contributed 37.4% and the brittlestar *Ophiothrix fragilis* 53.7%. The small sea urchin *Psammechinus miliaris* occurred at more than half of the

FIG 46. Colonies of the bryozoan *Alcyonidium diaphanum*, the largest c.15 cm high. (J. S. Ryland)

200 stations and ranked fourth in the maximum catch record, but comprised respectively only 3.4% and 1.0% of the echinoderm biomass at the east and west stations. The 24 echinoderm species recorded encompassed all trophic modes, suspension feeders, detritivores, scavengers and carnivores, and are of considerable ecological significance. While some species represent food for other benthic animals, the larger sea stars almost certainly compete for food with bottom-feeding fish. *Asterias rubens* and *Astropecten irregularis* (Fig. 76), in particular, consume large numbers of bivalves and might thus be considered to have an indirect economic significance. The populations of echinoderms sampled by this survey were enormous: in the northern Irish Sea *Asterias rubens* and *Ophiothrix fragilis* were trawled at rates equivalent to 1 tonne of biomass per hour; the largest single catch of *A. rubens* represented a haul of 40,000 animals per hour, and for *O. fragilis* 2.4 million per hour.

An interesting revelation provided by the CEFAS survey was the extraordinary predominance of the bryozoan *Alcyonidium diaphanum*, which would have

FIG 47. (a) Another familiar strandline object, Hornwrack, the bryozoan *Flustra foliacea*. (b) A young growing colony. (J. S. Porter)

FIG 48. (a) An abundant benthic predator, the starfish *Asterias rubens*. (b) A small individual attacking the clam *Chamelea striatula*.

TABLE 8. Echinoderms recorded in trawl surveys of the eastern English Channel, Bristol Channel and Irish Sea, as percentage occurrence (%O), and relative proportion by wet weight (%W) and number (%N), with maximum catch data for the most abundant species. Note the contrast between *Asterias rubens* and *Ophiothrix fragilis*. (From Ellis & Rogers, 2000)

Species	Eastern English Channel			Bristol Channel/Irish Sea			Maximum catch (numbers/ hour)
	%O	%W	%N	%O	%W	%N	
Antedon bifida	1.0	+	+	8.9	+	0.3	1,052
Astropecten irregularis	—	—	—	65.3	1.5	2.7	2,092
Luidia ciliaris	—	—	—	7.9	0.5	0.1	244
Luidia sarsi	—	—	—	3.0	+	0.1	546
Anseropoda placenta	13.1	+	+	7.9	0.1	0.1	350
Crossaster paposus	21.2	1.0	0.1	36.6	1.4	0.7	1,174
Solaster endeca	—	—	—	2.0	+	+	
Henricia oculata	18.2	0.1	0.1	28.8	0.1	0.1	234
Stichastrella rosea	—	—	—	3.0	+	+	
Asterias rubens	75.6	**37.4**	4.4	95.0	**86.4**	39.6	46,740
Leptasterias muelleri	—	—	—	1.0	+	+	
Marthasterias glacialis	—	—	—	10.9	0.4	0.1	470
Ophiura albida	29.3	+	0.3	22.8	+	0.3	8,690
Ophiura ophiura	22.25	0.1	0.1	70.3	2.7	42.1	114,112
Ophiocomina nigra	3.0	+	+	1.0	+	+	
Ophiothrix fragilis	23.2	**53.7**	93.0	27.7	**1.0**	11.1	2,404,320
Amphiura spp.	2.0	+	+	3.0	+	+	
Echinus esculentus	—	—	—	28.7	3.3	0.6	1,144
Psammechinus miliaris	60.6	3.4	1.7	51.5	1.0	1.4	24,360
Echinocardium cordatum	10.1	3.5	0.5	26.7	0.5	0.5	17,808
Spatangus purpureus	4.0	0.6	+	12.9	0.9	0.2	852
Cucumariidae	8.1	+	+	16.8	+	+	

comprised the greater part of the average 10.4% bryozoan biomass recorded per sample. It is familiar to strandline naturalists around most of southern Britain as fleshy, amber-coloured colonies thrown ashore in often considerable quantities following stormy interludes. They have a firm, rubbery consistency, with a lobed, knobbly or simply cylindrical form, up to 20 cm or more in length, flat or with a diameter up to 5 cm. It was described more than 200 years ago but beyond the fact that, like all bryozoans, it is a ciliary filter feeder, practically nothing is known about its biology or ecology. Its reproductive cycle is largely unknown, and the rate of growth and longevity of the substantial colonies washed ashore are quite unrecorded. Yet this enigmatic animal was present in 77–90% of the beam trawl samples in the eastern English Channel and southern North Sea, and in 53–73% of the Bristol Channel and Irish Sea trawls. In some habitats it apparently achieves an abundance unsuspected from even the largest strandings. The maximum recorded catch in the eastern CEFAS samples was equivalent to more than 1,400 kg per hour, and in the west 750 kg per hour, and that is an impressive biomass for a single bryozoan species.

Beam trawl surveys of epibenthic communities of most of the North Sea revealed similar patterns to those reported by the CEFAS study. Sampling at 152 stations in the northern part of the region, from 56° to 61° N, and at depths greater than 50 m, yielded 196 species of large epibenthic invertebrates, among which echinoderms were the most abundantly represented group (Basford et al., 1989). *Asterias rubens* and *Astropecten irregularis* (Fig. 49a,b) were the most frequently recorded of five widespread starfish species, and *Ophiothrix fragilis* was predominant among three common brittlestar species. The two bryozoans, *Alcyonidium diaphanum* and *Flustra foliacea*, were both common, and large crustaceans were abundantly represented by hermit crabs and spider crabs, especially *Pagurus bernhadus* and *Hyas coarctatus* (Fig. 49c,d), and the swimming crabs *Liocarcinus depurator* and *L. holsatus*.

A subsequent epifaunal survey, employing a 2 m beam trawl (Jennings et al., 1999), sampled quantitatively at fewer stations, 63, over a wider distributional range, 51° to 61° N, and with a shorter trawl duration, 5 minutes rather than 15–20, yet, probably as a result of greater taxonomic precision – i.e. identifying all of the smaller organisms – recorded a total of 334 species. Of 25 species of echinoderm, *Asterias rubens*, *Astropecten irregularis* and the brittlestar *Ophiura ophiura* were the most frequently recorded, in 82.5%, 65.17% and 61.9% of the samples respectively. *Alcyonidium diaphanum* and *Flustra foliacea* were also frequent, in 27% and 38.1% of the samples. The large spider crab *Hyas coarctatus* was present in 55.6% of the samples, and *Liocarcinus holsatus*, at 47.6%, was the most commonly occurring of four species of swimming crab. Hermit

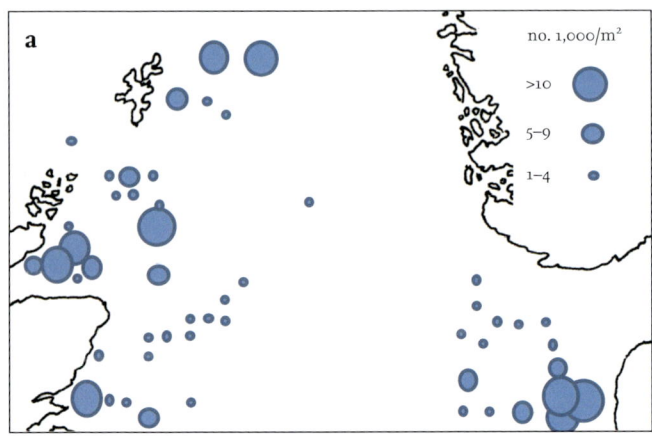

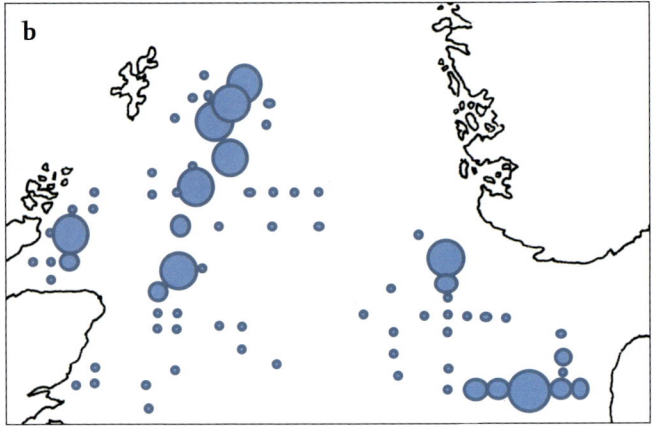

FIG 49a–b. Distribution and abundance of two common epifaunal species in the northern North Sea: the starfish (a) *Asterias rubens* and (b) *Astropecten irregularis*; (From Basford *et al.*, 1989)

crabs, in total, were probably the most abundant crustacean type found: seven species were recorded, with 79.4% occurrence for *Pagurus bernhardus*, 39.7% for *Anapagurus laevis* and 30.2% for *P. pubescens*. *Alcyonium digitatum*, with 28.4% occurrence, and *Metridium senile*, at 28.6%, were also common, but the most frequently encountered cnidarian was the sheet-encrusting hydroid, *Hydractinia echinata*, at 50.8% occurrence, a figure explained by its almost obligate association with hermit crabs (Fig. 50).

An especially extensive survey of North Sea epibenthos was conducted in the context of the International Bottom Trawl Survey (IBTS) by the five nations bordering the sea (Zühlke *et al.*, 2001). A total of 241 samples was collected, using standardised 2 m beam trawls, from 143 ICES rectangles, each measuring 0.5°

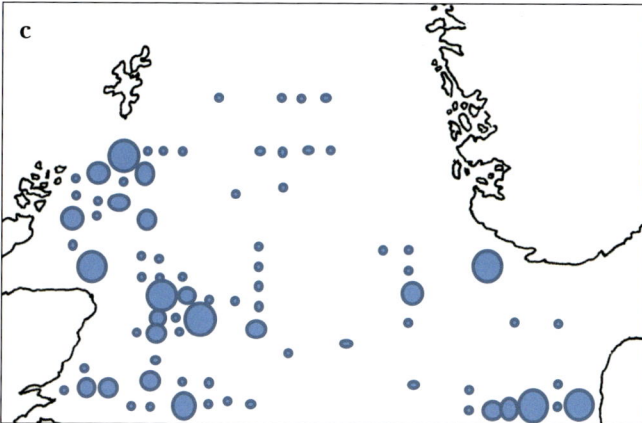
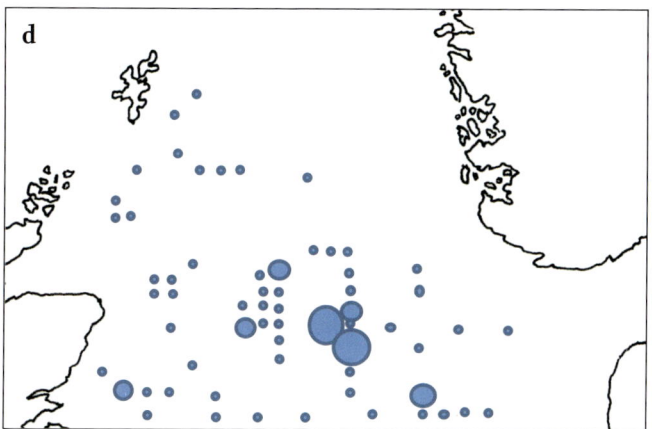

FIG 49c–d. Distribution and abundance of two common epifaunal species in the northern North Sea: (c) the hermit crab *Pagurus bernhardus*; (d) the spider crab *Hyas coarctatus*. (From Basford et al., 1989)

latitude by 1° longitude, between 51° 30' and 62° N, and 423 epibenthic invertebrate species were identified, with more or less the same dominance hierarchy noted previously. Echinoderm species totalled 21, with *Asterias rubens*, *Astropecten irregularis* and *Ophiura ophiura* most common, and the two bryozoans, the soft coral and *Hydractinia echinata* again abundantly represented. Among the crustaceans, *Hyas coarctatus* and *Liocarcinus holsatus* were still very common, with 42.3% and 50.6% occurrence, but two shrimp species, *Crangon allmani* (Fig. 51), at 53.9%, and *Pandalus montagui*, at 49%, were equally significant, while *Pagurus bernhardus* was the most frequently occurring crustacean, recorded from 88.8% of all samples.

These quantitative beam trawl surveys corroborated observations based on earlier investigations, that the epibenthic communities of the North Sea could

FIG 50. The hermit crab *Pagurus bernhardus*, with its symbiotic cnidarian *Hydractinia echinata*. (J. S. Porter)

FIG 51. The shrimp *Crangon allmani*. (J. E. Lancaster)

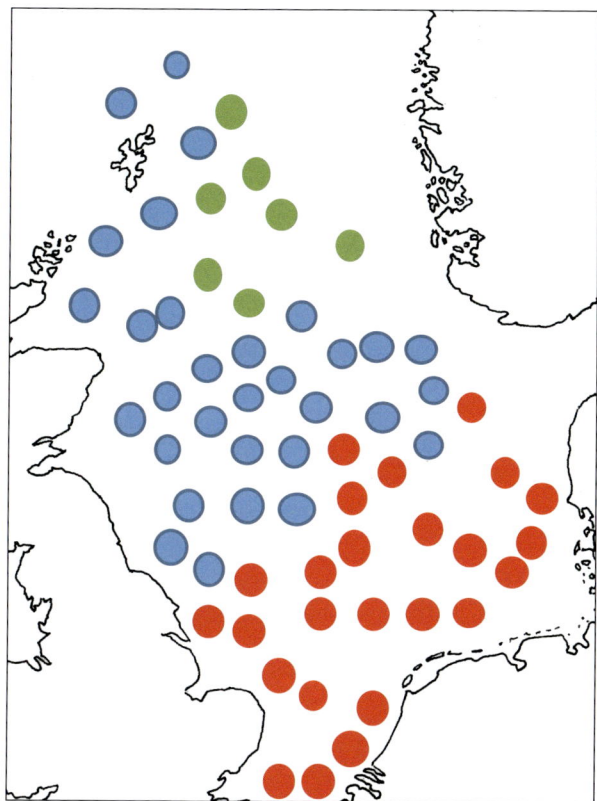

FIG 52. Occurrence of northern (green), central (blue) and southern (red) North Sea assemblages of free-living epibenthos. (After Jennings et al., 1999)

be divided into three distinct groups, the distributional boundaries of which coincided roughly with the 50 m, 100 m and 200 m isobaths (Fig. 52). Substratum is one factor that might influence diversity and composition of epifaunal communities, but the most significant factor correlated with depth is water temperature, specifically, mean bottom temperature and annual temperature range. Within the 50 m isobath these temperature parameters reflect physically unstable environments in the southern North Sea, with increasing stability northwards. The most significant consequence is that species richness increases northwards, and the proportion of individually abundant species in benthic communities also rises with increasing latitude. In the southern region the grid-based IBTS survey yielded an average 1–15 free-living epifaunal species per haul, with an additional 1–6 sessile species. In the central region the mean number of free-living taxa per haul ranged from 15 to 50, with up to 29 additional sessile species. This degree of diversity was not much greater in the deeper northern region, but the composition of the communities began to change; perhaps no species was limited to one depth zone, but the proportional representation of many changed markedly. For example, although, as noted, the hermit crab *Pagurus bernhardus* occurred in almost all of the hauls, it was most abundant in the southern and central North Sea (Fig. 53a). *Anapagurus laevis* was practically absent from the south but consistently more abundant than *P. bernhardus* beyond 50 m depth, while *P. pubescens* and *P. prideaux* (Fig. 53b) were most abundant in the northern sector. Subsequent surveys investigating diversity, community structure and biomass of the epibenthos, demersal fish assemblages and the infauna reinforce the distinction between the three divisions of the North Sea, and reveal some interesting patterns.

Sampling with both beam trawl and otter trawl at 270 stations, across 139 ICES rectangles, in 2000 yielded a total 456 epibenthic invertebrate species and 64 species of demersal fish (Fig. 54a) (Callaway *et al.*, 2002). Epibenthic

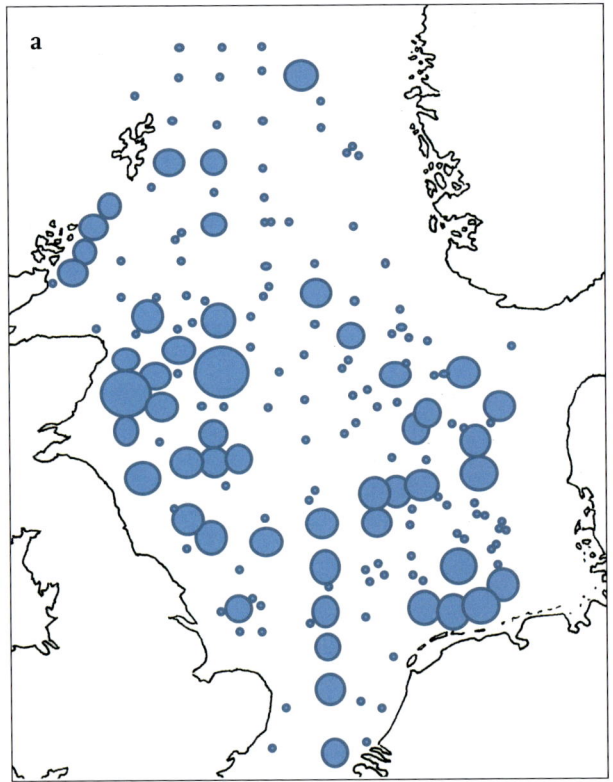

FIG 53a. Contrasting distributions of two hermit crab species recorded by the International Bottom Trawl Survey. (a) *Pagurus bernhardus* occurred in almost all samples, but was most abundant in the southern and central North Sea. (After Zühlke *et al.*, 2001)

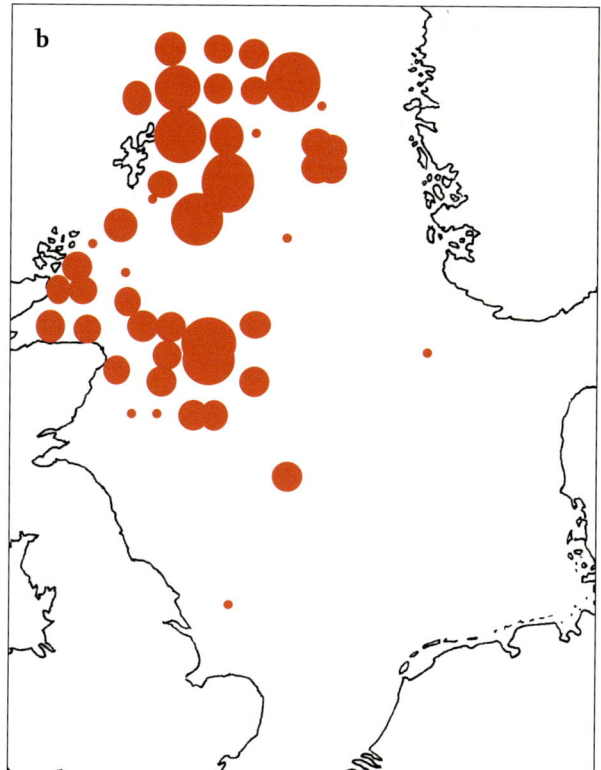

FIG 53b. Contrasting distributions of two hermit crab species recorded by the International Bottom Trawl Survey. (b) *Pagurus prideaux* was largely limited to, and most abundant in, the north. (After Zühlke *et al.*, 2001)

invertebrate diversity followed the usual gradient, from 5–19 species per station in the southern North Sea, within the 50 m isobath, to 19–58 per station towards the north, but with no difference between the < 100 m and < 200 m boundaries. An exception was an apparent transitional zone along the English coast, from the mouth of the Humber to the Wash, where diversity was as high as in the central North Sea. Invertebrate epibenthos biomass (Fig. 54b) showed no clear distributional pattern; it was greatest along the southern border of the region, from Denmark to the Netherlands, within the 50 m depth contour, and at depths greater than 50 m in the south and central North Sea. In the former areas biomass values were boosted by large populations of *Asterias rubens* and the brittlestars *Ophiura ophiura* and *O. albida*, while the maxima in the south/central region consisted largely of *Pagurus bernhardus* and *Astropecten irregularis*. High biomass values for a few stations in the northern sector were attributable to concentrations of large sea urchins, species of *Echinus*, and several species of

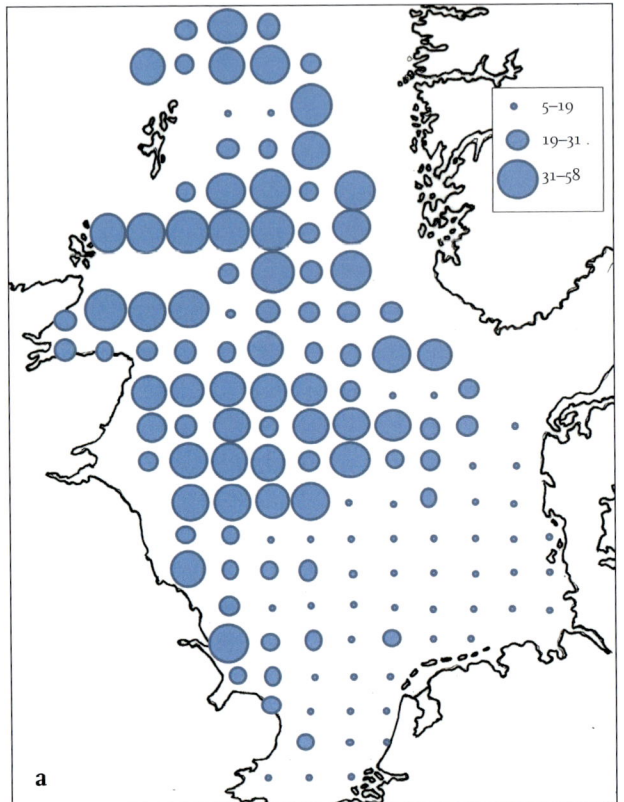

FIG 54a. Epibenthic invertebrates and fish recorded in the North Sea in 2000: numbers per ICES rectangle. (a) Epibenthic invertebrate species. (After Callaway et al., 2002)

hermit crab. Fish diversity showed contrasting patterns depending upon the sampling gear (Fig. 54c,d). The beam trawl tended to catch the smaller demersal fish, and for these samples diversity was greatest in the southern region, within the 50 m isobaths, and along the coasts of Scotland and Shetland, with 6–10 species recorded per station; it was lowest in the central region, with just 1–4 species per station. The otter trawl recorded consistently higher diversity, 16–24 species per station, in the northern region beyond the 100 m isobath, although similar diversities were found off the mouth of the Humber, the Wash and the Thames estuary, and offshore from major continental estuaries.

Distributional patterns of the most frequently occurring species, of both epibenthos and fish, accorded with those derived from previous surveys, and multivariate analysis of all species records, based on presence/absence data, demonstrated that statistically significantly distinct benthic communities

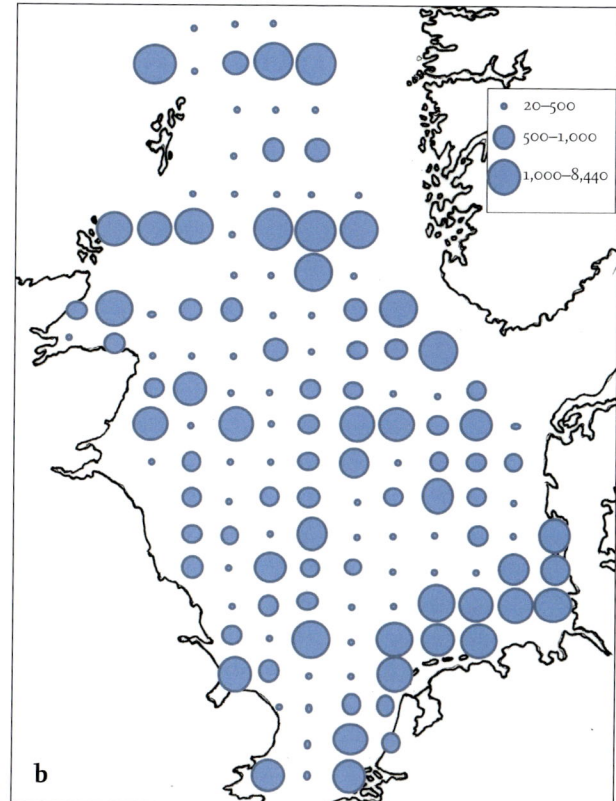

FIG 54b. Epibenthic invertebrates and fish recorded in the North Sea in 2000: numbers per ICES rectangle. (b) Biomass of epibenthic invertebrates (grams wet weight). (After Callaway et al., 2002)

could be recognised. Further, the 50 m isobath was a sharp boundary between two significantly different community groups: inshore of this boundary the epibenthos consisted primarily of free-living species, while beyond it sessile species were predominant. The most widely distributed epibenthic community in the southern North Sea was characterised by decapods, P. bernhardus, L. holsatus and Corystes cassivelaunus, and echinoderms, Asterias rubens, Astropecten irregularis, Ophiura ophiura, O. albida and Psammechinus miliaris. In the shallowest inshore waters the structure of this community changes as diversity decreases, and C. cassivelaunus and P. miliaris are much less abundant. Community structure also changes towards the Channel, where C. cassivelaunus and A. irregularis are less frequent and the shrimps Crangon crangon, C. allmani and Philoceras trispinosus are characteristic components. Although community types in the central, <100 m, and northern, < 200 m, regions were both quite distinct from those of the

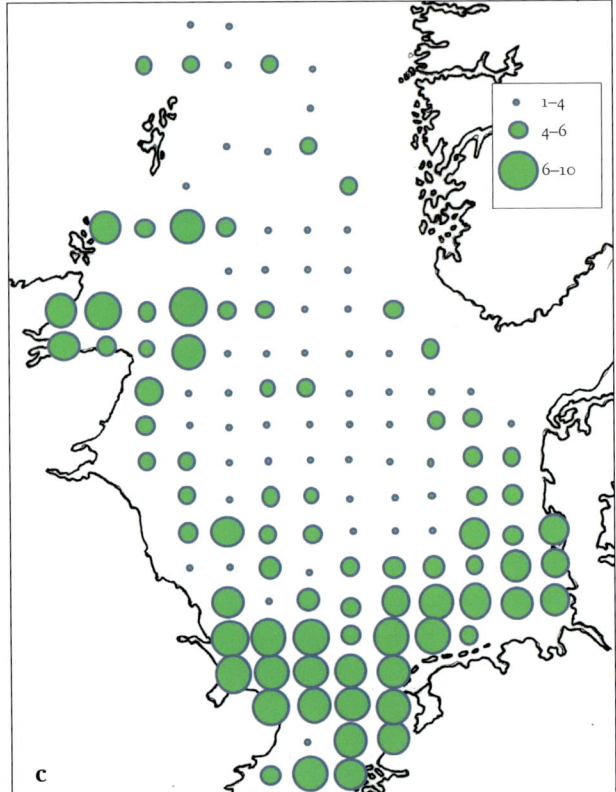

FIG 54C. Epibenthic invertebrates and fish recorded in the North Sea in 2000: numbers per ICES rectangle. (c) Fish species taken by 2 m beam trawl. (After Callaway et al., 2002)

shallow southern region, there was less difference between communities of the two deeper regions. Generally, benthic communities of the central region were characterised by numerous species not recorded within the 50 m isobath, such as the two whelks *Neptunea antiqua* (Fig. 55) and *Colus gracilis* (Fig. 56), and the decapods P. pubescens, A. laevis and H. coarctatus. Many of the species characteristic of these middle-sector communities were much less frequent beyond the 100 m isobath, while the deeper communities were especially characterised by species of *Echinus*, A. irregularis, the tubiculous polychaete *Hyalinoecia tubicula*, and a small burrowing anemone, *Hormathia digitata*.

The 50 m depth contour marked a significant boundary between demersal fish assemblages, and fish communities also differed significantly between the central and northern sectors of the North Sea. In the south, < 50 m depth, the assemblage most frequently recorded by beam trawling was characterised by small, mostly non-commercial species, including the Solenette, *Buglossidium*

FIG 54d. Epibenthic invertebrates and fish recorded in the North Sea in 2000: numbers per ICES rectangle. (d) Fish species taken by otter trawl. (After Callaway et al., 2002)

luteum, Dab, *Limanda limanda*, and Dragonet, *Callionymus lyra* (Fig. 57). The otter trawl records indicated a mosaic of two distinct assemblages, marked by particular proportions of Whiting, *Merlangius merlangus*, Grey Gurnard, *Eutrigla gurnardus*, Dab, and Scad, *Trachurus trachurus*. In the central North Sea, communities sampled by the beam trawl were characterised by abundant Dab and frequent Long Rough Dab, *Hippoglossoides platessoides*, while between 100 and 200 m *H. platessoides* and Hagfish, *Myxine glutinosa*, were characteristic. Analysis of the otter trawl samples found Haddock, *Melanogrammus aeglefinus*, Whiting, Herring, *Clupea harengus*, and Plaice, *Pleuronectes platessa*, to be characteristic of the central region communities, while Norway Pout, *Trisopterus esmarkii*, was the dominant species in the demersal fish assemblages of the northern sector.

To examine the distribution of community types in relation to environmental variables, the distributions of 280 free-living epibenthic

FIG 55. Shells of the Red Whelk, *Neptunea antiqua*. Scale bar: 5 cm.

FIG 56. Shell of the Slender Whelk, *Colus gracilis*. Scale bar: 5 cm.

invertebrates from 255 stations of the 2000 survey were analysed, together with records of 62 species of demersal fish from 316 stations sampled in the summer of 2000 and 489 infaunal species from 808 stations sampled over the period 1999–2002 (Reiss *et al.*, 2010). For all three faunal elements – epifauna, infauna and fish – the highest abundance values, as individuals per unit area or, for fish, catch per unit effort (CPUE), were found in the coastal regions of the southern North Sea, but while the epifauna and infauna were most abundant to the east, from the Netherlands into the German Bight, fish were most abundant along the eastern coast of England. In the central North Sea infauna were the most abundant faunal component and epifauna the least. Species richness of epifauna and infauna showed the now familiar gradient, increasing

FIG 57. The bottom-feeding Dragonet, *Callionymus lyra*. (J. S. Porter)

from south to north, and values for both were significantly correlated. For fish, species richness showed no clear pattern, and no correlation with either epifauna or infauna: maximum numbers of species were recorded from the Thames estuary along the east coast of England to Flamborough Head, and around the Shetland Isles. For each faunal element, community structures showed significant differences on either side of the 50 m isobath; for epifauna and infauna there were also significantly different communities between > 50 m and > 100 m depths, but there were no significant differences between fish communities for all stations deeper than 50 m. Within the 50 m depth contour, epifaunal communities and demersal fish assemblages of the eastern Channel region were significantly different from those of the southeast North Sea, but there was no difference between the infauna of these two areas, which occurred as a mosaic of distinct communities from the Dover Strait northwards along the eastern English coast. Diversity measures for the infauna and the epifauna were significantly correlated with environmental parameters, except for tidal stress and granulometry; the strongest correlations were with hydrographic variables rather than sediment, particularly temperature, salinity and depth. Fish diversity showed no significant correlation with any of the measured environmental parameters. Measures of similarity between communities also seemed to relate to hydrographic factors, for all three components of the benthos, the strongest correlation being with summer bottom temperature, stratification and salinity.

As the English Channel deepens westwards, tidal streams also increase markedly, and substratum has a significant effect on the composition and diversity of benthic assemblages, and coarse shell gravels in particular support rich epifaunal communities, including many encrusting species. A survey of the benthos of the English Channel over the years 1960 to 1962 sampled 311 stations from 1.5° E (just east of Dungeness) to 5° W (Lizard Point, Cornwall), with depths ranging from 40 m in the east to 100 m in the west (Holme, 1961, 1966). The gear employed, and appropriate for mixed, coarse substrata, was an anchor dredge, a rigid steel construction with a spade lip, that samples both epifauna and shallowly burrowed infauna. Samples were washed through a 2.2 mm sieve. The fauna recorded could not be considered entirely representative of the habitats sampled, given the methods employed, and the 207 species identified consisted mostly of shelled molluscs (130), echinoderms (35) and crustaceans (18). Species could be grouped into several categories, according to their distribution, with some, such as the Dog Cockle and the brittlestar *Ophiothrix fragilis*, common throughout the Channel, and others occurring predominantly in the east (the bryozoan *Flustra foliacea*) or the west (the brittlestar *Ophiocomina nigra*). Some were more abundant on the English side of the Channel, others were found principally on the French coasts. Tidal streams, and hence sediment grade, and depth, and hence bottom temperature, were probably important initial determinants of these distribution patterns. Much of the central Channel, the western Bay of the Seine and the Gulf of St Malo was shown to be dominated by gravel-associated communities, grading to boreal offshore sand communities (Table 1) towards the English coast, and extensive muddy sands between Portland Bill and Start Point. Examination of bryozoan species encrusting shell and other hard substrata from 102 of Holme's (1966) stations showed some significant correlations between bryozoan faunas and depth (Grant & Hayward, 1985). A total 121 species of Bryozoa was identified, from gravel associations; the number of species increased with depth, from fewer than ten at < 30 m depth, to a maximum of 62 at 90 m, and while seven species were shown to be uniformly distributed at all depths, 18 were positively associated with depth. Three distinct assemblages could be defined: a shallow-water assemblage, at mean depth 42 m, and a mean 15 species per station; an intermediate assemblage, with mean depth 51 m and a mean 30 species per station; and a deep-water assemblage at mean 86 m, with mean species number 46. There were significant associations between clusters of species, which appeared also to relate to the depth associations; for example, seven positively co-occurring species were also positively associated with the deep-water assemblage, and were considered to be at the northern limit of their geographical ranges in the western English Channel, while another cluster, of six species formerly regarded as rare, while showing no correlation with depth, occurred only on the smallest shell substrata (Fig. 58) and were all characterised by small individual colony size, and the small size of their constituent zooids.

Shelf habitats off the western coasts of the British Isles have not been surveyed as extensively as those of the North Sea, although they are now attracting increasing attention as the exploration and exploitation of benthic resources moves westwards. The total area of the western shelf is less than

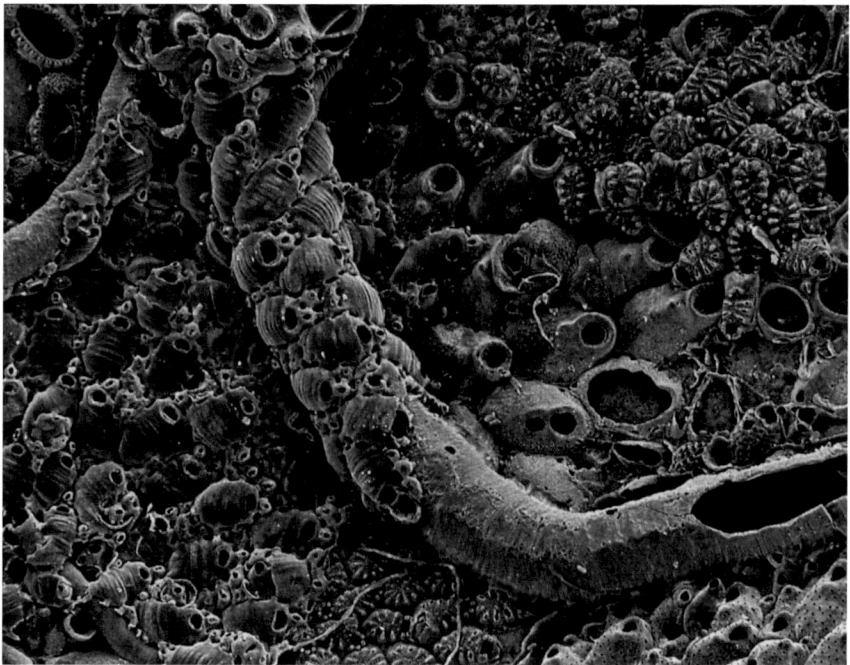

FIG 58. Scanning electron micrograph showing a cleaned, concave shell fragment, encrusted by a community of ten species of bryozoan. Scale bar: 1 mm.

half that of the epicontinental North Sea, and proportionately far less of it lies within the 50 m isobath. The summer thermocline is established across a greater proportion of it, and only the Bristol Channel and the Irish Sea are subject to the continual vertical mixing characteristic of the southern North Sea. Comparison of standardised beam trawl surveys of the southern North Sea, west Channel and Irish Sea (Rees et al., 1998) suggested an increase in species diversity from east to west, with 19–50 epifaunal taxa per sample in the east, and 46–75 in the west Channel and the Irish Sea. The reasons for these differences were not clear, but substratum seemed to be a contributory cause, as evidenced by the fact that coarse stony deposits in the eastern Irish Sea and off the coast of Dorset and the mouth of the Humber had very similar levels of diversity.

Beam trawl surveys of the western shelf, in the region of the Celtic Sea, showed that, similarly to the North Sea, epifaunal communities could be categorised into significantly distinct depth-related assemblages (Ellis et al., 2013). A total of 154 stations was sampled, from the western end of the English Channel, at 3° W, to the shelf edge, at 12° W, between 47.5° and 52.5° N, with a depth range of 35–480 m. More than 300 species of benthic invertebrate and fish were identified. These included a number of widely occurring species, present in more than 50% of the samples, such as the shrimp *Crangon allmani*, the hermit crabs *Pagurus prideaux* and *Anapagurus laevis*, and the echinoderms *Astropecten irregularis* and *Ophiura ophiura*, all familiar from the North Sea surveys. Six significantly distinct community types could be recognised (Fig. 59). Most widely distributed was an outer shelf assemblage, sampled at 67 stations, with depth range 49–175 m, in the western end of the English Channel, and from the Brittany coast northwards across the Western Approaches to the southern Irish Sea. It supported a rich benthic community, with 17–53 species (mean 34.2) recorded per haul. The most abundant animals were shrimps, *C. allmani* and *Pontophilus spinosus*, hermit crabs, *P. prideaux*, swimming crabs, *Liocarcinus depurator* and *L. holsatus*, starfish, *Asterias rubens* and *Astropecten irregularis*, and the Auger Shell, *Turritella communis*. South of 49° N, 19 stations, with depth range 132–232 m, yielded equally diverse faunas, with 16–49 (mean 31.1) species per haul, but with a distinctly different composition. While the hermit crab *P. prideaux* was the second most abundant species in these samples, the Devonshire Cup Coral, *Caryophyllia smithii*, ranked first, and the echinoderm species characterising the assemblage included several with predominantly western or southwestern distributions in the British Sea area, such as the Cushion Star, *Porania pulvillus* (Fig. 60), the Goosefoot Starfish, *Anseropoda placenta*, and the Feather Star, *Antedon bifida*. Four species of crab were especially typical of this assemblage: *Eurynome aspersa*, *Ebalia tuberosa*, *Macropodia tenuirostris* and *Macropipus tuberculatus*. Twelve inshore stations, principally off the south Wales coast, with

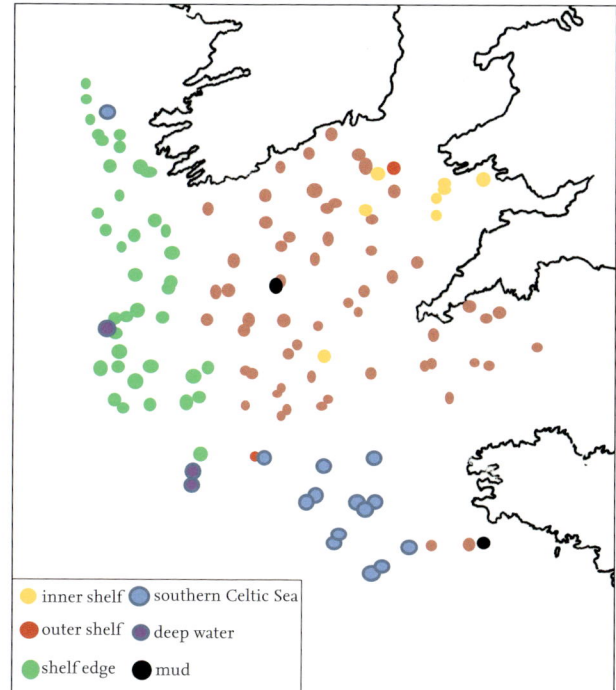

FIG 59. Epibenthic assemblages on the West European Shelf. (After Ellis et al., 2013)

FIG 60. The Cushion Star, *Porania pulvillus*.

depth range 35–130 m, yielded a much poorer fauna of 8–30 (mean 16.9) species per haul, and an assemblage dominated by *Asterias rubens*, *Ophiura ophiura*, three common hermit crab species, *Crangon allmani* and *Liocarcinus holsatus*, all common in North Sea communities. Mud assemblages were recorded at seven stations, with a depauperate fauna, 7–23 (mean 14.4 species) per haul, and *Nephrops norvegicus* and the nut shell *Nucula nitidosa* characteristic. The most westerly stations sampled were distributed in a narrow zone from 48.5° to 52.5° N, at 143–423 m depth, and revealed a shelf edge assemblage with a diversity of 9–54 (mean 30.8) species per haul. Most abundant were a sea anemone, *Actinauge richardi*, and its associated predator, *Pycnogonum littorale*, and *Caryophyllia smithii*. *Pagurus prideaux* and *Ophiura ophiura* were among the most abundant species in these samples, but *P. variabilis* was more abundant than the former, and another brittlestar, *Ophiothrix luetkeni*, ranked higher than the latter, while the sea pen *Pennatula phosphorea* and a stalked barnacle, *Scalpellum scalpellum*, were also characteristic of the assemblage. Three stations within this band, at depths of 394–480 m, and with 22–37 species recorded, were significantly distinct in composition from the rest, and could be defined as a deep-water assemblage.

POPULATIONS AND PATTERNS

Soft-sediment benthic communities are dynamic, their composition and structure determined by reproduction, migration and mortality of each constituent species, and reflecting differential responses to physical and biological stresses. The reproductive output of many benthic invertebrates may fluctuate between years, reflecting the nutritional state of the adult population: good food supplies through spring and summer result in a breeding population with high energy reserves surviving to breed the following spring; a poor food year leads to high winter mortality and a population with low reproductive reserves for the next breeding season. In other species egg production may vary little between years, but larval mortality may be high in years when sea temperatures are low and primary production depressed. Reproductive mode was formerly considered to be an especially significant factor: species with planktotrophic larval stages typically produce enormous numbers of larvae, while those that shed lecithotrophic larvae, or fully developed benthic juveniles, tend to produce far fewer offspring. In the latter cases juvenile survivorship is high, while in the former the huge output of pelagic, planktotrophic larvae is balanced by very high larval mortality. However, it is not at all clear that either case reflects reality.

While it has been generally assumed that benthic invertebrates suffer high mortality during the larval stage of their life cycles, in practice it is difficult to measure larval mortality rates. Thorson considered larval 'wastage' to be enormous, as a result of predation, offshore dispersal, and environmental stress, but it is not possible to estimate the effects of any of these factors, although their potential might be illustrated indirectly. For example, the Moon Jellyfish, *Aurelia aurita*, has a pelagic phase lasting about five months, through spring and summer. Like most scyphozoans (with the exception of the largest northeast Atlantic species, *Rhizostoma octopus*) it is carnivorous, feeding partly on holoplanktonic cladocerans and copepods, but particularly on the seasonal meroplankton, which consists largely of the larvae of benthic invertebrates. *Aurelia* has been estimated to consume 600 g wet weight of prey during its free-swimming phase, which equates to around a quarter of a million barnacle larvae, and with swarms of the jellyfish achieving densities of above 50 individuals per cubic metre in coastal waters, the loss of larval polychaetes, bivalves, crustaceans and echinoderms must be truly enormous. Over-dispersal, offshore and away from suitable habitat, might also be a serious source of larval loss, especially following prolonged stormy periods, although many species seem to have evolved reproductive strategies or larval behaviours that reduce the potential for loss.

The tubicolous polychaete *Owenia fusiformis* lives in dense populations in fine sands, and with an individual life span of about four years its populations tend to be persistent, and apparently stable. A study of *O. fusiformis* populations in the Bay of the Seine examined their structure in relation to the worm's breeding cycle (Thiébaut et al., 1994). *Owenia* spawns synchronously, and in the Bay of the Seine all populations spawn on a single day, around 12 May. The densities of *O. fusiformis* larvae recorded were impressive, ranging from 45 to almost 200,000 per cubic metre, with a record sample of one million. The larva has a pelagic phase lasting up to 28 days during which it passes through five developmental stages (Verdier-Bonnet et al., 1997). At about 17 days, halfway through the fourth stage, the larva becomes negatively buoyant, and begins to sink to the sea floor, and at day 23 the stage-5 larva is competent to settle. The cues inducing mass spawning in *O. fusiformis* are not known, but the result is that the populations in the Bay of the Seine produce their larvae at the beginning of a period of calm weather, when winds are on average west or southwesterly, and clockwise surface circulation on either side of the river mouth ensures that the pelagic larvae are retained within the bay, and above the optimum habitats occupied by the adult populations. Unseasonal weather and adverse winds during early summer in the Bay of the Seine thus might result in high mortality of larval *O. fusiformis*, carried offshore and away from optimum coastal habitat. That significant mortality in the early stages of the worm's life cycle must occur was suggested by the demographic structure of the adult populations: year class strength varied noticeably, with some populations dominated by a single year class resulting from one successful recruitment. An especially strong year class might remain the predominant proportion of the population for three years, while poor recruitment would result in a small year class that might not be detectable within the population for more than a year. Late-stage larvae lost from the Bay of the Seine are probably doomed, but it is possible that larvae dispersed early in their pelagic development might be carried into other areas of suitable habitat and either join other populations or found new colonies. Unless evidence can be found to suggest sharp genetic disjunction between similarly spaced populations of *Owenia*, intermixing through larval dispersal must be assumed to occur.

The more robust tube builder *Lagis koreni* provides an interesting contrast to *O. fusiformis* (Thiébaut et al., 1996). On the same area of the French coast this benthic polychaete has two breeding periods, in spring and summer, neither of which coincides with predictable weather conditions. It has a two-week pelagic phase during which larvae may be carried by tidal currents up to 40 nautical miles (74 km) from the originating population, and at settlement the juvenile worms are quite likely to find themselves in unsuitable habitats. However, unlike *O. fusiformis*, juvenile *L. koreni* are able to select their optimum substratum and to relocate through bottom drifting. Settled in unsuitable sediments, the juvenile *L. koreni* secrete mucus threads, up to 25 cm in length, that lift them off the bottom and into tidal currents. The density of drifting postlarval *L. koreni* has been measured at thousands per square metre of sea floor, and has been found to be greatest, by at least an order of magnitude, on flood tides, ensuring that they are carried to inshore sandy sediments.

Soft-sediment benthic communities are infinitely varied: species composition, total and relative abundance, total density and biomass all vary seasonally, annually and at longer intervals, and at many spatial scales. Populations of some species may display regular, seasonal or annual, fluctuations in density, or seem to change at apparently random, irregular intervals. Some species appear to reproduce successfully only infrequently, and the population may be sustained only for the life span of a single generation. Larval settlement and juvenile recruitment to the population are critical to its maintenance. Gravelly substrata in the Dover Strait are colonised by epifaunal assemblages, among which the brittlestar *Ophiothrix fragilis* (Fig. 61), is the predominant species, comprising more than 60% of the total community biomass, with densities of

FIG 61. The spiny-armed brittlestar *Ophiothrix fragilis*. (J. E. Lancaster)

up to 2,000 per square metre (Lefebvre & Davoult, 2000). The brittlestar has a protracted pelagic larval phase, encompassing four larval stages, and larvae occur in the local plankton from early June to late September. Total larval abundance and the relative frequency of each larval stage show great seasonal and spatial variation, and patterns and extent of dispersal vary in response to tidal currents and weather systems. Adult populations persist because they do not depend solely upon local recruitment. Early larval stages present in the plankton in June, and as late-stage larvae in September, are the product of local reproductive output, while late-stage larvae present in June and early-stage larvae appearing in August are immigrants dispersed from other populations, the breeding periods of which are progressively earlier to the southwest and later to the northeast than the Dover Strait populations.

Two species especially characteristic of muddy sand habitats across the southern North Sea are the brittlestar *Amphiura filiformis*, one of Petersen's original indicator species (Fig. 62), and the small bivalve *Kurtiella bidentata* (Fig. 63). They frequently live in association, although for neither is it an obligative mutualism. Juvenile *A. filiformis*, measuring less than 2 mm across the central disc, are predators of the meiofauna and the meiofaunal stages of infaunal deposit feeders, but adults, with disc diameters up to 10 mm, are captorial suspension feeders, selecting individual food particles from the sediment surface. They occupy branching galleries around 5 cm deep in the sediment, with the tips of

FIG 62. *Amphiura filiformis*. The slender arms of this fragile brittlestar are invariably damaged in benthic sampling. (J. R. Ellis (CEFAS))

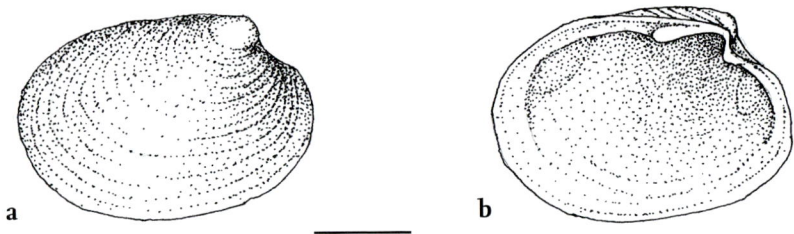

FIG 63. *Kurtiella bidentata*, a frequent associate of the brittlestar *Amphiura filiformis*. (a) Exterior of right shell valve. (b) Interior of left shell valve. Scale bar: 1 mm.

two or three arms protruding above the surface. *Kurtiella bidentata* lives within these burrows, feeding from the particle stream provided by *A. filiformis*, and both species may occur at extraordinary densities. Benthic surveys northeast of the Dogger Bank showed that these two animals were among the top five most abundant benthic invertebrate species, both with densities commonly ranging from 500 to more than 1,000 per square metre, but with different population structures (Künitzer, 1989). Size/frequency plots for the *A. filiformis* populations were bimodal, with the greatest proportion being adults of disc diameter 4–6 mm, and the secondary peak consisting of juveniles less than 2 mm diameter. The bivalve populations had the reverse structure, the bulk consisting of juveniles less than 1.5 mm shell length; in most populations, frequencies of adults, to 3 mm length, showed no consistent mode, although small peaks at around 2 mm were apparent in some samples. In these cold, seasonally layered waters growth was slow in both species, and growth rate decreased with age. The age structure of the *A. filiformis* populations might be related to density: competition for space and food might result in high juvenile mortality and thus explain the asymmetry of the frequency distributions, but the densities of the bivalve populations are correlated with those of the brittlestar and have no effect on juvenile mortality rates. An alternative explanation for the brittlestar population structure is suggested by some experimental data demonstrating that in dense populations food shortage leads to migration (Rosenberg *et al.*, 1997), and it might be that small animals are competitively inferior and simply move away. In shallower regions south of the Dogger Bank, with less mud, greater annual temperature fluctuations and no seasonal thermocline, sparse populations of *A. filiformis* may show a peak in juvenile densities following recruitment, but no stable adult component; the *K. bidentata* populations, also with much lower densities, achieve faster growth rates, and include higher proportions of larger individuals.

Amphiura filiformis populations in both sandy and muddy habitats on the west coast of Sweden showed fluctuations in density and biomass over a 12-year period, but with no particular pattern (Smith, 1995). The proportion of juveniles in the population varied most sharply in sandy habitats, with good recruitment in years when increased riverine outflow led to enhanced organic enrichment of the environment, but the populations were otherwise stable. However, benthic research in the North Sea since the last decade of the twentieth century has revealed that invertebrate assemblages can change radically and abruptly, a phenomenon that has been termed 'marine benthic regime shift', and such a shift profoundly affected populations of *A. filiformis* north of the Frisian Islands (van Nes *et al.*, 2007). Ecologists sampling benthic assemblages in the boundary zone, or front, between shallow, mixed coastal waters and deep, seasonally stratified waters to the north recorded a sharp switch in their structure. The Frisian frontal zone is characterised, as expected, by very high primary production, and consequently high sedimentation of fresh phytoplankton. *Amphiura filiformis* was the dominant component of a persistent, stable community, with densities up to 2000 per square metre, but in the late 1990s its populations collapsed, to densities scarcely equivalent to 10% of their previous levels. Concurrently, populations of the mud shrimp *Callianassa subterranea* boomed, from average densities of 50 to more than 400 per square metre (Fig. 64). The shift seems to have been

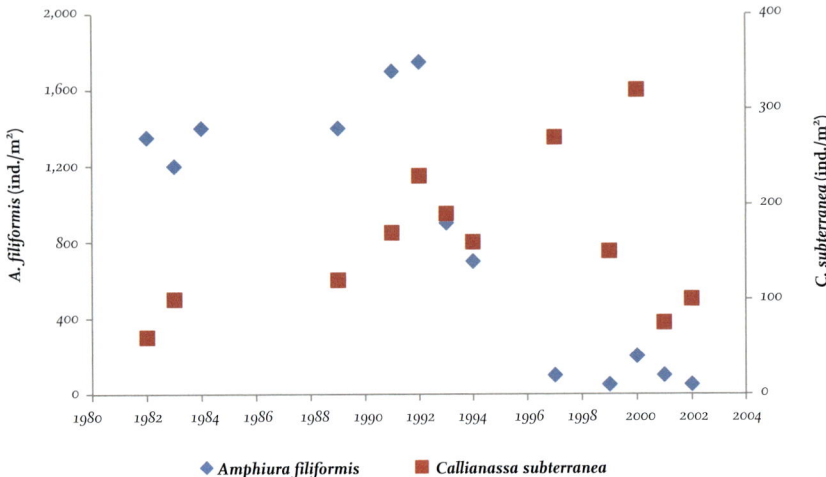

FIG 64. Average densities of *Amphiura filiformis* and *Callianassa subterranea* at the Frisian frontal zone (53° 70' N, 4° 30' E), 1982–2002. (After van Nes et al., 2007)

permanent, as 15 years later C. *subterranea* remained the dominant organism, while A. *filiformis* populations had failed to recover. The researchers proposed that the A. *filiformis* assemblage and the C. *subterranea* assemblage represented alternative stable communities, or regimes, but what provoked the abrupt switch from one to the other is not clear. The two species appear to have similar diets, but A. *filiformis* collects food particles at the sediment/water interface, while C. *subterranea* is a subsurface deposit feeder. Both are significant bioengineers, but the activities of the brittlestar act to increase sediment stability, while those of the mud shrimp result in the continual re-suspension of fine material. The feeding efficiency of A. *filiformis* might be impaired by re-suspended material, and recruits might be damaged or displaced by the burrowing mud shrimps. It is also possible that successive recruitment failure over a number of years might have resulted in an ageing and vulnerable adult brittlestar population. It is likely that the C. *subterranea* regime may prevent re-establishment of A. *filiformis* populations, but it was probably not the cause of their initial collapse. A plausible explanation of the dramatic regime shift in the area of the Frisian Front may be that intensive beam trawling simply destroyed the integrity of the shallow-burrowed brittlestar populations, enabling the deeper-burrowing C. *subterranea* to predominate. Regime shift is now recognised as occurring widely over large areas of the northwest European shelf, and may be a consequence of increasing anthropogenic disturbance.

FOOD WEBS AND ENERGY FLOW

Food for the benthos is provided by primary and secondary production in the water column, and the energy it supplies is distributed among the five basic functional groups, suspension and deposit feeders, infaunal and epifaunal predators, and the meiofauna. Categorising feeding modes is not always straightforward. Many species may function as both suspension and deposit feeders, particularly among the brittlestars, and some starfish and echinoids switch between deposit feeding and predation. Diets may be narrow or broad, in many cases unknown or difficult to determine, and a large number of benthic invertebrates appear to be omnivores. Gut contents provide evidence of the types of material ingested, but may not necessarily indicate which components of the diet are nutritionally significant. Stable isotope analysis is an effective technique for defining the types of food absorbed, but cannot definitively identify prey species. With the exception of parasites and a few instances of narrowly defined mutualism, there are no strict trophic specialisms among the benthos. The two starfish *Astropecten irregularis* and *Asterias rubens* are predators of bivalves, each consuming prey within a particular size range but neither restricted to a particular prey species. At a higher trophic level, the large starfish *Luidia ciliaris* and *L. sarsi* feed on their smaller relatives, but without predilection for any one species.

Marine food webs are extremely complex, with multiple trophic pathways between infaunal, epifaunal, hyperbenthic and pelagic communities. Food webs illustrate the multitude of links between successive trophic levels, but flow diagrams are more informative in emphasising the most important energy transfers between marine functional groups. Ecologists tend to express energy transfers in units of energy, formerly calories but now more usually joules, but for the marine biologist lacking the perspective of the theoretical ecologist, energy flow expressed in terms of units of biomass or carbon enables a simpler appreciation of the significance and magnitude of each transfer. Energy pathways from pelagic production to benthic consumption have been revised a great deal over the past few decades. Formerly, flow diagrams linking pelagic and benthic communities emphasised intervening bacterial communities, implying that primary production passed to benthic consumers through a bacterial decomposition chain. This relationship is far too simplified, and it is now accepted that energy sources for benthic suspension and deposit feeders are far more diverse. Some species may be largely reliant upon bacterial biomass, and for others bacterial production doubtless supplies a proportion of their energy requirements, but it is also apparent that for many species the most significant source of nutrition is provided by fresh, undecayed phytoplankton.

Researchers at the Dove Marine Laboratory, Newcastle upon Tyne, established a permanent sampling station at 80 m depth off the Northumberland coast in 1971, beginning the longest series of continuous sampling for any British coastal location (Frid et al., 1996, 2009). The habitat is a medium-grade, silty sand with a macroinvertebrate fauna of 100–140 species, dominated by the echinoid *Brissopsis lyrifera* and the brittlestar *Amphiura chiajei*. During the first ten years, spring and autumn samples revealed an interesting pattern in annual macrofaunal densities: low March densities were followed by high September densities and high densities the following spring, but low values were recorded in the next September and March samples (Buchanan & Moore, 1986). That seemed simple to explain: high winter mortality results in low spring densities, but good summer recruitment leads to high autumn densities; winter mortality is density-dependent, but there is high survivorship and thus high spring densities. However, there is then increased competition, resulting in poor recruitment and low autumn values that are further depressed by winter mortality. This was a straightforward biennial cycle, but it became particularly interesting when compared with spring phytoplankton cycles for the same area: the benthic cycle matched the pelagic cycle, but with a two-year lag, partly attributable to sampling procedures. A good spring bloom supported a well-fed benthic community, with a high winter survival, and an enhanced reproductive output the following spring that became apparent as 0.5 mm sieve recruits in the next spring: phytoplankton production and benthic secondary production were thus demonstrated to be tightly linked, or coupled.

Since that first ten-year programme off the Northumberland coast, bentho-pelagic coupling has been established as a significant ecological phenomenon in all temperate shallow-shelf environments, and in many soft sediment macrofaunal species cycles of growth and reproduction are closely attuned to local phytoplankton cycles. An especially convincing example was revealed by a study of the annual cycling of phytoplankton pigments in the sediments and the sediment/water interface in the Frisian Front area of the southern North Sea (Boon et al., 1998). The spring phytoplankton bloom is marked by peak concentrations in surface waters of chlorophyll *a*, the most important of the photosynthetic pigments, and correspondingly high values for bottom waters. Highest concentrations are recorded in May and June, with a lesser peak in November marking the brief autumn bloom. The most important components of the bloom are diatoms, the largest and heaviest phytoplankton cells, with consequently fast sinking rates, which are thus the principal source of undecayed chlorophyll *a*. Once sedimented, chlorophyll *a* has a half-life of nine days, and 90% of it decays within three weeks. Through late summer the chlorophyll *a* content of the sediment declines sharply while the products of decay, phaeopigments, accumulate, and the ratio of fresh to decayed chlorophyll *a* increases towards autumn. Growth and reproduction of the bivalve *Kurtiella bidentata* are phased to this cycle. Growth rate accelerates through June, reaches a peak in July, and then drops again (Fig. 65); the lipid content of the bivalves rises sharply as they accumulate reserves for gamete growth and maturation. Through August and September the biological oxygen demand of the sediment is high, and is attributable to increasing bacterial biomass following the period of fast growth and reproductive activity of the bivalve. This suggests that rather than *K. bidentata* populations depending upon bacterial communities for energy during this critical time, the reverse may be true, as the bioturbatory activities of actively feeding macrofauna aerate the sediment and allow bacterial communities to exploit the increasing content of phaeopigments.

That many benthic species phase their annual growth and reproduction to the spring bloom is apparent from the huge values for benthic biomass increase along frontal zones. The border between the Kattegat, the strait separating Denmark and Sweden, and the bight of the Skagerrak, between Denmark and Norway, is a highly productive frontal zone (Josefson & Conley, 1997). It has the characteristics of a persistent plume front, with low-salinity Baltic water (25 psu) flowing northwards into the Skagerrak, and augmented by river outflows, and

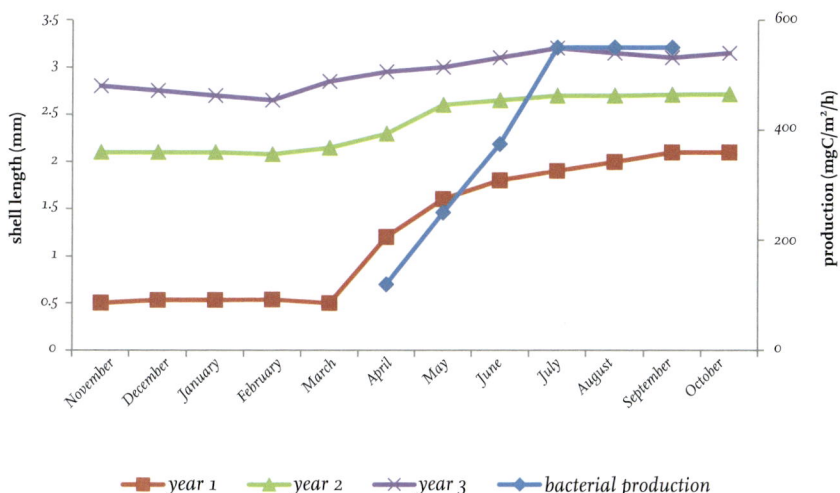

FIG 65. Growth curves for three year classes of the bivalve *Kurtiella bidentata*, and bacterial production, in a fine muddy sand sediment at the Frisian Front. (After Boon *et al.*, 1998)

high-salinity (> 32 psu) water of North Atlantic and North Sea origin flowing northeastwards along the west coast of Denmark. Transects across the frontal zone showed that benthic biomass was greatest along the front, and on either side of it decreased sharply in relation to decreasing ratios of fresh to decayed phytoplankton (Fig. 66). Here, maximum recorded densities of *Kurtiella bidentata* exceeded 10,000/m^2, and of its host, *Amphiura filiformis*, were above 3,000/m^2. Total biomass across the front varied by more than an order of magnitude, and total abundance by a factor of more than 50. The densities of *A. filiformis* and *K. bidentata* affected the distribution of functional groups across the zone, with numbers of infaunal deposit feeders increasing as the two dominant species increased, while surface deposit feeders decreased. At its greatest density, the metabolic demand of the *A. filiformis* population was estimated to be 140 g of carbon per square metre per year, compared to a primary production of 190 g per square metre per year. Density of the brittlestar population seemed to be limited eventually by the space required for each burrow, but the bivalve suffered no such limitation as the number inhabiting each host burrow rose significantly in the areas of maximum chlorophyll *a* concentration.

Deep-water habitats in the northerly limb of the Baltic, the region between Sweden and Finland known as the Gulf of Bothnia, are characterised by harsh physical environmental conditions that are subject to only small seasonal fluctuations. At 125 m depth salinity is low and constant at around 6 psu and temperature ranges from a winter minimum of 2 °C to a maximum of 5 °C in late autumn. The spring phytoplankton bloom is short, lasting just five or six weeks, and sedimentation peaks in late May or early June, approximately two weeks after the bloom. Less than one-fifth of primary production reaches the benthos, equivalent to 8.1 and 13.9 g of carbon per square metre in two years monitored by Finnish biologists (Lehtonen & Andersin, 1998). The macroinfauna at the study site consisted of just four species – a polychaete, an isopod and two amphipods – and of the latter the deposit-feeding *Monoporeia affinis* comprised 94–99% of the total faunal density and 45–99% of the total biomass. The populations of *M. affinis* show strong annual fluctuations in both biomass (Fig. 67) and density (Fig. 68); long-term fluctuations recorded between 1965 and 1995 had a period of about seven years between peaks and troughs. These cycles could not be related to fluctuations in environmental factors, such as temperature or salinity, or to hydrodynamic effects, and did not seem to correlate with observed fluctuations in phytoplankton production. Instead, population density and structure in these deep-water populations of *M. affinis* seem to be regulated by benthic food levels, and interannual fluctuations appear to reflect density-dependent mortality, probably attributable to intraspecific competition for food (Wenngren

MUD, SAND AND GRAVEL · 127

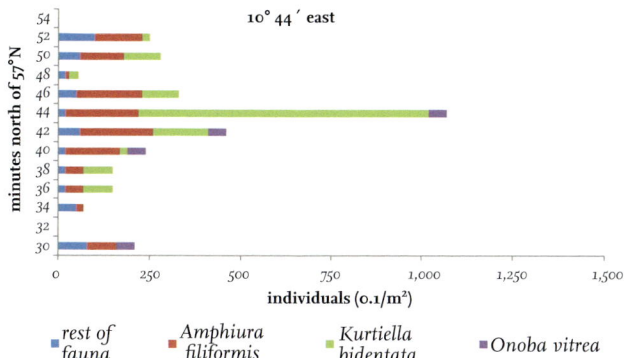
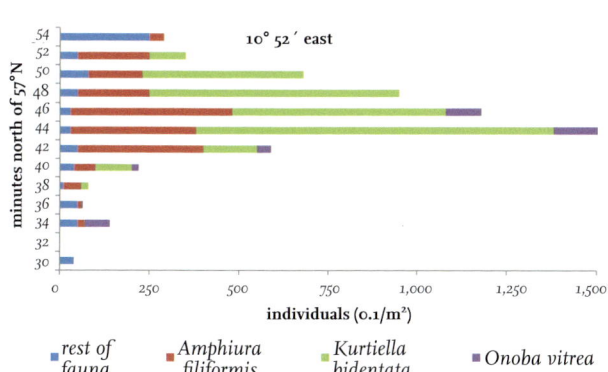
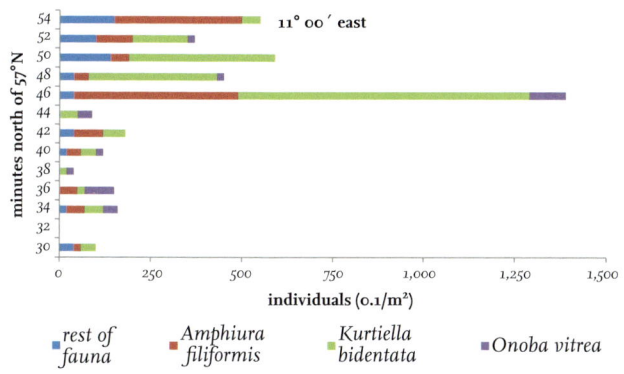

FIG 66. Macroinfaunal density along three longitudinal transects through the Skagerrak-Kattegat frontal zone. (After Josefson & Conley, 1997)

128 · SHALLOW SEAS

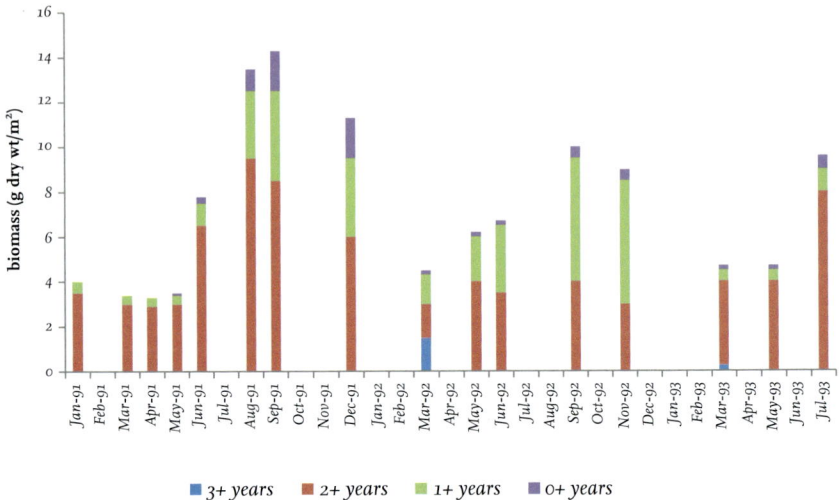

FIG 67. Biomass of a population of the amphipod *Monoporeia affinis*: proportional representation of four cohorts through three years. (After Lehtonen & Andersin, 1998)

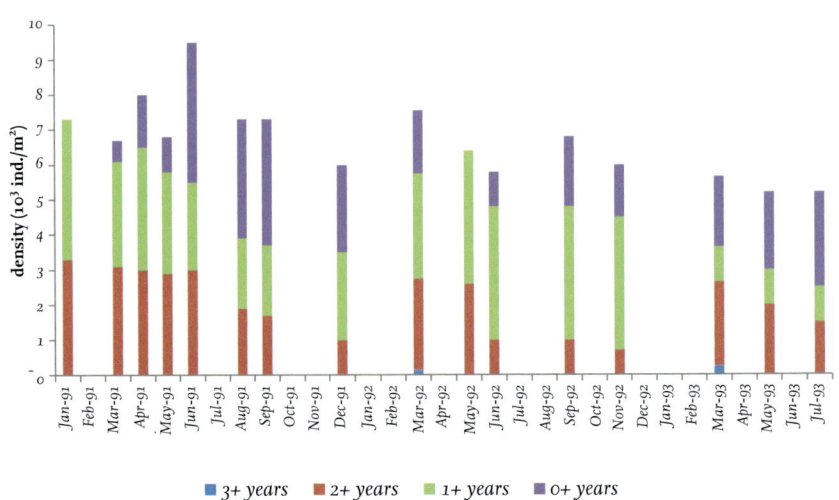

FIG 68. Population density of the amphipod *Monoporeia affinis*: proportions of four cohorts through three years. (After Lehtonen & Andersin, 1998)

& Ólafsson, 2002). *Monoporeia affinis* has a maximum life span of four years. It reaches sexual maturity in its third autumn, and it is semelparous; males die in late autumn, following summer mating, and the females after releasing a single brood of juveniles in March or April, and thus three year classes are usually apparent in the population. Growth and reproduction are both phased to the annual sedimentation, which may vary in quality, from large diatoms rich in chlorophyll *a* to small cells with low chlorophyll *a*, as well as quantity. All age classes undergo a period of rapid, almost exponential, growth between May and June, and then a long period of no, or negative, growth until the following May (Fig. 69). The brief period of rapid growth follows the peak in sedimentation as the animals accumulate lipid stores for the next cycle, and density, biomass, individual size and fecundity all vary in relation to food quality and quantity. Growth in adult *M. affinis* is density-dependent when food supplies are high, but the total biomass may decrease because individual body weight is depressed at high densities. Fecundity is related to body weight, and cohorts of small-sized adults have a reduced reproductive output. Growth of juvenile *M. affinis* is depressed at high adult densities, and juvenile mortality rises with increasing density of both adults and juveniles. The long-term fluctuations of biomass and density in *M. affinis* in these food-limited habitats in the Gulf of Bothnia appear to result from competition for food between and within age classes. As densities

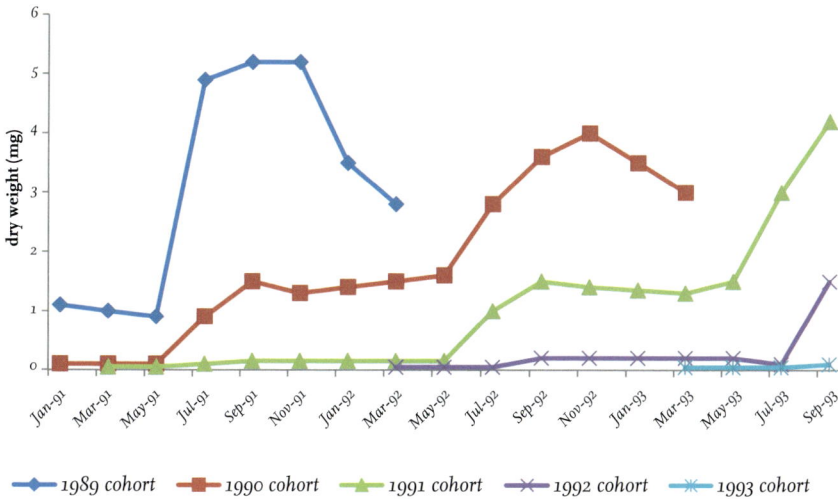

FIG 69. Mean individual dry weight of the amphipod *Monoporeia affinis*, at seasonal intervals through three years, for five cohorts. (After Lehtonen & Andersin, 1998)

peak, strong competition leads to decreased body size, lowered reproductive output and increased mortality, and declining population densities.

The coupling of benthic and pelagic processes may be even more complex than the examples considered so far. It is known that during periods of bloom some suspension feeders have the capacity to deplete phytoplankton in the lower layers of the water column, and under some circumstances phytoplankton blooms may be limited by benthic suspension feeders. In very sheltered habitats, where the water body is exchanged very slowly, over a period of several months, benthic suspension feeders will control phytoplankton biomass, affecting both the concentration of chlorophyll a in the water column, and the succession of phytoplankton species through the duration of the bloom. Especially complex interactions between phytoplankton communities and benthic suspension feeders was revealed by French ecologists in the Bay of Brest, a semi-enclosed ria on the northeast corner of Brittany, with an area of 180 km^2, but an average depth of just 8 m (Chauvaud et al., 2000). Five rivers flow into the bay carrying nutrient-rich terrestrial runoff, the nitrate concentration of which has increased tenfold through the twentieth century, to an annual input of 8,000 tonnes by 2000. Such significant eutrophication of the bay might have been expected to have dramatic effects on phytoplankton communities, increasing primary production and phytoplankton biomass, leading to oxygen depletion and changes in species composition. An increasing incidence of toxic blooms, composed of species of the dinoflagellates *Dinophysis* and *Gymnodinium*, is also characteristic of high levels of eutrophication. However, although toxic blooms began to occur sporadically from 1993, the only measurable eutrophication effect on the pelagic ecology of the Bay of Brest was a dip in the concentration of the very first spring bloom each year, and other effects appeared to be checked by the suspension-feeding benthos, the dominant species of which, ironically, was the invasive American Slipper Limpet, *Crepidula fornicata* (Fig. 70). It was first recorded in the bay in 1950, and by 1995 had colonised half of the area, and had an estimated wet weight biomass of 12,878 tonnes. Slipper limpets have a curious lifestyle. They are quite sedentary, and aggregative, forming size-graduated stacks of 10–15 individuals. The largest, invariably a female, clamps tightly to a hard surface, either a pebble or shell, or a large bivalve, especially oysters and mussels; several large females support a chain of successively smaller intersexes, and finally small males. Such stacks are stable, resisting movement by current, individuals do not disperse, and populations develop as dense patches. Slipper limpet populations flourished in every sedimentary habitat type in the Bay of Brest, from mud and sand, to shells, gravel and maerl, with profound effects on the original benthic assemblages, and the dynamics of the phytoplankton were also subtly affected. Suspension feeders were

FIG 70. The invasive American Slipper Limpet, *Crepidula fornicata*. The largest individuals in the clump are female, the smallest are male. Apertural view of a large female shown on right.

unable to exert control of the phytoplankton blooms because the tidal regime of the Bay of Brest ensured a maximum water residence time of 25 days, but that was long enough for the vigorous pumping of millions of slipper limpets to depress chlorophyll *a* concentrations during the first period of the bloom. A further consequence was that suspension feeding by *C. fornicata* led to biodeposition at up to four times the natural sedimentation rate, and continuous tidal mixing led to rapid recycling of nutrients and silica. The continuing availability of silica allowed diatom blooms to extend into late spring and early summer, delaying the usual succession to non-siliceous flagellate communities, expected in semi-enclosed eutrophic environments such as the Bay of Brest.

In the Netherlands Wadden Sea, in the 1980s, rising eutrophication promoted rising primary production, which doubled from c.150 g of carbon per square metre per year in 1970 to ca 300 g in the early 1980s (Beukema & Cadée, 1991). The biomass and productivity of the benthic consumers of the zoobenthos also doubled in response, and populations of one predominant infaunal bivalve, *Macoma balthica*, appeared to show a corresponding increase in growth rate. However, the composition of the plankton community began to change, and by 1989 diatom production had decreased noticeably, and the aggregating flagellate, *Phaeocystis pouchetii*, became a predominant element of the plankton blooms. *Macoma balthica* selects food particles from a fairly wide range of sizes, and functions as both a suspension feeder and a surface deposit feeder, and it grows faster, and builds better reserves, in years of diatom abundance. It is able to feed and grow on a diet of *P. pouchetii*, but as the latter increased in abundance the growth rate of the bivalves dropped, because at dense concentrations the small cells of the flagellate form bulky, mucilaginous colonies that are too large for *M. balthica* to ingest. Whether abundant slipper limpet populations have the potential to limit *Phaeocystis* blooms is an interesting question.

Benthic suspension feeders and particulate deposit feeders, both reliant on pelagic food sources, are the secondary producers, or primary consumers, in shelf ecosystems. Predators and scavengers constitute the next trophic level, the secondary consumers, and fall into several disparate categories. Probably a majority of meiofaunal animals are predatory, feeding upon other meiofauna, and on the eggs, larvae and meiofaunal stages of macrofaunal species. The meiofaunal predation bottleneck contributes importantly to post-settlement mortality of soft-sediment invertebrates (page 53): the permanent meiofauna includes many predatory species, particularly among the nematodes and turbellarian flatworms, that prey upon the recruits of macrofaunal species that constitute the temporary meiofauna. Meiofaunal predation is rarely quantifiable through field sampling, but was demonstrated in a study of a fine sand habitat, at 10 m depth, off the northwest coast of Italy (Danovaro et al., 1995). The meiofaunal community achieved peak summer densities of 3,463 individuals per 10 cm^2, 75% of which were nematodes, part of the permanent meiofauna, and on average 39% of these (10.6% in March to 58.6% in November) were predators (Fig. 71a). The temporary meiofauna consisted of 30 species of polychaete and six bivalve species (Fig. 71b), with combined densities comprising 1% of the total meiofauna. The spring recruitment of polychaetes was correlated with an increase in turbellarian and nematode predators, which reached peak densities as the second, summer, polychaete recruitment began (Fig. 71). About 98% of the mortality of polychaete recruits was attributed to meiofaunal predators,

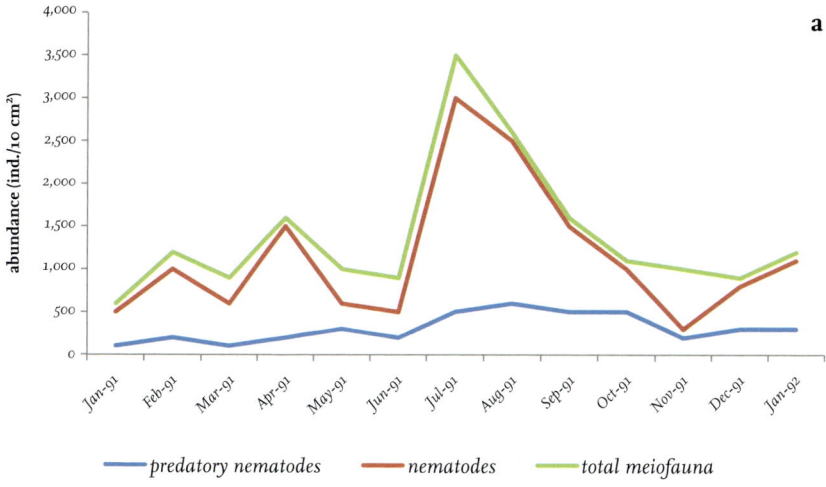

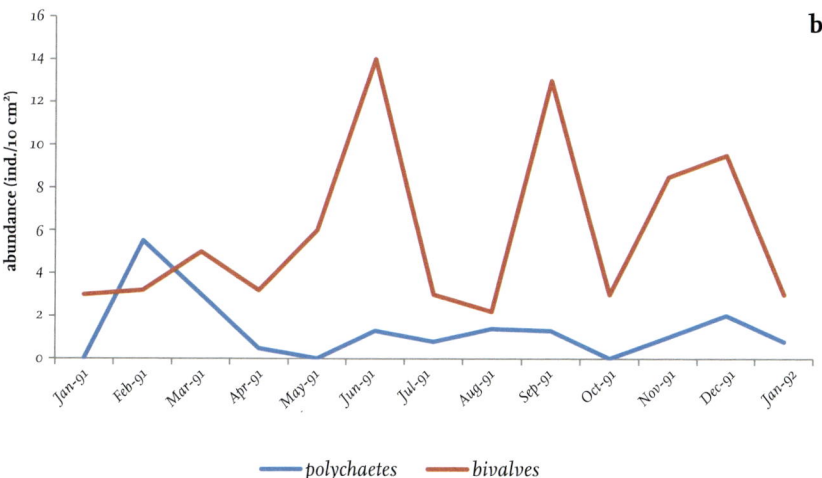

FIG 71. Seasonal abundances of (a) nematodes and total meiofauna, and (b) temporary meiofauna (polychaetes and bivalves), individuals per 10 cm², in a fine sand habitat. (After Danovaro et al., 1995)

but predation pressure seemed to lessen in late summer as juvenile nematodes switched feeding mode. In contrast, mortality rates of the bivalve temporary meiofauna were unrelated to meiofaunal predator densities, and perhaps reflected macrofaunal predation.

Macroinfaunal predators include large species of nematodes, some species of decapod, but especially large polychaetes, in particular among the Glyceridae and Nephtyidae, and for these there are some good data demonstrating their role as hunting predators. *Nephtys hombergii* grows to a length of 20 cm and is an important primary consumer in coastal and estuarine muddy sands, where it preys upon smaller polychaetes. The annual consumption of an adult *N. hombergii* has been estimated as equivalent to 300–700 g AFDW of prey, and at natural densities, in optimum habitats, of 80–100 individuals per square metre it is a significant factor regulating populations of other species of polychaete, particularly *Hediste diversicolor*, *Scoloplos armiger* and *Heteromastus filiformis*.

Infaunal predators, though motile, are resident, but many epifaunal carnivores and scavengers are peripatetic, and predation pressure on the benthos may be episodic, and often seasonal, in nature, and vary in intensity in relation to size of both predator and prey. Each size class of each predator species is likely to select and consume prey within an optimum size range. Demersal fishes provide good examples: most species spend their first few months of life in coastal nursery habitats, feeding initially on small prey items, and switching to successively larger prey as they grow, or in response to competition from other juvenile predators. O-group Plaice, juveniles in their first year of life and just 1–2 cm long, feed on postlarvae of spring-breeding polychaetes, and on the siphons of adult bivalves, later switching to newly recruited spat of inshore bivalves, especially Cockles, *Cerastoderma edule*, and the Thin Tellin, *Tellina tenuis*, which become abundant in the summer months. Juvenile gobies, *Pomatoschistus* species (Fig. 72), and Sole, *Solea solea*, have similar diets to o-group Plaice, and may switch preference from cockle spat to polychaetes in response to competition from them. Most crab predators also display prey size selection. The optimum prey size for each size class of crab is that which yields the best nutritional return for the effort entailed in cracking the shell and extracting the flesh: large crabs do not necessarily select the largest prey items available.

The Brown Shrimp, *Crangon crangon* (Fig. 73), is the most abundant benthic decapod in shallow soft-sediment habitats of the northwest European shelf, and is probably the most significant trophic link in shallow benthic food webs. Its distribution extends from the western Baltic to the Mediterranean, but it occurs most abundantly in the southern North Sea and around the coasts of England and Wales, where it is a commercially important species. *Crangon crangon* populations are mostly confined within the 20 m isobath, and most dense in proximity to large estuarine systems, including the Wash, the Thames, the Bristol Channel, Morecambe Bay and the Solway Firth, with smaller populations in smaller bays and estuaries between. Brown Shrimp are

FIG 72. Two sand gobies, *Pomatoschistus* sp. (P. E. J. Dyrynda)

FIG 73. The Brown Shrimp, *Crangon crangon*, a significant trophic stage in estuarine coastal food webs. (J. E. Lancaster)

omnivorous predators; they feed especially on polychaetes, bivalve spat and small arthropods, and are probably significant consumers of meiofaunal biomass; they may be effective predators of bivalve spat, but only during brief periods in the spring and summer. Spat larger than 2 mm in diameter are too large for the shrimps to handle and are safe from predation, which has the greatest impact when the smallest bivalve recruits coincide with the largest shrimps. The overall contribution of *C. crangon* populations to energy flow to higher trophic levels must be considerable. Brown Shrimp populations feeding on 40 km² of mudflat in Bridgwater Bay were estimated at 1–100 million individuals, which, with a mean of 10 million (Henderson *et al.*, 2006), and with a mean individual weight of 1 g, equates to a mean biomass of 10,000 kg; the booming *C. crangon* population after 1985 (page 300) comprised an estimated mean biomass 26 times higher than that figure, which would be 260 tonnes of shrimp!

Predation is a powerful biological factor affecting the abundance, composition and distribution of soft-sediment benthic communities. Baltic populations of the tellin *Macoma balthica* are limited by the amphipod *Monoporeia affinis* (page 54), and coastal populations of many species of bivalve suspension feeders are similarly constrained by predation pressure. Larvae of Cockles, Sand Gaper clams, *Mya arenaria*, and Blue Mussels, *Mytilus edulis*, settle in summer at densities measured in tens of thousands per square metre of shallow sandy habitat along the edges of the southern North Sea, coinciding with peak densities of 0-group Green Crabs, *Carcinus maenas*. In one study the juvenile crabs, with carapace width < 1 cm, were found to consume huge numbers of juvenile Cockles, shell length 4 mm, accounting for around 80% of all recruits in just two summer months. Green Crabs have their greatest impact in intertidal and shallow coastal habitats, their distribution offshore being limited by larger predators, and in subtidal habitats the most important macroinvertebrate predators are starfish. Fish and decapods are the most familiar epibenthic predators, and echinoderms do not come to mind as significant predatory components of the benthos, but starfish and brittlestars, in particular, have impacts on the composition and structure of benthic components just as striking as those resulting from predation by fish and crabs. On the Pacific coasts of North America starfish are regarded as keystone predators in rocky intertidal habitats, serving to maintain community diversity through selective predation on potentially spatially dominant sessile species. They do not fulfil the same role on temperate northwest European seashores, but in sublittoral habitats, within the 50 m isobath, they undoubtedly have a key function in structuring benthic communities. The most abundant temperate European starfish, *Asterias rubens* (Fig. 48), has been recorded in huge swarms covering many square kilometres of sea floor, at densities exceeding 15–20 per square metre, which move slowly along coastlines, presumably impelled by depleting food resources, in particular infaunal and epifaunal bivalves. Such swarms seem often to consist largely of individuals all of a similar size, perhaps a single year class, representing a single successful recruitment; but larval development, settlement and recruitment, and growth rates of starfish are all sensitive to temperature, and it is not possible to determine the age of the individual easily. New recruits, at around 1 mm diameter, may grow to 2–5 mm within two months, and may be distinguished from the previous year class, with diameters above 20 mm, but growth rates fall throughout the winter, and also vary in response to food availability; beyond two years separate year classes may not be discernible, and discontinuities in size frequency plots may represent years in which recruitment failed.

Echinoderm predation may influence the composition and structure of benthic invertebrate communities, and this was demonstrated by a study of epifaunal species in the Bay of Douarnenez, northwest Brittany, a semi-enclosed marine embayment, 20 × 15 km, floored with unconsolidated deposits, sorted by tidal currents into discrete sedimentary habitats (Guillou, 1990). Seven species were common, comprising three trophic categories, and occurring in different proportions on each sediment type. The suspension-feeding brittlestar *Ophiothrix fragilis* (Fig. 61) was restricted to mixed muddy sands, and where these graded towards muddy gravels, small numbers of another suspension feeder, *Ophiocomina nigra* (Fig. 74) were also present. Two small carnivores, the brittlestar *Ophiura ophiura* (Fig. 75) and the starfish *Astropecten irregularis* (Fig. 76), were predominant in fine-sand habitats, where they preyed upon small infaunal bivalves, while the larger *Asterias rubens* occurred across the whole spectrum of habitat types, feeding on large epifaunal or shallow-burrowed bivalves. Another large predator, *Marthasterias glacialis* (Fig. 77), was found only in muddy sands, and was most common on coarser, mixed deposits. The top predator was *Luidia*

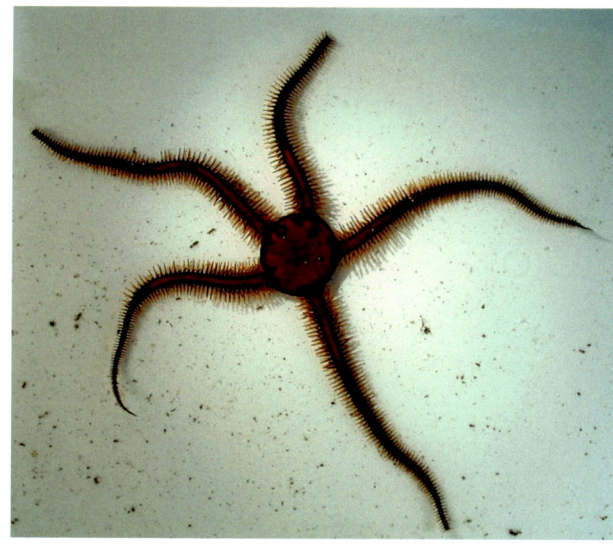

FIG 74. The suspension-feeding brittlestar *Ophiocomina nigra*. (J. E. Lancaster)

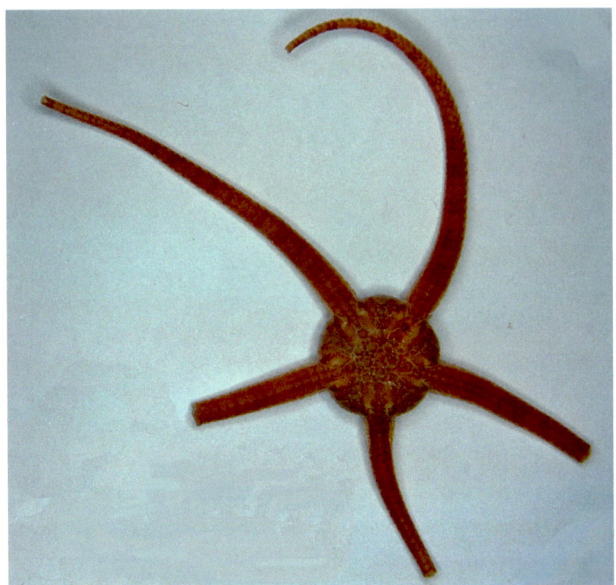

FIG 75. The brittlestar *Ophiura ophiura*, an epifaunal carnivore. (J. E. Lancaster)

FIG 76. The small starfish *Astropecten irregularis*, a predator especially of small bivalves.

FIG 77. The large, predatory Spiny Starfish, *Marthasterias glacialis*.

FIG 78. The top starfish predator, *Luidia sarsi*.

sarsi (Fig. 78), which was found only on mixed, muddy sediments, and fed upon *O. fragilis*, *A. rubens* and *M. glacialis*.

Populations were sampled every autumn over an eight-year period; total echinoderm biomass reached a peak halfway through the period, after which it dropped sharply, all species disappearing completely at one site after five years. Fine-sand habitats (Fig. 79a,b) were initially dominated by *O. ophiura* and *A. irregularis*, which fed upon small bivalves such as *Chamelea striatula*, but the proportion of the two small predators decreased and *A. rubens* became the numerical dominant. As total echinoderm biomass peaked, annual food demand was calculated as equivalent to 1.2 and 1.1 g of carbon per square metre at two fine-sand sites, against an estimated bivalve production of 0.66 and 1.05 g of carbon per square metre. *Asterias rubens* preyed upon large bivalves such as Cockles, but as densities of preferred prey decreased it began to feed on smaller

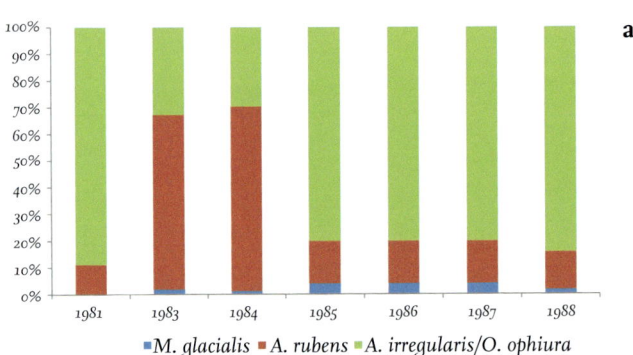

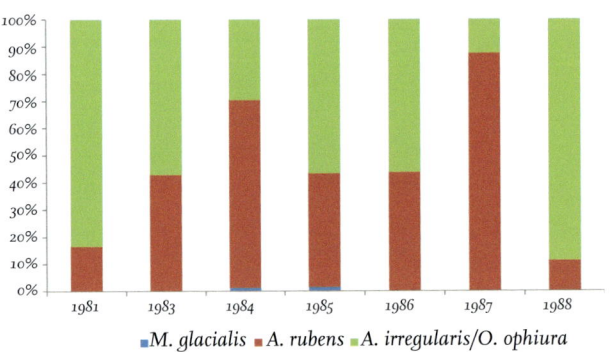

LEFT AND OPPOSITE: FIG 79. Relative densities of epifaunal ophiuroids and asteroids over an eight-year period at four soft-sediment sites. (Data from Guillou, 1990)

species, in effective competition with O. ophiura and A. irregularis. Subsequently, the dominance of the latter two species in one of the fine-sand habitats was restored as numbers of A. rubens dropped by almost 70% (Fig. 79a). This was probably a result of migration out of the area by the hungry starfish, but might also have reflected competitive pressure by the larger and more efficient predator, Marthasterias glacialis. At the second site (Fig. 79b) the proportion of A. rubens remained at around 50% of the echinoderm community after its initial increase, and only declined, to below its original level, in the final sample; here, the shortfall in bivalve production was smaller than at the first site, and competitive pressures were probably lower, but eventually O. ophiura and A. irregularis became the numerical dominants as the population of A. rubens slumped.

On muddy sand O. ophiura and A. irregularis were the numerical dominants for the first three years of the survey (Fig. 79c), but as A. rubens increased in

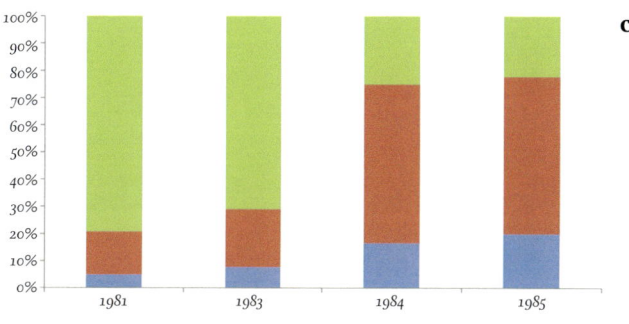

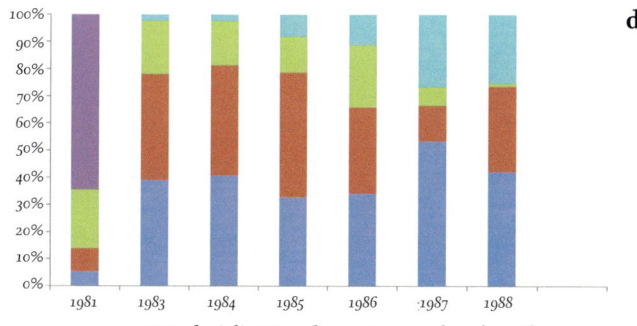

numbers, the population of *M. glacialis* also grew, and by the fifth year the two larger predators each comprised about 40% of the total number of individuals in the community. Echinoderm biomass peaked at a level three times higher than in the fine-sand habitats; this was attributable to *M. glacialis*, a large starfish, which contributed 75% of the total wet weight. Annual bivalve production was estimated at 1.57 g of carbon per square metre and food demand at 0.95 g, but the increasing biomass of large starfish outran its resources and vanished.

The mixed muddy-sand habitat (Fig. 85d) was initially dominated by a bed of *O. fragilis*, the sole suspension feeder, but this had completely gone by the second sampling occasion, perhaps dispersed following disruption of the bed's structure by physical disturbance, or even by predation. The two small predators then increased, but had decreased by the next year in the face of competition from *A. rubens* and *M. glacialis*. Bivalve production was high, at 3.04 g of carbon per square metre, while food demand of the bivalve predators was only 0.5 g C/m^2, but the most abundant bivalve was the free-living scallop *Aequipecten opercularis*; its motility provided an ability to escape predation, but also gave the competitive advantage to *M. glacialis* over the more slow-moving *A. rubens*. Populations of both starfish were limited by the top predator, *Luidia sarsi*, but their production, at 0.21 g C/m^2, was less than half of the energy requirements of the *L. sarsi* population, and total echinoderm biomass crashed.

Competition between predatory starfish was seen as one of the factors determining benthic community structure in the Bay of Douarnenez. Starfish locate their prey through chemosensation; there is some evidence that species of *Astropecten* employ the same means to select the most energetically rewarding prey items, but as they ingest them whole, choice is constrained by mouth diameter. *Asterias rubens* feeds by everting its stomach, engulfing and partially digesting the prey before ingestion. It is able to tackle larger species of bivalve than *Astropecten irregularis*, but whether or not it selects prey on the basis of size, energetic content or frequency of encounter, the choice broadens as larger species are depleted, to include the smaller species favoured by *A. irregularis*.

There are many more species of predatory fish than starfish, and some of them, under natural conditions, might perhaps occur as abundantly as *A. rubens* and have equally strong impacts on benthic communities. However, in some habitats it is probable that bottom-feeding fish might reduce inter- and intraspecific competition by partitioning food resources between species and size classes. The International Bottom Trawl Survey of the North Sea (Callaway et al., 2002; Reiss et al., 2010) recorded 64 species of fish, from 270 beam trawl samples. The total included a small number of pelagic species that feed in the water column, such as Herring, Sprat, *Sprattus sprattus*, Mackerel, *Scomber scombrus*, and

Scad, but the majority was essentially demersal, and covered a broad taxonomic range, from dogfish and skates, to gadoids (the cod family) such as Cod, *Gadus morhua*, Haddock and Whiting, to pipefish, gurnards, dragonets, gobies and flatfish. As with the invertebrate epibenthos, three distinct assemblages could be distinguished: in the southern North Sea, within the 50 m isobath, Solenette, Dragonet and the goby *Pomatoschistus minutus* were the most abundant species; in the central area, between 50 and 100 m, Dab and Long Rough Dab were characteristic, while below 100 m the latter species was co-dominant with the Hagfish.

In contrast to the other epibenthos, fish communities were most diverse in the shallow southern region, and along the east coast of the British Isles, with up to 24 species recorded per hour of trawling, but dropped to a maximum of 12 per hour in the central area, although increasing again beyond 100 m depth, to the north. Thirteen species of flatfish were collected, six of which were among the ten most frequently occurring species. The Dab was most frequent, occurring in 57.8% of all stations, followed by the Long Rough Dab (40.4%) and Solenette (37.1%). Most flatfish are strictly demersal in habit, feeding upon infaunal and epifaunal prey, particularly polychaetes, bivalves and small species of Crustacea, but Dragonets (28.1% frequency) and Whiting (24.4%) also take suprabenthic prey. Juvenile Whiting, 18–33 cm long, in the southern North Sea were found to feed during the summer on small crustaceans (Singh-Renton & Bromley, 1999), especially Brown Shrimp, prawns (*Pandalus* spp.), hermit crabs (*Pagurus* spp.) and swimming crabs (*Liocarcinus* spp.), and small fish, including sand eels (*Ammodytes* spp.), gobies (*Pomatoschistus minutus*) and smaller Whiting. Diets probably reflected prey availability, and varied with habitat, location and time: *C. crangon* formed 39–40%, by weight, of stomach contents at two sites, while *Liocarcinus* (53%) and *Ammodytes* (53%) were the predominant diets at two others. Elsewhere in the North Sea, in summer months, young Whiting have been shown to feed almost exclusively on fish. Like Whiting, Cod and Haddock are demersal in the sense that they are bottom feeders, but they are a trophic step above the flatfish, and the smaller species such as the Solenette, characteristic of the southern North Sea benthos, may be important food sources for the larger size classes of these predatory fish.

The diets of nine species of flatfish common in the Irish Sea have been shown to vary seasonally, and in relation to the benthic habitats they occupy (Amezcua *et al.*, 2003). Intensity of feeding was low in March and high in October. In early spring fish fed on different prey items on muddy, sandy and mixed coarse substrata, but the diet of each species was very similar, suggesting that there was little competition between them. In early autumn dietary composition displayed some interesting patterns. Most species seemed to have a characteristic diet that might vary in relation to sediment type, or between size classes of each species, indicating that both inter- and intraspecific competition was limited through food resource partitioning; the range of food items eaten by all fish was broad, but the diet of each size class, and each species, was characterised by a particular combination. The smallest sizes of Dab, for example, ate the smallest infaunal crustacean species; medium-sized individuals consumed amphipods and brittlestars, while the largest fed on hermit crabs, the stout brittlestar *Ophiura ophiura*, sea anemones and small fish, depending on the sediment type occupied, and thus the availability of different prey items. Large Lemon Sole, *Microstomus kitt*, had a consistent diet of sedentary and tube-dwelling polychaetes, and amphipods, regardless of the sediment it lived upon, and the diet of all sizes of Plaice was especially characterised by bivalves and species of *Nephtys*. For some species, the diet of a particular size class might be very similar to the equivalent size class of other species, perhaps reflecting mouth or body shape: the smallest Dab, Long Rough Dab and Solenette, for example, had very similar diets. That result notwithstanding, most size classes of most species of flatfish appeared to have a distinct diet that enabled co-occurrence through resource partitioning and reduced competition.

Demersal fish assemblages in large estuarine systems bordering the southern North Sea, the English Channel and western Britain all seem remarkably similar, dominated by mostly juvenile cohorts of just a few species of gobies, gadoids and flatfish, for all of which *C. crangon* is an important prey species. Brown Shrimp populations fluctuate both seasonally and annually, but in most habitats, above salinities of 16 psu, they are present throughout the year. In Bridgwater Bay juvenile Whiting are present from September to April (Henderson *et al.*, 1992). Adult Sprat overwinter in the bay through December and January; they do not feed during these months, but contribute to energy flow as a major prey item for the Whiting. Brown Shrimp are the principal food of Whiting prior to the arrival of the Sprat, remain part of its diet through December and January, and become its principal energy source again once the Sprat have left. Whiting abundance drops sharply in the summer months, but then *C. crangon* become an increasingly important food source for small gadoids, particularly Pout, *Trisopterus luscus*, and Poor Cod, *T. minutus*, and for summer-migrant Flounder, *Platichthys flesus*.

CHAPTER 4

Hard Grounds

Hard substrata provide habitat for sedentary and sessile organisms, and the density and diversity of these epifaunal communities are determined by the persistence of the habitat in relation to prevailing hydrodynamic regimes. In low-energy environments accumulations of dead shell on soft, muddy sea floors may support clumps of large hydroids or sea squirts, tolerant of silty, turbid waters. Offshore shell gravels, in higher-energy environments, may be habitat for diverse communities of encrusting animals, but these will have a sheet-like, or otherwise low-profile, form, able to withstand a high degree of current scour. Mussel beds and banks of coralline algae, maerl, may be the bases of high-diversity assemblages, with numerous encrusting species, an associated fauna of small, free-living species, and a variety of predators, large and small. However, although biogenic habitats may persist for decades, they are not truly permanent: mussels and coralline algae grow and die, though at dramatically different rates, and the structure of the habitats they provide, and the faunal assemblages they support, are continually changing. Boulder habitats are also less stable than they appear. They are characteristic of turbulent hydrodynamic environments, often scoured by sand and gravel, and sporadically overturned, with a frequency related to the size of each boulder. Bedrock provides the only natural, stable, hard substratum habitat, and while rock is subject to slow solution, abrasion and occasional catastrophic collapse, in comparison with other hard substrata it is effectively permanent. Offshore hard grounds, consisting of bedrock interspersed with patches of unconsolidated coarse sands and gravels, cobbles and boulders, and shell debris, extend across much of the shelf to the northwest, populated by sessile and motile epifaunas.

A degree of inclination often results in a change in faunal assemblages, and steep to vertical rock faces may be carpeted by sessile animals, benefiting from a refuge from competitors, predators and physical scour.

Hard-ground benthos consists of an encrusting epifauna of bivalves, tubeworms, large, solitary sea squirts (or ascidians) and, especially, colonial, or modular, animals (Box 3), including sponges, cnidarians, bryozoans and small ascidians, and non-modular clonal species, such as sea anemones. There is also a motile fauna of small, non-colonial or unitary organisms, particularly polychaetes, small crustaceans, molluscs and echinoderms, that varies in density and diversity in relation to water flow and the incline of the rock substratum. Many of these will build and occupy fine silt tubes, weaving between the primary space occupiers, and will be suspension feeders or detritivores; others will be free-roving detritivores or micropredators. There may also be guilds of larger predators, each species often with a narrowly defined array of preferred prey.

The composition, diversity and density of hard-ground benthic communities is initially defined by physical factors, and may change sharply with depth, water flow, rock type and inclination, and the abundance and diversity of the motile component depend largely upon the shelter and food provided by the attached epifauna. Gradient is important: flat areas in low-energy environments tend to accumulate silt, the finest silt collecting in the stillest water, and silt is inimical to many suspension feeders. Captorial suspension feeders, such as some of the larger hydroids, may tolerate considerable siltation, but filter-feeding bryozoans cannot, and the aquiferous systems of sponges are soon clogged. On sloping and vertical rock surfaces silt does not persist, and as water flow increases towards more exposed habitats,

---

Box 3. Unitary, colonial and clonal animals

Unitary animals exist as morphologically discrete individuals, each of which may be the product of either sexual (gametic) or non-sexual (agametic) reproduction.

Colonial animals exist as structurally linked and functionally integrated units, or modules, developed by continuous replication from a single founding module, which may be the product of either gametic or agametic reproduction. Thus, also described as **modular**.

Clones are groups of genetically identical individuals originating by agametic reproduction, through budding, fragmentation or fission. Clonal organisms may be unitary, as in the case of many sea anemones, or modular, as in the colonial sea squirts.

sand scour has another structuring effect on sessile benthos. The bushy bryozoan *Flustra foliacea* (Fig. 47) lives in aggregations on current-swept sea floors. Its colonies are largely chitinous, and flexible, and the orifice of each constituent module, or autozooid, through which the feeding lophophore of tentacles is exserted, is ringed by a palisade of thick, chitinous spines. *Flustra foliacea* seems highly resistant to sand scour; it is long-lived – individual colony ages are known to exceed 15 years – and while its populations may increase by larval settlement and recruitment, colonies are also able to regenerate though new growth from the base. On steep and vertical faces, above the turbulent bottom layers of the water column, *F. foliacea* may be displaced by populations of the gregarious tubeworm *Sabellaria spinulosa* and the clonal anemone *Metridium senile* (Fig. 80). This apparent zonation may reflect not only the

FIG 80. The Plumose Anemone, *Metridium senile*. (J. S. Porter)

tolerance of each species to physical stress, but perhaps also feeding functions. Anemones collect large food items from strong water currents, and sabellid worms require finer materials, while bryozoans are microphagous filter feeders.

SPATIAL COMPETITION

For hard-substratum sessile communities, space, to which to attach, and on which to grow and expand, is a limiting resource, and competition for space is constant and intense. On well-lit surfaces, in clear water, algae will prevail and encrusting invertebrates will be restricted to impermanent refuges on stipes and holdfasts of some of the larger brown seaweeds. In shaded habitats, and in deeper or turbid water, algal populations decline, and hard substrata support communities of encrusting invertebrates (Fig. 81). New space on crowded surfaces becomes available following the death of an occupant, through senescence, predation or random accident, and is soon colonised by new settlers. Unitary organisms, such as serpulid tubeworms, saddle oysters and solitary ascidians, can only retain living space by resisting overgrowth or dislodgement by spatial competitors, and some species may be only transitory components of the

FIG 81. An epilithic community of sessile, suspension-feeding, modular animals, with bryozoans especially predominant. (J. S. Porter)

photographic recording over a five-year period, were dominated by two solitary ascidians, *Dendrodoa grossularia* and *Ciona intestinalis*, and a tubeworm, *Pomatoceros triqueter*, which also constituted the bulk of the biomass as well as the greatest coverage; the population of *C. intestinalis* (Fig. 84) a soft-bodied animal that may grow to 15 cm in length, exceeded 400 individuals per square metre (Gulliksen, 1980). Yet suction sampling of the habitat yielded more than 150 additional species, mostly consisting of tube-building amphipods and polychaetes, and small molluscs. Dense populations of *C. intestinalis* accumulate large quantities of fine silt, drawn by the combined inhalant currents of the animals, which becomes packed between them, and augmented by faecal pellets. This creates a turbid environment that results in decreasing densities and diversity of other filter-feeding species, but provides perfect conditions for the small detritivorous amphipods and polychaetes.

Rocky habitats, at around 17 m depth, in Kinsale Harbour on the south coast of Ireland were found to be populated by sparsely distributed kelps, *Laminaria hyperborea*, and beds of the brittlestar *Ophiothrix fragilis* (Ball *et al.*, 1995). Vertical surfaces bore dense assemblages of sessile animals, with no apparent spatial dominants; large ascidians, in particular, seemed rare, but the community consisted of 116 species, of which around 100 were modular organisms. Some of these were perhaps encrusting kelp holdfasts, but none was strictly epiphytic in habit, and the sessile faunas of the holdfasts would have developed through the settlement of larvae shed by rock-encrusting species in immediate proximity to

FIG 84. A group of the solitary ascidian *Ciona intestinalis*. (J. S. Ryland)

the kelps. More than half of the total number of sessile species recorded from natural rock habitats were bryozoans, 57 species, indicating how extraordinarily diverse bryozoan assemblages may be in the absence of large, superior spatial competitors. The potential diversity of the epilithic faunas of this area was suggested by an additional 44 species, including 16 bryozoans, that were recovered from artificial settlement panels.

POPULATIONS AND COMMUNITIES

Some of the larger spatial dominants in sessile benthic communities may be relatively short-lived. *Ciona intestinalis*, for example, has a maximum life span of two years, and in sparse populations individuals probably do not survive even that long, being outgrown by competitive superiors or succumbing to predation. A strong year class, following from high reproductive output and good recruitment, may establish a dense population by late summer that persists not through competitive superiority but rather through the blanketing effect of the silt and faecal material it generates. As well as discouraging other suspension feeders, an example of amensalism (in which the life mode of one species precludes co-occurrence with others, as opposed to commensalism), the inhalant and exhalant streams of the ascidian population prevent settlement of prospecting larvae, which are either engulfed or swept away. Community structure may change as a consequence: larvae of *C. intestinalis* suffer the same fate as all others, the demography of the population does not change, and it collapses and disappears at the termination of the two-year life cycle (Fig. 85).

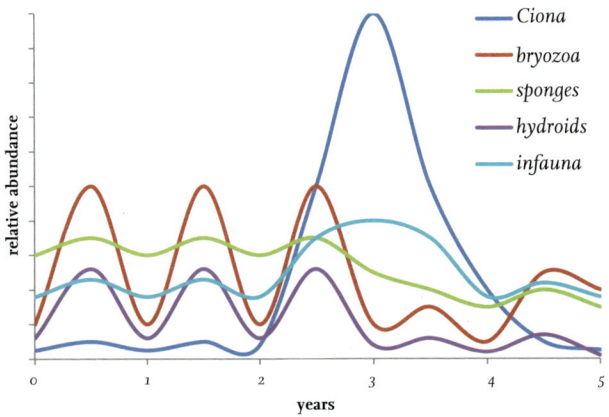

FIG 85. Hypothetical fluctuations in the population densities of components of an epilithic community colonised by *Ciona intestinalis*. (After Gulliksen, 1980)

Such exceptions aside, most spatially dominant sessile animals tend to be long-lived; clones of *Metridium senile*, for example, seem to persist for decades. The Plumose Anemone is a large animal, with presumably a high energy demand; it is reliant upon water flow for its food supply, and is common in high-energy environments, tolerating current velocities up to 75 cm per second. Turbulent water flow and the cnidocyte-laden tentacles of the anemones usually result in a high-density but low-diversity assemblage. The upright, perennial colonies of the large bryozoans *Flustra foliacea* (Fig. 47) and *Pentapora foliacea* (Fig. 86) will also be found

FIG 86. (a) The Ross Coral, *Pentapora foliacea*. (b) Detail showing the growing edge and feeding lophophores. (J. S. Porter)

in high-energy habitats, while the solitary ascidian *Ascidia mentula* favours a quieter milieu, with current velocities no greater than 18 cm per second, but all three have a characteristic associated fauna of both sessile and vagile species, and constitute high-density mesocosms within the broader hard-substratum habitat. The firm outer test of *A. mentula* is often encrusted with bryozoans, hydroids and small tubicolous polychaetes, and as it grows it gathers a coating of small red algae. In the field these epizootic communities effectively camouflage the ascidian, which is sometimes only recognised by its open siphons. The capacious branchial chamber of *A. mentula*, through which flows a constant stream of food-laden, oxygen-rich water, provides a secure environment for a number of endosymbiotic species, especially small crustaceans. These include several species of cyclopoid copepod within the genus *Notodelphys*, the systematic affinities of which are patent (Fig. 87a), and the extravagantly dimorphic *Notopterophorus papilio* (Fig. 87b), which may not be immediately recognised as a copepod at all. A large amphipod, *Leucothoe spinicarpa*, growing to almost 2 cm in length and with distinctively chelate gnathopods (Fig. 87c), is also likely to be found in the branchial sac of *A. mentula*, as well as in other large, solitary ascidians, and in the aquiferous chambers of sponges.

An especially interesting symbiont of *A. mentula* is a small mussel, *Modiolarca subpicta* (Fig. 88), which can be found embedded within the outer surface of the test, its siphons communicating to the exterior through a small opening. *Modiolarca subpicta* is not restricted to *A. mentula*, but can be found in association with several species of solitary ascidian, and its biology has been investigated in a population commensal with *Ascidiella aspersa* (Fig. 89), a smaller species than *Ascidia mentula* with a maximum length of 12 cm and a life span of no more than two years

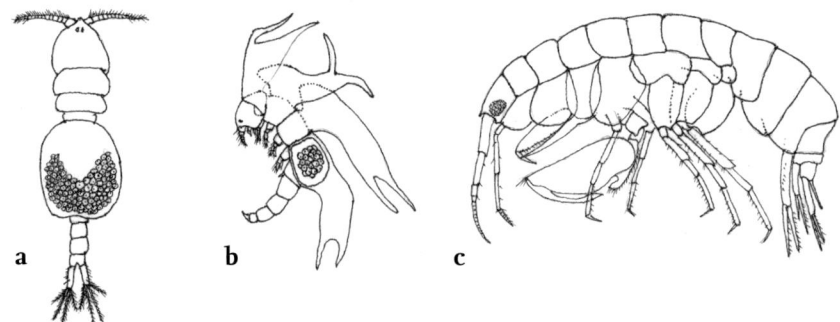

FIG 87. Some common endosymbionts of *Ascidia mentula*: the cyclopoid copepods (a) *Notodelphys allmani* and (b) *Notopterophorus papilio*, and (c) the amphipod *Leucothoe spinicarpa*, with distinctive chelate (clawed) ganathopods (first limbs). Not to scale.

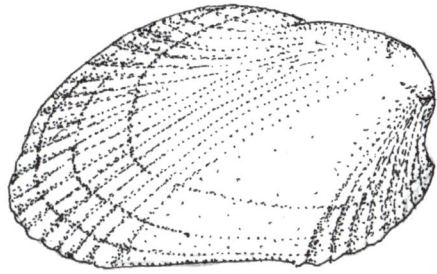

FIG 88. The small mussel *Modiolarca subpicta*. Scale bar: 2 cm.

FIG 89. Two individuals of the solitary ascidian *Ascidiella aspersa* with (left) cushion-shaped colonies of the modular species *Polyclinum aurantium* and (right) a laminar colony of *Botryllus schlosseri*. (J. S. Ryland)

(Morton & Dinesen, 2011). *Ascidiella aspersa* is a protandrous hermaphrodite; its lecithotrophic larvae settle in midsummer following a brief free-swimming period and limited dispersal. *Modiolarca subpicta* has a single short spawning period in summer; planktotrophic larvae develop during a protracted pelagic phase of up to six weeks, before settling and metamorphosing on the ascidian host. Newly settled spat are first apparent on *A. aspersa* individuals of 10–25 mm in length; adult mussels are most frequently found on ascidians 35–60 mm long, and are most abundant, to a recorded maximum of 11, in those in the 45 mm size class. It does not burrow into the outer tunic of the host, contrary to previous conceptions. Instead the newly settled mussel uses its elongate foot to attach byssal threads on its dorsal, hinge-bearing side, securing it to the host tunic before turning onto its dorsal side so that the ventral shell gape and the mantle siphons face upward and outward. As the host grows, folds of tunic enclose and envelop the mussel, so that only its siphons protrude to the exterior. The life cycles of host and symbiont seem synchronised: both breed during a short summer period, and the smallest mussels occur on the smaller size classes of the ascidian. The association terminates with the death of the host. *Ascidiella aspersa* breeds once in its second summer, and then dies in its second winter, and it is possible that the mussel also has just a two-year life span, determined by that of its host. Yet individuals of *M. subpicta* have been recorded from interstices of kelp holdfasts and attached beneath shells and stones by byssal threads, so it is equally possible that it survives the death of the initial host, and is perhaps able to seek and attach to a second host. An investigation into size frequencies of *M. subpicta* populations on longer-lived hosts, such as *A. mentula*, would be an interesting preliminary test of this possibility.

Two perennial bryozoans, *Flustra foliacea* and *Pentapora foliacea*, contribute importantly to habitat heterogeneity. *Flustra foliacea* populations occur as patches on level coarse grounds and on inclined rock surfaces, often forming permanent shrubberies over several square metres. Its flexible, horn-coloured colonies, from which it derives its vernacular name, Hornwrack, have been aged at up to 12 years, and provide ideal surfaces for small, encrusting organisms, and the flabellate fronds offer protection from scour. Encrusting bryozoans are particularly frequent on *F. foliacea* colonies, together with small hydroids and a suite of motile predators. More than 50 species of bryozoan have been recorded encrusting *F. foliacea*, across a geographical range encompassing the entire shallow shelf region of northwest Europe. An ecological study of a population of Hornwrack at 15–20 m depth on the south Wales coast reported an associated fauna of 32 sessile invertebrate species, and noted, unfortunately without detail, large numbers of the small, suspension-feeding porcelain crab *Pisidia longicornis*, and the regular echinoid *Psammechinus miliaris*, and 'thousands' of caprellid amphipods, which were assumed to be living amongst the *F. foiliacea* fronds (Stebbing, 1971). Colony form of the bryozoan species reported from *F. foliacea* encompasses most known morphologies, with the exception of heavily calcified, branching, rigid forms. Apart from a few ubiquitous species, such as *Electra pilosa* (Fig. 90), the most

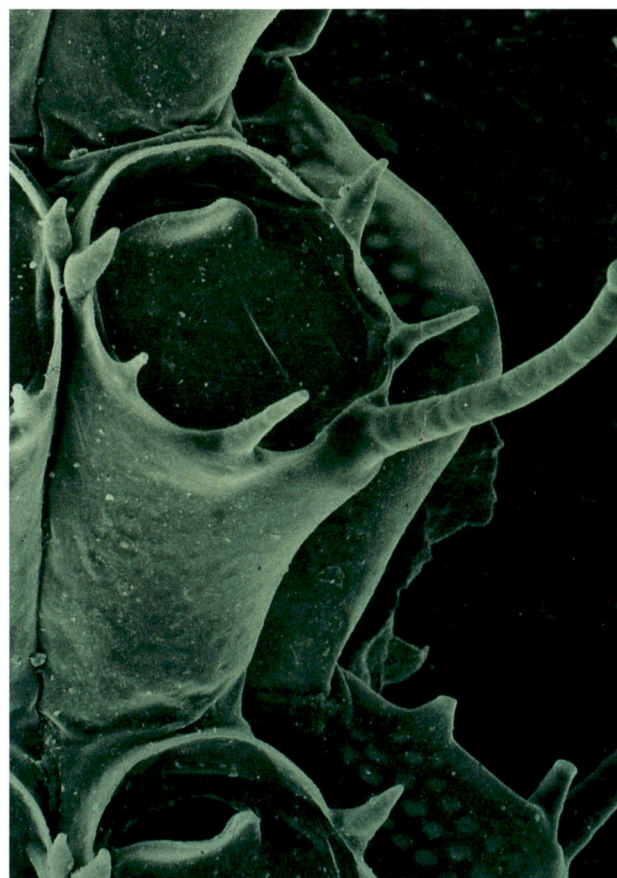

FIG 90. Scanning electron micrograph showing autozooids (modules) of the bryozoan *Electra pilosa*. For scale, the elongate frontal spine is approximately 0.25 mm long.

frequently occurring bryozoan associates of Hornwrack are characterised by delicate, flexible, branching colonies. Predominant among these are stiff, white colonies of *Crisia eburnea*, species of *Scrupocellaria* (Fig. 91) with open, fan-like colonies, and flabellate fronds of *Dendrobeania* and *Bugula* species. *Crisia eburnea* and *Scrupocellaria reptans* were abundant in the south Wales population, with 260 and 210 individual colonies, respectively, on a single 10.5 cm high *F. foliacea* colony. *Bugula flabellata* (Fig. 92) was the third most abundant bryozoan species found, and had the most intimate association with its host. The branching rootlets – rhizoids – that attach each *B. flabellata* colony to its substratum penetrated the autozooids of the *Flustra* colony, and ramified throughout the frond. This phenomenon is not unique, however, as *B. flabellata* can be found

FIG 91. Colonies of the erect bryozoan *Scrupocellaria*. (J. S. Porter)

in similarly close association with many other species of bryozoan, and none of the sessile species recorded from Hornwrack is restricted to it. There is no characteristic, recurring sessile community associated with *F. foliacea*, probably because its populations tend to be patchily distributed, and the colonising fauna is determined by local larval supply.

Colonies of *Pentapora foliacea*, commonly referred to as the Ross Coral (Fig. 86), are heavily calcified and rigid, built from bilaminar plates of autozooids that divide and anastomose along the growing edges. They frequently exceed diameters of 20 cm, for a height of 10 cm, and some huge specimens have

FIG 92. A luxuriant colony of the bushy bryozoan *Bugula flabellata*. (J. S. Porter)

been reported in the past, during times when the benthic environment was less disturbed by human activity. The frequency of division and fusion of the growing edges is perhaps determined by local environmental factors, such as hydrodynamic regime. Colonies may have an open lattice structure, with large interstices between the plates of autozooids, while in others the growing edge may be closely folded, with small spaces in a honeycomb-like structure; the latter growth form might be an adaptation to especially turbulent conditions, but the possibility has not been investigated. The biology and ecology of *P. foliacea* have been little studied. It is present on coarse grounds and in rocky habitats along the south coast of England, sparsely from the Solent eastwards to West Sussex, increasingly common westwards, and abundant around Cornwall; it is found off Pembrokeshire, the Isle of Man and north Wales, but is otherwise absent from most of the Irish Sea, and from the southern and western coasts of Ireland northwards to the Hebrides. It seems to be scarce around Orkney and Shetland, where it perhaps reaches its northern limit. Population densities, colony growth rates, individual maximum sizes and longevity have not been documented with any precision. Oxygen isotope analysis has suggested growth rates of around 2 cm vertical extension per year; actual age has not been established for any specimen, although a minimum age of three years has been estimated for a colony from Pembrokeshire (Pätzold *et al.*, 1987). The colony is attached to the substratum by an encrusting sheet of zooids; erect plates may be produced during the first year's growth, or perhaps commence in the second, and it is probable that *P. foliacea* has the same facility for repair and regeneration as most cheilostome bryozoans, so that accurate ageing of any individual is likely to be very difficult to achieve. *Pentapora foliacea* colonies contribute significantly to hard-ground habitat heterogeneity, providing secure refuges for an abundance of other organisms. As in the case of *Flustra foliacea*, it supports a diverse community of encrusting bryozoans, although no similar inventory of species associated with it has been made. *Amphiblestrum flemingii*, *Callopora dumerilii*, *Membraniporella nitida* and *Smittoidea reticulata*, all developing encrusting sheets, are commonly found on *P. foliacea* colonies, and possibly represent a perennial community, regenerating annually from overwintered fragments. Erect, bushy bryozoans seem to be uncommon, perhaps because of turbulent water flow, or because the rigid structure of the *Pentapora* colony supports an assemblage of small browsers, grazers and predators, as well as relatively large suspension feeders. Sedentary bivalves, especially small species of scallop and mussel, may be permanent residents within the chambers defined by the anastomosed plates of the *Pentapora* colony; suspension-feeding decapods, such as *Pisidia longicornis*, may also be abundant, but they are probably outnumbered by hosts of lesser crustaceans, especially amphipods. By analogy with coralliform bryozoan biotopes described from other regions, including the eastern Mediterranean and New Zealand, dense populations of Ross Coral are probably utilised for food and shelter by juveniles of many species of large decapod and fish, and represent foci of high biomass and high density in moderate- to high-energy habitats.

The upper portions, and the plate edges, of live and actively growing *P. foliacea* colonies have a deep orange-red coloration, and are generally free of epibionts; the colour fades towards the older, basal regions of the colony, which are usually encrusted with sessile organisms. The clean growing edges might indicate a mismatch between the period of larval settlement and recruitment of epibiotic species, and the growth period of the host, but might also indicate that the vulnerable growing edge is a source of bioactive compounds that deter settlement of larvae and spores. However, microbial communities show an opposite pattern

of distribution, with thick layers of bacteria present on the growing edges of the
colony, and thin, diffuse bacterial growth on the older regions. *Flustra foliacea*
(Fig. 47) and *Alcyonidium diaphanum* (Fig. 46) are known to generate metabolic
by-products that may have toxic or irritant properties. Trawler fishermen
constantly exposed to *A. diaphanum* through cleaning nets may become sensitised
to an alkaloid metabolite released by the bryozoan, and develop a painful allergic
dermatitis, termed 'Dogger Bank itch' (Pathmanaban et al., 2005); notably, *A.
diaphanum* colonies are usually free of epibionts. Fresh, live colonies of *F. foliacea*
have a lemon odour, and also contain alkaloid secondary metabolites, some
of which have been found to disrupt functions of bacterial colonies. Bacteria
were found to be differentially distributed on the surface of *F. foliacea* zooids,
suggesting suppression of bacterial growth by secondary metabolites (Peters et al.,
2003), yet colonies typically support rich communities of encrusting animals, with
the larvae of some species settling preferentially along the tips of the growing
fronds. The roles of secondary metabolites in the biology of bryozoans are thus
still quite unclear, but the possibility that in some species their interaction
with bacterial communities, and bacterial products, may have consequences for
macrofouling communities is intriguing.

Spiny-armed brittlestars are abundant in hard-ground habitats, from coarse
gravels to rock, and brittlestar beds, dominated by *Ophiothrix fragilis* (Fig. 61)
but often with a small proportion of *Ophiocomina nigra* (Fig. 74), constitute a
biotope characteristic of all moderately exposed northwest European coasts.
Ophiothrix fragilis was the most common species recorded by the CEFAS surveys
in the eastern English Channel, the Bristol Channel and the southern Irish Sea
(page 94), and it ranges, at equivalent densities, along the western coastlines
of the British Isles and Ireland. *Ophiothrix fragilis* can be found in the low
intertidal on rocky shores, and offshore to the shelf edge; in coastal habitats
it lives in small groups among kelp holdfasts, and on inclined, even vertical,
rock faces, but beyond the limit of wave surge, below 15 m, its populations form
dense aggregations, in discrete, sharply demarcated patches (Warner, 1971). The
brittlestars actively associate: removed from a patch and placed alone, regardless
of the type of substratum a solitary individual will begin to walk in a straight
line, changing direction at intervals, and continue to walk until it encounters a
group of conspecifics. A patch, or bed, of *O. fragilis* has a minimum population of
around 100 individuals. Smaller groups tend to continue moving and eventually
coalesce with other groups, and the most extensive may cover thousands of
square metres of sea floor. The largest patches consist mostly of the largest
adults, with disc diameter 8–12 mm, together with the smallest recruits, disc
diameter < 1 mm, which cling to the arms and arm spines of the adults. At disc

diameters greater than 2 mm juveniles appear to migrate, congregating as small groups in adjacent habitats, such as clumps of large hydroids, and returning to the main concentration as they grow. This behaviour perhaps serves to mitigate competition for food between adults and juveniles. Patch stability is partly related to hydrodynamic factors: patches form at current speeds of around 10 cm per second and achieve stability at an optimum speed of 25 cm per second as the closely packed individuals interlink their arms, but are scattered at speeds above 50 cm per second. Within the largest beds densities of *O. fragilis* are often in the range 1,500–2,000 individuals per square metre; biomass estimated by wet weight suggested a value of 340 g/m^2 for a population density of 309/m^2, and mean biomass in terms of ash-free dry weight (AFDW) has been measured at 210 g/m^2 for a population density of 1500/m^2.

TROPHIC INTERACTIONS

Ophiothrix fragilis is principally a particulate suspension feeder, holding two or more arms vertically into the water column, trapping and binding fine food items with mucus-bearing tube feet, and passing packages of material along a tube-foot conveyor to the mouth. One advantage of aggregated populations may be that they result in increased feeding success, as the dense thickets of upraised arms probably slow boundary-layer current flow and enhance particle capture. Phytoplankton provides the most significant source of nutrition through much of the year, but this is probably supplemented by detrital material of various origins. Optimal current speed for suspension feeding was found to be 20 cm per second, and duration of feeding was determined by tidal currents: in the Dover Strait *O. fragilis* populations were able to feed through 86% of a neap tidal cycle, but only for 37% of a spring cycle (Davoult & Gounin, 1995). Brittlestar aggregations must be a significant element of the benthic ecosystem, contributing substantially to organic carbon input and the cycling of calcium carbonate. In the Dover Strait and the eastern English Channel *O. fragilis* beds extend over 6,000 km^2, and the Dover Strait populations have been estimated to produce 682 g $CaCO_3$ per square metre per year, significantly boosting levels of dissolved CO_2. However, their further ecological roles are still only incompletely documented, and the importance of brittlestars in marine food webs are generally poorly known. High-density *O. fragilis* populations affect the composition of hard-ground benthic communities, partly through resource competition with other suspension feeders, and almost certainly through the exclusion of potential spatial competitors. *Ophiothrix fragilis* is preyed upon by bottom-feeding fish, especially gadoids and flatfish, and by several species of decapod, but it is a principal prey item only for the large starfish *Luidia ciliaris* and *L. sarsi*.

Hard-ground sessile benthos, with its epibionts and symbionts and its associated complement of small tube builders, and the blankets of brittlestars comprise the primary consumers and detritivores that support a dependent community of secondary consumers. The most important of these are bottom-feeding fish, large decapods, numerous species of gastropod, and a number of echinoderm species. Some are scavengers, including most whelk and many decapod species, while others are carnivorous predators, such as the cottid fish *Myoxocephalus scorpius* and *Taurulus bubalis* (Fig. 93) and most starfish, or broadly omnivorous, like the epifaunal sea urchins. Echinoderms are as important to the ecology of hard-ground habitats as they are to that of soft sediments. *Asterias rubens* swarms can have the same devastating effect in both environments, and on rocky grounds are probably only limited by greater water turbulence, which will eventually disperse them. The Edible Sea Urchin *Echinus esculentus* (Fig. 94) browses kelp fronds, with a predilection for those encrusted with the bryozoan *Membranipora membranacea* (Fig. 95), and below the kelp forest feeds on smaller

FIG 93. The bottom-feeding Sea Scorpion, *Taurulus bubalis*. (J. S. Porter)

FIG 94. The Edible Sea Urchin, *Echinus esculentus*, browsing the stipe of *Laminaria hyperborea*. (J. S. Porter)

red and brown seaweeds, as well as on barnacles, bryozoans and other encrusting invertebrates. At high densities *E. esculentus* has a considerable impact on sessile macroalgal and invertebrate assemblages; the animals are very motile, seem to move offshore during the winter months, and migrate in response to changing food resources.

Submarine cliffs and rock outcrops, offshore installations, and of course wrecks, all increase habitat complexity in hard-ground environments, and support a correspondingly broad diversity of scavengers and predators. Blennies and gobies, the small flatfish *Zeugopterus punctatus*, known as the Topknot, the Tadpole Fish, *Raniceps raninus*, and the Bull Rout, *Myoxocephalus scorpius*, are often permanent residents, venturing out from holes and crevices to capture prey. Individuals appear to show a degree of home-range fidelity, defending a foraging area against conspecific intruders, though not always against other species, and display rhythmic activity cycles (Nickell & Sayer, 1998). Other fish species may be only temporarily resident, seeking refuge as juveniles, visiting to feed as adults,

FIG 95. A young colony of the bryozoan *Membranipora membranacea*, on the Serrated Wrack, *Fucus serratus*. (J. S. Porter)

or as seasonal visitors establishing breeding territories during part of the year. Diets of these benthic-feeding fish are not known in detail, but most seem to take small prey items, and to show little selectivity. Crabs, squat lobsters, amphipods and other minor crustaceans, molluscs and polychaetes all form part of the diet of most species. In Lough Hyne, southern Ireland, five species of small wrasse were shown to establish breeding territories on a vertical rock face, constructing and guarding nests in which their eggs develop (Bell *et al.*, 2003). Some were observed foraging among the sea squirts *Ascidia mentula* (Fig. 83), *Ascidiella scabra* and *Ascidiella aspersa* (Fig. 89). The latter was the most abundant of the three species and attracted the greatest attention from the wrasse, and in one year 95% of its population was consumed. The detached tests of the sea squirts, collected from the foot of the cliff, had been eviscerated, and holes in the outer tunic showed where the fish had also extracted the bivalve symbiont *Modiolarca subpicta* (Fig. 88). The bulk of the biological material raining down from the cliff consisted of sponges; 22 species were recognised, but these had probably been dislodged by wrasse in the course of nest construction, or by other cumbersome foragers, particularly echinoids and spider crabs, rather than through predation.

Most of the benthic secondary consumers are generalists, consuming a wide variety of food items and displaying little selectivity. However, hard-ground benthos also provides habitat for a very diverse suite of specialist predators, some with diets so narrow that their life cycles are spent in association with just one or very few prey species. Many small arthropods fall into this category, but most familiar, and conspicuous, among them are species of nudibranch sea slug (Picton & Morrow, 1994). *Tritonia hombergii* is especially striking (Fig. 96): it is the largest species of sea slug found in northwest European seas, a bulky animal growing to 20 cm in length, recognised by a lobed and fringed oral veil that overhangs the head, and thick-stemmed arborescent gills in pairs along its dorsal surface. This huge sea slug feeds exclusively on *Alcyonium digitatum*, reportedly biting off and ingesting large pieces of its prey.

FIG 96. The large predatory sea slug *Tritonia hombergii*. (J. S. Ryland)

Dendronotus frondosus (Fig. 97) is a rather similar species, but with a maximum length of only 10 cm, with a more slender body and proportionately longer, delicately branching gills. Its diet is less restricted than that of *T. hombergii*. Juveniles feed on several species of small hydroid, while adults browse clumps of the larger *Tubularia indivisa* and *Ectopleura* (formerly *Tubularia*) *larynx* (Fig. 98).

Goniodoris and *Onchidoris* are two genera of medium-sized sea slugs, species of which are mostly less than 4 cm long. All have firm, oval and domed bodies, the dorsal surface typically covered with rounded tubercles, with an anterior pair of sensory, club-shaped rhinophores, and a posterior ring of contractile gills. *Goniodoris castanea* and *G. nodosa* feed on encrusting bryozoans as juveniles, and on colonial ascidians such as *Dendrodoa grossularia* when mature. There are numerous species of *Onchidoris* in European shelf habitats, all of which appear to feed on encrusting, calcified bryozoans, although whether each species has a particular dietary preference is not known. Specialist sponge predators are found mostly in the superfamily Doridacea, and many dorid species seem to have quite narrowly limited diets.

FIG 97. (a) The sea slug *Dendronotus frondosus* browsing on a hydroid colony. (b) A larger individual with especially arborescent gills. (J. S. Porter)

The most attractive of the nudibranch predators are the many species of *Doto* (Fig. 99), a genus usually classified with *Tritonia* and *Dendronotus* in the Dendronotacea. They are all tiny animals, usually less than 2 cm in length and many not exceeding 1 cm, with often prettily patterned bodies bearing pairs of

FIG 98. A colony of the hydroid *Ectopleura larynx*. (J. S. Ryland)

FIG 99. The sea slug *Doto fragilis* – recognised by its knobbly, brown cerata – on its prey, the hydroid *Nemertesia antennina*, with its spiralled, white spawn ribbon. (J. S. Porter)

clubbed, tuberculate processes – cerata – dorsally. They are an interesting contrast to larger carnivorous sea slugs in that their prey is also their habitat. *Doto* species live among clumps of hydroids, on which they deposit their spawn, and to which their larvae must be attracted at settlement. At least 18 species of *Doto* have been described from northwest European seas, and more perhaps remain to be recognised. They are not easily distinguished from each other by size, colour or morphology, and it is probable that apparently widely occurring species, reported from several different species of hydroid, may each represent a number of superficially similar species, each with only a limited diet, and perhaps restricted to one, or very few, hydroid host species. Cryptic species groups have already been revealed among some *Doto* populations through molecular genetic studies (Morrow *et al.*, 1992)

SETTLEMENT AND SUCCESSION

Newly available substratum soon becomes encrusted with assemblages of sessile organisms. Fresh surfaces provided by rock falls, boulder overturn, coastal constructions and offshore installations become coated with bacterial films, diatoms and microalgae, and then colonised by macroalgal propagules and invertebrate larvae. Settlement may seem to show succession, with hydroids, bryozoans and serpulid tubeworms often among the early colonists, and ascidians and anemones arriving later. The composition of the community thus changes with time, but there is no predictable climax community. Development of the community is determined by the season in which the new habitat is created, its depth, larval supply, and the growth rates and longevity of each colonising species. The species founding the new community might be barnacles and other winter breeders, in the case of habitat newly available in spring, while hydroids, tubeworms and bryozoans might be the pioneers in summer colonisations, and these early successional species might be displaced by late successional invaders, or may persist as the spatial dominants, partly depending on the season in which the community begins to develop. Spatial competition between the filter-feeding sessile component will further mediate development of the assemblage, and biomass and diversity will fluctuate accordingly. Complexity increases as vagile suspension feeders, scavengers and predators begin to colonise.

Where hard-substratum benthic assemblages form on anthropogenic surfaces they may become an expensive nuisance. They may develop an especial luxuriance, as the conveniently provided habitat may benefit from enhanced water flow, increased volumes of particulate food, more light and higher water temperatures, and even decreased competitive pressure and a much reduced risk of predation. Such assemblages, on docks and pontoons, on and within structures associated with coastal power stations and cooling systems, on the hulls of inshore pleasure craft and oceanic shipping, and on oil and gas rigs, are referred to as 'fouling communities' (Fig. 100). Marine biofouling may have severe economic impacts, obstructing cooling-water intakes, modifying water

FIG 100. (a) The invasive tubeworm *Ficopotamus enigmaticus* in a dense band on a pontoon float, Swansea marina. (b) Detail to show regenerated tubes intergrown with small barnacles.

flow regimes and creating additional frictional drag; on ships' hulls fouling may affect speeds, and result in increased fuel consumption, while on stationary structures, such as offshore rigs, it may result in increased stress, and in both cases encrusting communities may conceal potentially damaging phenomena, such as cracks and corrosion. Marine biologists have researched the ecology of fouling communities for decades, through field observation and manipulation, and through the use of artificial settlement panels to record the settlement and succession of epifaunal assemblages on different substrata and under differing environmental conditions. These approaches still have importance for understanding the structure and development of fouling assemblages, and in order to devise methods for control and eradication, but now have an added significance in relation to environmental management and conservation. Information on the settlement, succession and development of sessile benthic communities may be useful in evaluating the sensitivity of habitats to disturbance, and for estimating rates of recovery following disturbance. Further, such data might be important when considering the creation of artificial reefs, which may prove to be useful means of increasing habitat heterogeneity, and hence biodiversity.

The composition of epilithic communities varies according to the type of rock they encrust. Carboniferous limestone, with frequent bedding planes and numerous fissures, and pocked by solution, biogenic boring and physical erosion, offers a greater range of niches than hard schists and soft shales. Communities established on artificial substrata are often structurally different from those on naturally occurring hard substrata, and it has been suggested that they will only approximate to a 'natural' state if the artificial substratum encompasses the same degree of physical heterogeneity as natural substrata. Submerged wooden pilings, concrete, smooth steel and corroded iron surfaces each comprise distinct, new microhabitats, and comparisons between benthic communities on different natural and artificial hard substrata must be carefully drawn. Nevertheless, research into fouling communities on artificial substrata supplies some interesting insights into the ecology of natural sessile benthos. Hydroids are among the earliest invertebrate colonisers of newly available substrata. In March 2004 an artificial reef was created in Whitesand Bay, south Cornwall, when the decommissioned frigate HMS *Scylla* was scuttled, and over a number of years successive settlements of algae and invertebrates were recorded by diving biologists (Hiscock et al., 2010). Barnacles (*Balanus crenatus*), tubeworms (*Pomatoceros triqueter*) and the hydroid *Obelia dichotoma* settled within a few weeks of the establishment of the reef, but all three species seemed to have disappeared by the end of the first summer. The first two remained as inconspicuous, and probably insignificant, components of the developing assemblage, but by summer the most abundant species of hydroid were *Ectopleura larynx* (Fig. 98) and *Tubularia indivisa* (Fig. 101), both of which had also first appeared soon after the ship had been sunk, and continued to show brief annual peaks in abundance during the five-year duration of the monitoring programme.

These early successional species have sometimes been termed 'pioneer' species, expanding rapidly into often extensive populations that are eventually displaced, though not entirely obliterated, by competitively superior, late-successional species that may come to constitute the spatial dominants of the community. Fifty-nine taxa of algae and invertebrates had colonised *Scylla* by the end of 2004, and the number increased annually to a total of 148 by the end of the five-year survey. These consisted mostly of 'conspicuous' species that could be identified with confidence in the field, under water, together with species identified from a number of samples scraped from upper surfaces of the ship, on a single occasion halfway through the survey. It is probable that the actual diversity of the new reef's epifauna was underestimated, and that regular (but, alas, expensive) sampling would have greatly increased the number of species recorded. The ascidian *Ciona intestinalis* was an early colonist, increased in abundance in the second year of the survey, and thereafter maintained a more

FIG 101. A clump of the hydroid *Tubularia indivisa*. (J. S. Ryland)

or less prominent presence in the communities. *Metridium senile* and *Ascidiella aspersa* were not recorded until the year following the creation of the reef, but increased in abundance in the third year, and as potential spatial dominants would probably have established a permanent presence by year five. The authors of this study noted that an artificial reef of concrete blocks installed in Poole Bay attracted a settlement of 90 species within a year, with the total number of taxa recorded rising to 220 during the second year and to 240 in the third year following the submersion of the blocks. They also noted that natural rock habitats in the vicinity of *Scylla* supported a community of 122 conspicuous species, 39 of which had not been recorded on the artificial reef during the five years it was monitored. Three different types of hard substrata, one natural and two artificial, appeared to support three different benthic assemblages.

As a pioneer species, *Tubularia indivisa* seems to have a role in facilitating the settlement of newcomers, and thereby promoting the diversity of the developing community. This may be in part a passive process: *T. indivisa* is a non-modular hydroid, but its solitary polyps grow in clumps, as a result of larval settlement behaviour. Its intertwined hydrocauli ('stalks'), each often in excess of 15 cm high, provide stable frameworks for tube-building amphipods, and habitat for caprellid amphipods. Species of *Jassa*, in particular, may be abundant within *Tubularia* clumps, and caprellids such as *Pseudoprotella phasma* find ideal vantages from which to seize small crustacean prey items or, as effective kleptoparasites, to steal food items from suspension-feeding amphipods, or even from the tentacles of the host hydroid. There is also a possibility that the larvae of some ascidians, *Ciona intestinalis* and *Ascidiella aspersa* in particular, may respond positively to a cue originating from the chemical constituents of the outer chitinous covering of the hydroid, the perisarc, when exploring settlement sites. *Tubularia indivisa* has a life span of around one year (Hughes, 1983). It breeds more or less continuously, with reproductive rate perhaps regulated by ambient temperature, but recruitment of juveniles shows a large peak in the summer months, May and June, with a lesser peak in late September. It does not have a pelagic larval stage. Instead it releases a creeping, tentaculate larva resembling a tiny polyp, termed an actinula, which settles on and attaches to the perisarc of the adult. However, the larva must have sufficient dispersal potential to enable the species to colonise new substrata so swiftly. It develops attached to the hydranth (the tentacle-bearing distal end of the polyp) of the adult, at the top of the hydrocaulus (polyp stem), and is released into the water column up to 20 cm above the substratum, With an estimated sinking rate of 1 mm per second, the actinula might achieve significant dispersal in an average tidal current flow of 10 mm per second. There are no data available on the duration of the free larva phase of *T. indivisa*, but it must be reckoned in days, or even weeks, as it colonises offshore structures as readily as coastal buoys and pontoons. Prospecting actinulae appear to settle preferentially on adult perisarc, but as a general rule marine invertebrate larvae show decreased discrimination in choosing a settlement site with time, as competence to settle also decreases, and this explains their colonisation of new habitat following perhaps extended dispersal.

The biomass of *T. indivisa* populations fluctuates annually and seasonally, and usually shows an early peak, reflecting high adult growth rates, and a high proportion of juveniles. Belgian scientists studied *T. indivisa* populations on a shipwreck 17 km offshore and at a depth of 30 m (Zintzen et al., 2008). Over a two-year sampling programme the total biomass of the epifauna on the wreck ranged from an October minimum of 9 g AFDW/m^2 to a summer maximum of 1,106 g AFDW/m^2. This range largely reflected seasonal and annual variation in the biomass of *T. indivisa*, which comprised from 59% to 82% of the total biomass. Mean biomass of *T. indivisa* ranged from 60 to 324 g AFDW/m^2 in late winter, rising to 362–912 g AFDW/m^2 in early summer, but dropping to only 5–14 g AFDW/m^2 in October. The sharp decline in biomass in late summer and autumn, probably characteristic of all populations of the species, may have had several causes. Predation is one important factor: pycnogonids, polychaetes and several species of small nudibranch are known to feed on *T. indivisa*, mostly as partial predators, but the large sea slug *Dendronotus frondosus* will ingest entire hydranths as well as a proportion of the supporting hydrocaulus. A further cause may be a sharp decline in the number of recruits in late summer as the adult clumps become crowded with a dense associated fauna, and perisarc becomes encrusted with sessile taxa, especially bryozoans, preventing settlement of actinulae. Tube-dwelling suspension feeders may disperse or disable prospecting larvae, and other symbionts may even eat them.

Much of the southern North Sea is floored with mixed sands, and wrecks are islands of otherwise scarce hard substrata. On these *T. indivisa* forms a constant component of an epifaunal assemblage, in which it is dominant for part of the year, providing habitat for a diverse associated fauna. Numbers of individuals inhabiting *T. indivisa* clumps fluctuated seasonally, in concert with varying biomass of the host, from 6,500 ± 56 individuals per square metre in October to 445,800 ± 189,800 in July; the most abundant species was the amphipod *Jassa herdmani*, with a minimum 1,000 ± 385 individuals per square metre in October and a July maximum of 398,000 ± 189,000. The number of species recorded was 102, with a sample total of 15 in October and 42 in August, and for 24 species their abundance was positively correlated with the biomass of *T. indivisa*. The caprellid amphipods *Phtisica marina* and *Caprella tuberculata* and the porcelain crab *Pisidia longicornis* were the next most abundant crustaceans in the community, while cnidarians, annelids and molluscs showed maximum abundances of 6,100, 1,500 and 5,000 individuals per square metre respectively. These communities were considered unusual in that pioneer *T. indivisa* was the main structural element every year, while elsewhere it seems to be eventually displaced by competitively superior late-successional species. It is not clear how these Belgian wreck communities are maintained, but it is possible that environmental pressures, such as current flow, turbidity and abrasion by suspended material, modify community succession, preventing settlement of late colonisers, and larval supply must also be an important factor where the wrecks are remote from natural hard-substratum habitats.

Offshore drilling rigs provide good opportunities for observing colonisation and succession of sessile biota, in essentially simplified systems isolated from natural, coastal benthos. Two oil-drilling rigs in the northern sector of the North Sea, in waters 158 m and 167 m deep, and two in the central sector in 80 m and 85 m, were monitored over an 11-year period (Whomersley & Picken, 2003). High-resolution video cameras were employed to record fouling communities on clamps securing the oil riser pipe, carrying oil from the well head to the surface, to the drilling rig. Composition of the communities varied with depth and over time. Riser clamps on all rigs were settled by mussels, *Mytilus edulis*, and green algae, forming a narrow zone, close to the surface, in the upper few metres of the water column, that persisted through the entire survey period. Below the mussel zone, hydroids and tubeworms occupied all areas monitored, at all depths, but were displaced within 3–6 years by the Plumose Anemone, *Metridium senile* (Fig. 80). On the two northerly rigs *M. senile* initially settled on the lower riser clamps, but gradually extended its range upwards, outcompeting and displacing the hydroid assemblages, and subsequently-settled colonies of the soft coral *Alcyonium digitatum* (Fig. 82). On the rigs in the central sector hydroids persisted longest, but after six years they were eventually crowded out by overlapping zones of *M. senile*, which was dominant at 20–59.9 m, and occurred as deep as 79.9 m, and *A. digitatum*, which was the spatial dominant between 60 m and 89.9 m. The early-successional, essentially *r*-selected species were outcompeted by *K*-selected species, and the developing communities were defined by competition for space and food. The mussel community, in the upper reaches of the water column, might also be considered to be *K*-selected, but the most important pressure determining their community structure was predation. On the *Scylla* reef mussels settled in the first summer of its creation but were obliterated by the end of the year, and absent thereafter. The wave-beaten environment of the offshore oil rigs was inimical to the most significant predators of *Mytilus edulis*, namely *Asterias rubens* and *Carcinus maenas*, and the mussels flourished in a predation refuge.

The Stone Coral, *Lophelia pertusa* (page 264), was an unexpected colonist of several oil-drilling rigs in the northern North Sea, sited in deep water beyond the 50 m isobath (Gass & Roberts, 2006). *Lophelia pertusa* is adapted to cold oceanic waters, requiring an optimal temperature within the range 4–12 °C and a minimum salinity of 33 psu. In the northeast Atlantic it is found below the thermocline, but was not known to occur in the deep shelf areas of the North Sea. It is present off Shetland, and in deep, cold Norwegian fjords, but had not been recorded from the North Sea until colonies were found on the decommissioned Brent Spar oil storage buoy in the 1990s. Subsequently it was discovered encrusting 13 out of 14 drilling platforms installed in the northern North Sea between 1975 and 1988, in waters deeper than 100 m. Two platforms studied in detail were found to bear a total of 947 colonies of *L. pertusa*, within depth ranges of 59–122 m and 62–118 m, and at densities of 0.02–0.39 and 1.23–2.8 colonies per square metre. The period of time that each platform had been established was known, and on the assumption that the largest colonies represented the earliest colonists, growth rates, in terms of radial expansion, were estimated at 6–26 mm per year and 24–33 mm per year for populations on each of the two platforms. This extension of the coral's range southwards into the deep North Sea was facilitated by the provision of new hard substratum that the founding planula larvae require. Recruitment into a particular rig community would depend upon its distance from the nearest source of larvae, and its proximity to south-flowing cold currents, and would probably have been infrequent and sparse. However, size frequency distributions of the two populations sampled were bimodal, suggesting that the initial colonists had begun to reproduce, and that the new populations had begun to increase through self-recruitment.

CHAPTER 5

Some Key Habitats: Kelps and Seagrasses

Some benthic habitats are sufficiently singular to warrant separate, individual and detailed consideration. They are discrete habitats, usually with clearly marked boundaries differentiating them from their surroundings, and they are habitats engineered by their most significant living components. These are the habitats provided by kelps and seagrasses, by huge beds of Horse Mussels, *Modiolus modiolus*, by stiff blankets of coralline algae, or maerl, and on the outer shelf by the enormous banks of the Stone Coral, *Lophelia pertusa*. They are 'key' habitats within the wider benthic realm in providing a high degree of substratum and habitat heterogeneity, and the broadest range of microenvironments for occupation and exploitation. These habitats are consequently foci of very high biological diversity, supporting rich and dense populations of resident species, which offer food and refuge for adult and juvenile stages of the widest selection of epibenthic and pelagic species. The physical framework of each such habitat consists of one or more living species, and they are characterised by extremely high productivity. In the case of kelp forests and seagrass beds, the respective habitat engineers are significant primary producers, the kelps in particular providing much of the energy that fuels the entire coastal system.

KELPS

Kelps are large brown seaweeds belonging to the order Laminariales and are characteristic of cool temperate and polar waters in both hemispheres (Kain,

1979). Their geographical distribution is limited by the 20 °C summer isotherm; in the North Atlantic this largely confines them to the coasts of northwest Europe, from northern Spain to the Kara Sea, Svalbard, Iceland, southern Greenland, and the coasts of eastern Canada and New England. In the North Pacific kelps flourish from eastern Japan northwards to the Kamchatka Peninsula. South of the equator species of kelp are found around New Zealand, southeast Australia, the west coast of South Africa, the shores of Patagonia and many of the subantarctic islands. On the western coasts of the Americas upwelling cold currents enable kelp species to achieve their greatest latitudinal spread, ranging from Tierra del Fuego almost to the equator in the south, and from Alaska almost to the tropic of Cancer in the north.

The kelps of northwestern Europe comprise eight species in three families. The Alariaceae accommodates the small kelp *Alaria esculenta*, a seemingly delicate species growing no longer than 1.5 m, but found only on the most wave-battered rocky headlands, and the Asian exotic *Undaria pinnatifida*, a recent introduction that seems likely to spread steadily through European coastal habitats. The Phyllariaceae include the largest kelp species found on British coasts, the 4.5 m *Saccorhiza polyschides*, which is, nonetheless, a short-lived opportunistic species that achieves its remarkable individual size in just 10 months' growth. The Laminariaceae comprises the bulk of the northwestern European kelps, in all senses, with five species in the genus *Laminaria*, of which three constitute the overwhelming majority of European macroalgal biomass.

Kelps are found on rocky coasts, from the lowest level of the tide seawards to depths of around 20 m. Their lowest limit is determined by the lowest depth to which blue light penetrates, and will be greatest in clear oceanic waters and least in turbid coastal waters. The band of kelp fringing a coastline will be narrower or broader depending on coastal geomorphology and the gradient of the sea floor, and populations will often be extremely dense. This kelp fringe is often referred to as 'kelp forest', its density and productivity being compared to those of terrestrial forests. However, the analogy is otherwise inappropriate, and the term not particularly informative. Dense populations of large macroalgae certainly contribute significantly to the energy budget of coastal ecosystems, and provide shelter and habitat for diverse and abundant communities of plants and animals. But they are not rooted into their substratum, they do not provide or sustain a soil, and they are not directly grazed to any appreciable extent. Seaweeds derive the inorganic nutrients required for photosynthesis and growth from the surrounding seawater, and their production is returned to the environment in the form of mucus, which contributes to the ambient pool of dissolved organic material, as reproductive products, and as finely comminuted fragments of decayed tissue, termed particulate organic material (POM). While fresh kelp is eaten by very few animals, the decayed POM supports many species of small, detrital-feeding animals, often in very large numbers. However, in a final contrast to terrestrial forest habitats, none of the many species relying on kelp production forms a symbiotic association with the seaweed, or, with the exception of the Blue-rayed Limpet, *Patella pellucida*, is completely dependent upon it.

Three perennial species of *Laminaria* constitute the majority of the kelp biomass around British coasts, *L. hyperborea*, *L. digitata* and *L. saccharina*. They all display the same basic morphology, with a holdfast of stiff, branching haptera that attaches the plant to the substratum, a short stipe, or stalk, and an elongate blade – frond or lamina – that is shed annually. The gigantic *Saccorhiza polyschides* may be found sporadically among all or any of the three *Laminaria* species, and is distinguished from them by its knobbly, bulbous, hollow holdfast, its huge crinkly frond, and by its life cycle, which is limited to a span of less than a year. *Laminaria hyperborea* (Fig. 102a) is characterised by its stiff, round-sectioned stipe, tapering slightly towards the frond, which is rough-surfaced and typically supports a diverse community of other, small, seaweeds. Its holdfast, when grown uncrowded, is a tall, flat-topped cone with regular whorls of haptera, each complete whorl representing a single year's growth increment. The stipe of *L. digitata* (Fig. 102b) is smooth and flexible, offering little purchase for epiphytic communities; its holdfast has a low profile and a very variable shape, and the smooth-surfaced haptera are frequently intergrown with those of neighbouring plants. *Laminaria saccharina*, the Sugar Kelp (Fig. 102c), is distinguished by its dimpled, wavy-edged frond, quite distinct from the flat, smooth blades of the other two species, and it has a very short stipe and a smooth holdfast of tightly knotted haptera.

Laminaria hyperborea occurs in the temperate northeast Atlantic, and is abundant along the western coasts of Britain, Ireland and Scandinavia; it ranges northwards to the North Cape, Norway, and reaches its southern limit on the coast of Portugal, but does not extend into the Baltic. *Laminaria digitata* is distributed on both sides of the North Atlantic; in the east it does not occur quite as far to the south as *L. hyperborea*, its populations declining just south of the Brittany coast, but it has a far more extensive northerly distribution, beyond the Arctic Circle to Svalbard and the Kara Sea, and to Iceland and east Greenland. The geographical distribution of *L. saccharina* seems to shadow that of *L. digitata*, but is not clearly defined to the far north, where it is perhaps confused with other, Arctic, species of the genus. The warm-water kelp *L. ochroleuca* is widely distributed in the Mediterranean, and ranges south along the Atlantic coast of Morocco, and north to the Isles of Scilly and southwest England, where it has

FIG 102. Kelps at extreme low water of spring tides. (a) *Laminaria hyperborea* is recognised by its stiffly erect stipes. (b) *Laminaria digitata*, showing the characteristic divided frond, and smooth, flexible stipe. (c) Crinkly fronds of the Sugar Kelp, *Laminaria saccharina*.

become established only within the last century. It was first recorded on the English coast in 1949, had spread east as far as Salcombe by 1969 (John, 1969), and by 2006 could be found along the southern and southwestern coasts from the Isle of Wight to north Devon, and on the French coast east to the Cotentin Peninsula.

Laminaria digitata is the predominant kelp species in the low intertidal; it will tolerate only a minimum of tidal emersion and is uncovered only at extreme low water of spring tides (ELWS), but occurs commonly in large tide pools, and on exposed coasts in even the smallest rock pools. Its lower limit is variable, but *L. digitata* forms a narrow band extending 1–2 m below the lowest tidal level around much of the rocky coast of Britain and Ireland. It occurs at 3–4 m depth on the Brittany coast, at 5 m in Milford Haven; it grows at 1–2 m depth along the open sea coast of Norway but at 10–20 m in the fjords, and even deeper to the north of its range, to more than 20 m at Svalbard. Water clarity is an important factor determining the depth range of all kelps, but variation in the vertical distribution of *L. digitata* across its geographical range also reflects competitive interaction with *L. hyperborea*. Where they occur together, *L. hyperborea* will outcompete *L. digitata*; its stiff, rigid stipe allows it to overtop and shade the laminae of the latter, and ultimately to outgrow and displace it. However, *L. digitata* has a compensatory advantage in its slender, flexible stipe. At low spring tides the stipe of *L. digitata* allows the plant to lie flat, reducing the extent and duration of tidal emersion, while that of *L. hyperborea* projects from the water, exposing the vulnerable basal part of the blade to dehydration. In areas of vigorous water movement, on exposed coasts and in areas of strong tidal current, *L. digitata* has the advantage, its flexibility reducing the potential for physical damage. *Laminaria saccarhina* has a competitive edge over both of the other species in especially sheltered habitats, and is also able to flourish on more unstable substrata, attaching readily to cobbles, while *L. hyperborea* requires firm rock.

Reproduction and growth

Kelps have complex biphasic life cycles, alternating between an asexual sporophyte phase and a sexual gametophyte. The sporophytes are the enormous seaweeds cloaking rocky coasts, while the gametophytes are microscopic, consisting of just a few cells and undetectable in nature by an unaided eye. During a short breeding season the sporophyte sheds millions of single-celled haploid meiospores which settle among the small understorey seaweeds in the kelp bed and develop into male and female gametophytes. The males release free-swimming gametes which locate female gametophytes, and fuse with the still-attached female gametes to form diploid zygotes. These detach soon after and attach to the substratum as microscopic sporophytes that eventually grow into juvenile kelps beneath the blanketing canopy of the adults. The microscopic stages of the kelp life cycle are exposed to numerous hazards, and of the millions, or billions, of meiospores, gametes and zygotes released over every square metre of kelp bed during the reproductive season, only the smallest proportion will survive as juvenile plants. This massive reproductive output is the first contribution of *Laminaria* beds to the energy budget of coastal ecosystems, and it supports countless numbers of microherbivores. Juvenile kelps must survive dislodgement by the sweeping fronds of understorey algae, and achieve a size beyond which they are safe from the attentions of small grazers, and will then grow slowly beneath the dense adult canopy. As with other large brown macroalgae, the size frequency distribution of a kelp population is always strongly skewed to the left (Kain, 1979): there are always more small plants than large ones. Populations of *L. hyperborea* on the Norwegian coast comprised as many as 14 adult, canopy-forming plants per square metre of rock substratum, and from 14 to 16 at sites along the west coast of Scotland, with a mean of eight per square metre across the whole of that range. The number of juveniles was not recorded in those instances, but at a site off southwest Ireland, a square metre of rock within a *L. hyperborea* bed, at three metres depth, supported 61 one-year-old plants.

In all species of *Laminaria*, growth of the sporophyte is strictly seasonal, and occurs in distinct phases. The holdfast and stipe are perennial, but the lamina is replaced annually, the new tissue growing in a zone of proliferation just above the stipe, termed the intercalary meristem. The tip is thus the oldest part of the frond. In *L. saccharina*, over much of its geographical range, the tip of the lamina is continuously eroded and the boundary between new and old tissue is not always evident. The lamina of *L. digitata* tends to split longitudinally (thus becoming digitate or fingerlike), and the border between old and new tissue, again, is not usually distinct. Growth of the sporophyte displays a circa-annnual rhythmicity, with a period of fast growth during late winter and spring, and a period of slow or no growth during summer and autumn. The lamina produced during the first, fast period of growth is long and wide, and that developing during slow growth is short and narrow, and forms a boundary, visible especially in *L. hyperborea* as a narrow 'collar', between the two fast-growth increments. This rhythm is an ecological adaptation, enabling the kelp to adjust its growth and reproduction to nutrient cycles, and it is an internal, or endogenous, rhythm, cued and synchronised by daylength. The sporophyte commences growth in late winter when inorganic nutrients, most importantly nitrate, are at high ambient levels. The old blade, originating from the previous winter/spring growth phase, acts as a reserve providing about 70% of the carbon needed for the synthesis of new tissue. In spring the old frond is eventually shed, detaching along a line of weakness marked by the slow-grown part of the lamina. This reserve is replenished in the next summer slow-growth period, before the plant enters an autumn resting phase.

Experimental cultivation of sporophytes of *L. hyperborea* and *L. digitata*, with controlled temperature and light regimes, demonstrated that the seasonal rhythm of growth in both species is cued by daylength (Schaffelke & Lüning, 1994). Exposed to 12 hours of daylight, *L. hyperborea* displayed a predicted growth cycle rhythm with an average period of 36 weeks: growth rate of the lamina increased steadily for the first 20 weeks of the experiment, simulating the natural winter/spring phase of fast growth, but thereafter decreased sharply to a minimum. With eight hours of daylight the plants embarked on continuous growth with little variation in growth rate and no apparent rhythm. A daylength of 16 hours resulted in a decline or cessation of growth, simulating the autumn resting phase, while continuous light resulted in no growth at all. Sporophytes of *L. digitata* subjected to two light regimes displayed rhythmic growth with eight hours of light daily, but with 16 hours showed continuous slow growth with no discernible rhythm. These growth patterns are controlled by endogenous physiological mechanisms, commonly termed 'biological clocks'. Compressing the annual cycle of daylength to six months, and then to three months, by shortening proportionately the experimental day, resulted in the experimental plants displaying compressed growth cycles. However, it is difficult to explain why, for each species, one daylength was critical in cueing the growth cycle, particularly when both occur over a considerable latitudinal range, although the fact that critical daylength was shorter for *L. digitata* than for *L. hyperborea* may be related to the greater latitudinal range of the former. At the northern ends of their respective geographical ranges, *L. digitata*, which extends far into Arctic regions, experiences shorter winter daylengths and longer summer daylengths than *L. hyperborea*, and a swift change from short daylength to long daylength in spring. Physiological adaptations permitting initiation of rapid growth at a short, eight-hour daylength, and greater light tolerance allowing continuing growth at long daylength, enable *L. digitata* to flourish in high Arctic habitats. *Laminaria saccharina*, which has as broad a geographical range as *L. digitata*, must possess similar adaptations. Greenland populations of this species begin the rapid growth phase long before the summer ice melt, when less than 1% of surface sunlight penetrates to the sea floor (Borum *et al.*, 2002). Under these conditions the carbon reserves of the old frond are especially important, and unlike populations in temperate British waters, Arctic *L. saccharina* does not shed its fast-grown frond annually but retains it for up to two further years, and the lamina is thus, effectively, perennial.

Growth rates of *Laminaria* plants are easily monitored by punching holes in the midline of the frond, just above the intercalary meristem, and measuring their rate of progression as new tissue is added behind them (Kain, 1979). In all temperate northeast Atlantic species, rapid growth proceeds through early spring, peaking in May in *L. saccharina* at a mean elongation of 1.7 cm per day. *Laminaria digitata* grows more slowly, with a peak, also in May, of 1.1 cm daily, but its rate of growth declines only gradually from then to August, and begins to increase again from November. Growth rate in *L. saccharina* continues to drop sharply from May to November, when it also begins to pick up once more. *Laminaria hyperborea* achieves its maximum growth rate of around 0.9 cm per day in April, but growth slows sharply thereafter, and is negligible between June and November. Expressed as annual production, kelp growth is certainly significant: Norwegian kelp beds contribute 1,000–1,500 g of carbon per square metre per year to marine food webs. However, such bald figures only become impressive when related to the actual tonnage of kelp represented, its biomass. The annually grown lamina can be considered to represent the total annual growth of the plant, and thus the wet weight per unit area is a shorthand measure of annual production which emphasises the immense importance of kelp production to coastal ecosystems. *Laminaria hyperborea* covers at least 5,000 km^2 of sea floor along the coasts of Norway; size and growth rate of individual plants vary with depth and temperature, and thus latitude, but it has been estimated that each square metre produces 3–12 kg wet weight of kelp annually (Fredriksen, 2003). Taking an average of 10 kg/m^2/year, not unreasonable given a measured maximum of 40 kg, this equates to 50 million tonnes of *L. hyperborea* annually, which is a great deal of seaweed. No similar figures have been derived for *L. digitata*, which occupies an enormously larger geographical range in the northeast Atlantic than *L. hyperborea*, but *L. saccharina* biomass in the eastern English Channel was measured at 5.2 kg/m^2 at the end of the fast growth period (Gevaert *et al.*, 2001).

Kelp communities

Kelp beds provide shelter, habitat and food. On rocky coasts the dense canopy of fronds moderates water movement, providing calm refuge for pelagic fish and surfaces for settlement and exploitation for a small number of invertebrate species. The stipes of *L. hyperborea* offer abundant surface area for colonisation by small seaweeds, especially rhodophytes (red algae), which in turn provide habitat for many species of sessile and sedentary animals. The perennial holdfasts of all species of *Laminaria* (Fig. 103) contribute significantly to benthic habitat complexity and are largely responsible for the high taxonomic diversity of kelp beds. The kelp-associated fauna is extremely diverse, in terms of both

FIG 103. The perennial holdfast of *Laminaria hyperborea*, with a rich associated community of small red seaweeds, pink coralline algae and encrusting invertebrates. (J. S. Porter)

numbers of species and modes of life, and the number of species and individuals supported per unit area is amazingly high. A survey of *L. hyperborea* beds at four localities along 1,000 km of the Norwegian coast yielded a total of 238 species of invertebrate macrofauna, as well as more than 40 species of small epiphytic seaweeds (Christie *et al.*, 2003). The macrofauna consisted almost entirely of motile species, few encrusting species were identified, and meiofaunal records were beyond the resources of the survey, so total taxonomic diversity remained unknown. From a total of 56 mature plants carefully bagged and collected with their entire communities, almost half a million invertebrate individuals were collected. This represented an average of around 8,000 individuals per plant, and an average density of 10 mature, canopy-forming plants per square metre yielded an average of 100,000 individuals per square metre of kelp bed. At one site, the total counted computed as equivalent to one million macrofaunal invertebrates per square metre. There have been no such detailed surveys of kelp habitats dominated by the other species of *Laminaria*, but 15 specimens each of *L. digitata* and *L. ochroleuca* from a single site at Plymouth yielded a total of 130 species, including 12 species of algae and 33 species of sessile organisms that had not been considered in the Norwegian survey (Blight & Thompson, 2008). Nevertheless, the diversity of motile macrofauna was sufficient, perhaps, to suggest that over an equivalent latitudinal range the species richness of *L. digitata* and *L. ochroleuca* communities might be as great as that recorded for *L. hyperborea*.

The various invertebrate taxonomic groups are unequally represented in kelp habitats, and both numbers of species and numbers of individuals may be disproportionately large for some groups. The taxonomic composition of the kelp fauna and the relative abundance of species are also likely to vary seasonally, and according to the objectives and taxonomic expertise of the investigating biologists. Small crustaceans, gastropods and polychaete worms are usually the most abundant and taxonomically diverse faunal categories. The Norwegian study found 60 species of amphipod alone, totaling almost 200,000 individuals; gastropods comprised 48 species, with a total of more than 100,000 individuals, while 36 species of polychaetes yielded a more modest 18,000 individual specimens (Christie *et al.*, 2003). Of the gastropod total, 11,000 consisted of unidentified juveniles of a single family, the Rissoidae, which, together with a substantial majority of the 93,000 individual bivalves recorded, might have reflected a seasonal peak of post-settlement juveniles that might eventually disperse to other habitats, such as rocky substrata adjacent to the kelp bed.

Kelp habitat is conveniently divided into three vertical zones, each characterised by particular environmental conditions and particular faunas. The canopy constitutes the greater part of the kelp habitat but demands the greatest adaptability. Kelp laminae are fast growing, flexible and slippery, and while the canopy provides an enormous area for colonisation and exploitation, its surface exudes bioactive compounds that protect the most vulnerable parts of the frond. Kelp laminae support few sessile epiphytic organisms, few grazers, and practically no large browsing organisms, but those that are adapted to the silky-surfaced kelp fronds are especially interesting. The lightly calcified bryozoan *Membranipora membranacea* is the most constant and conspicuous sessile epiphyte associated with kelp laminae. Small colonies of *M. membranacea* can be found on *L. saccharina* and *Saccorhiza polyschides*, and even on low-shore fucoids such as *Fucus serratus* (Fig. 95), but it is especially adapted to the broad, flat laminae of *L. digitata* and *L. hyperborea*, on which it forms extensive, brilliantly white, lacy sheets. Following a prolonged planktonic existence the distinctive cyphonautes larva (Fig. 26b) of *M. membranacea*, a brown triangle with ciliated base and apex, resembling a microscopic, gyrating samosa, attaches to the kelp lamina and metamorphoses into the founding zooid of a new colony. It grows by rapid peripheral budding, and growth is directional, the expanding colony spreading onto the new tissue at the base of the lamina, and away from the fraying and eventually discarded tissue at the frond's tip. By late summer much of the fast-grown frond of *L. hyperborea* and *L. digitata* is covered with a delicate

FIG 104. A late-summer frond of *Laminaria digitata*, extensively overgrown by the bryozoan *Membranipora membranacea*.

tracery of *Membranipora* (Fig. 104); it may not be continuous, as it provides prey for a number of small predators, especially the sea slug *Polycera quadrilineata*, but it is perennial, and fragmented colonies will regenerate in spring, spreading onto the new season's growth of lamina. *Membranipora* is a suspension feeder; as with most bryozoans its diet consists of a wide range of fine particulate material, although what part of it has any nutritional value is not known. However, it has been shown to absorb kelp mucus directly from the immediate environment, and it is likely that this may supply a perhaps significant proportion of the bryozoan's energy requirements. In the northeast Atlantic there is no evidence that the kelp suffers any disadvantage from its coating of *Membranipora* colonies, and perhaps the rate, and period, of growth of the kelp lamina, coupled with predation pressure on the bryozoan, allow for sufficient unshaded, photosynthesising tissue to support the late-summer and autumn growth and respiration of the seaweed. In contrast, along the coasts of New England kelp beds seem to have been disastrously affected by anthropogenic import of *M. membranacea* (Lambert et al., 1992). Perhaps because of a lack of natural predators to check it, the bryozoan spread rapidly southwards, colonising and overwhelming the kelp beds, which died as rapidly. It is possible that this reflected not only a lack of predation pressure on the bryozoan, but also a mismatch between growth periods of the kelp and its unexpected new epiphyte.

The crinkly fronds of *Laminaria saccharina* and *Saccorhiza polyschides* are not suited to the broad, spreading sheets of *Membranipora*, and the annual life cycle of *S. polyschides* is unsuited to a perennial epiphyte. However, both support a taxonomically wider range of sessile epiphytes, including several species of encrusting and erect, tufted bryozoans, tubeworms of the genus *Spirorbis* and a variety of small hydroids, and these in turn support a number of associated predators, especially small sea spiders and sea slugs, leading to communities much more diverse than those supported by the smooth fronds of *L. digitata* and *L. hyperborea*.

There are very few herbivores, either invertebrates or fish, which feed directly on fresh kelp, but these include several that have significant effects on kelp communities. Sea urchins are among the most important of these, although for reasons that are quite unclear they have little or no impact on northeast Atlantic kelps. On the northwest Atlantic coasts of Canada *Strongylocentrotus droebachiensis* has the most significant effects on kelp ecology. It is omnivorous, grazing detritus as well as algae, but at a critical population density it will switch to feeding solely on fresh kelp. This diet switch coincides with the onset of an aggregative behaviour leading to 'urchin fronts', dense bands of sea urchins that move across the sea floor transforming the habitat from one dominated by kelp beds to one dominated by encrusting, coralline algae. These new habitats are termed 'urchin barrens', and they are stable, maintained by continuous grazing by the urchins, now primarily detritivorous again. It has been noted that there is a cycle between mature kelp bed and urchin barrens, with a period variously asserted to be from three to four or 15 to 20 years, but what drives the cycle is still unclear. It is certain that kelp beds can only begin to re-establish following a great mortality of sea urchins; it is not clear what may cause a mass die-off, but one possibility is an increasing incidence of parasitic disease (Hagen, 1995). It is even more uncertain what causes the dramatic increase in sea urchin density at the beginning of a cycle. Natural factors might include an exceptionally strong recruitment, and most echinoderm species are known to show good larval settlement, survival and recruitment in benign years. The switch in feeding mode, from omnivory to herbivory, and in behaviour, from dispersed and sedentary to aggregated and migratory populations, are also components of the change, but what drives them is quite unknown. In northeast Pacific kelp habitats anthropogenic effects, such as the removal of competitors, grazing shellfish such as abalones, or predators such as lobsters and sea otters, have also been implicated, but the answer is still to be found. It has been suggested that a boom in sea urchin populations in northern Norway in the 1980s reflected profound changes from a benthic ecosystem characterised by long-term persistence to one characterised by cyclical phenomena (Hagen, 1995), in which urchin-dominated communities alternated with immature kelp assembages. What is most curious is that over its northwest Atlantic and northeast Pacific ranges *S. droebachiensis* controls and engineers kelp habitats, and further south along the Californian coast its congener, *S. franciscanus*, has an equally powerful influence on all kelp environments. Yet, although *S. droebachiensis* occurs in the northeast Atlantic from Arctic Norway to west Scotland and northeast England, population outbreaks are rare and infrequently documented events, and the species appears to have little effect on northwest European kelp habitats. *Echinus esculentus* (Fig. 94), the common Edible Sea Urchin, is another, larger, grazer that is also known to display both solitary and aggregative behaviour, and both omnivorous and determinedly herbivorous feeding. It has been proposed as an important agent limiting the depth range of *L. hyperborea* across part of the urchin's range, but appears to play no significant role in ecological cycling.

Two small molluscan grazers are especially characteristic of the fronds of *L. hyperborea* and *L. digitata*, the specialist Blue-rayed Limpet, *Patella pellucida* (Fig. 105), and the generalist Banded Chink Shell, *Lacuna vincta* (Fig. 106). *Patella pellucida* can be found occasionally on large plants of *Fucus serratus*, but it is otherwise limited to the two kelp species on which it spends its entire life cycle. Newly settled juvenile limpets are often seen clumped together close to the base of the frond, and larger individuals are more widely distributed. The limpets graze the surface of the lamina, leaving pale patches of abraded surface, and appear to show no preference for new or old tissue. In autumn adult limpets migrate basally, a significant proportion surviving the winter to breed in the spring, and dying before the next generation settles from the plankton. In some populations, however, single, large, old individuals eat into the underside of the stipe, at the top of the holdfast, excavating a deep cavity which weakens the plant, making it vulnerable to snapping in winter storms (Fig. 107). The Blue-rayed Limpet occupies a narrow vertical range, from ELWS to around 10 m depth, within which it may impose an effective limit on the life span of *L. hyperborea*.

FIG 105. Blue-rayed Limpets, *Patella pellucida*, on *Laminaria digitata*.

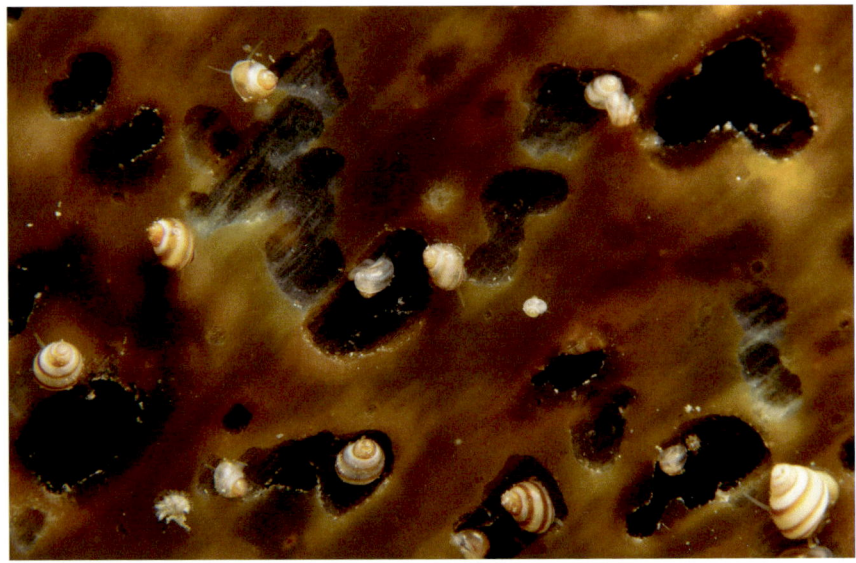

FIG 106. A group of the herbivorous gastropod *Lacuna vincta* grazing a kelp frond. (J. S. Porter)

FIG 107. An adult Blue-rayed Limpet in the cavity it has created in the underside of a *Laminaria digitata* holdfast.

Lacuna vincta lives on most species of large lower-shore and subtidal macroalgae; it has a prodigous reproductive output, with mature females producing tens of thousands of eggs in a short breeding period, and in 'plague years', when larval settlement and recruitment are overwhelmingly high, it can have a dramatic effect on whatever seaweed it settles upon. In contrast to *P. pellucida*, *L. vincta* was shown experimentally to feed by preference on newly grown lamina tissue close to the meristem, and thus in high numbers it is far more damaging to its host plant (Toth & Pavia, 2002a).

The stipe and holdfast of the kelp plant together support the majority of the kelp-associated fauna in terms of species, and the overwhelming bulk of the community in terms of numbers of individuals. However, although they may have many species in common, stipe and holdfast display distinct ecological differences, and the distinction between the two is particularly sharp in *L. hyperborea*. The stiff, rough-surfaced stipe of *L. hyperborea* provides a firm support for attachment and colonisation by sessile organisms (Fig. 108), and supports a diverse fauna of sponges, cnidarians, bryozoans and greyish granular sheets of colonial ascidians. It is also especially important as habitat for epiphytic

FIG 108. The epiphytic community of a *Laminaria hyperborea* stipe. (J. S. Porter)

seaweeds, which derive a competitive advantage in growing above the dense understorey algae. In turn the often luxuriant ruffs of epiphytic algae contribute to the complexity of the stipe habitat, providing shelter and a massively increased surface for colonisation. The range of morphologies of the different seaweeds flourishing on the kelp stipe leads to maximum structural complexity and a corresponding magnification of microhabitats, with the greatest potential for faunal diversity.

While more than 40 species of epiphytic algae were recorded from Norwegian *L. hyperborea* beds, the total biomass, measured as wet weight, would have been dominated by just a few species. In temperate northwestern European waters the stipe epiphytes of *L. hyperborea* consist in the large part of just four species of rhodophyte: *Palmaria palmata*, *Membranoptera alata*, *Phycodrys rubens* and *Ptilota gunneri*. A classic study of kelp epiphytes at St Abbs Head, Northumberland, found that these four species comprised more than 95% of the total stipe epiphyte biomass (Whittick, 1983). They have four contrasting morphologies: *Palmaria palmata* has broad, flat, deeply lobed fronds up to 30 cm long, while *Membranoptera alata*, also up to 30 cm, develops narrow, dichotomously branching fronds. *Phycodrys rubens* resembles bunches of beech leaves, with sharply lobed edges, arising from narrow branching stalks, and *Ptilota gunneri* has a densely branching, feather-like form. The numerous other species, principally rhodophytes, comprising the minority of *L. hyperborea* stipe epiphyte biomass might include stiff, chalky clumps of *Corallina officinalis*, long, strap-like fronds of *Delesseria sanguinea*, and many smaller species growing as feathery clumps or short, lobed blades.

There is competition between the seaweeds for space, and access to light, and the dominant four species tend to be zoned with respect to both water depth and position on the kelp stipe. *Palmaria palmata* occurs only at the top of the vertical range of *L. hyperborea*, at 1–2 m below ELWS, where it may comprise as much as 97% of the total epiphyte biomass. It is adapted to high light levels and flourishes on the upper portion of the stipe, and attaches readily to the smooth young tissue immediately below the blade, a capacity which also enables it to colonise the slippery stipe of *L. digitata* (Fig. 109). *Phycodrys rubens* constitutes around 80% of the total epiphyte load at the deepest levels to which *L. hyperborea* ranges, which at St Abbs Head was 12 m. *Membranoptera alata* and *Ptilota gunneri* occupied the middle regions of the stipe at intermediate depths, competing with *P. palmata* in shallow water, and with *P. rubens* at greater depth, and thus occurred higher on the stipe as water depth increased. The zonation of the four species, and competitive interactions between them, and between each and the rest of the seaweed epiphyte community, will also vary seasonally, in relation to the growth cycles of each species, and to winter loss of fronds.

The total biomass of epiphytic seaweeds reaches a maximum between June and September, depending on the depth at which they grow, and drops abruptly in November and December, with minimum values, as wet weight or volume, recorded in January. Most of these small red seaweeds have annual life cycles, or overwinter as inconspicuous holdfasts. The winter loss of foliage may represent 75% of the epiphyte biomass at some depths, or may be practically total at others. The invertebrate fauna associated with the stipe epiphytes must be adapted to this cycle of growth and loss, and indeed many species have annual life cycles coincident with, or overlapping, those of their host algae. Sessile animals attached to the fronds of the epiphytes – sponges, hydroids and bryozoans – must recolonise through the reproduction of individuals overwintering on perennial substrata, but motile animals have the option to move. Norwegian *L. hyperborea* (Christie et al., 2003) supported an average 42 motile species and 5,728 individuals for a sample of 56 stipes, while the 56 holdfasts sheltered an average

FIG 109. A *Laminaria digitata* plant with a ruff of the rhodophyte *Palmaria palmata*.

63 species and 1,780 individuals. Small gastropods, such as *Lacuna vincta* and species of *Rissoa* (Fig. 110), were predominant in the stipe habitat, but bivalves and tube-building amphipods were also abundant. In winter much of this fauna is doubtless lost through wholesale removal of the epiphyte burden by wave surge, but many of the species may be rafted on detached algae, or drift from mucus threads, or simply crawl into the holdfast, or onto any adjacent rocky habitats. That there is significant winter migration of stipe fauna is demonstrated by the fact that the taxonomic composition of the holdfast fauna changes as a result of an influx of stipe-associated species.

In contrast to the stipe, with its annually replaced epiphyte assemblage, the holdfast is a perennial habitat that increases in size, and hence in habitat volume

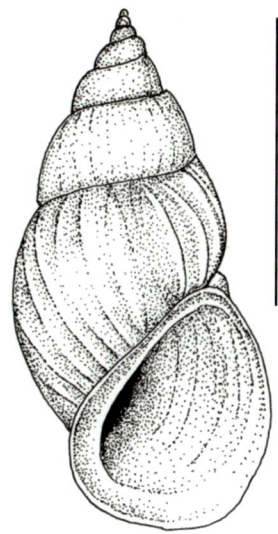

FIG 110. A common epiphytic gastropod, *Rissoa membranacea*. Scale bar: 5 mm

and complexity, by annual addition of a ring of anchoring haptera (Norderhaug *et al.*, 2002). This increase continues through the five- to ten-year life span of the individual kelp. The holdfast acts as a trap, accumulating silt and organic detritus, and thus providing food as well as habitat space. The fauna may not be as abundant as that supported by the stipe, but it is generally much more diverse. The motile fauna is dominated by polychaetes, and by a wide range of small crustaceans, including isopods and tanaids as well as amphipods. Many of the species may be relatively sedentary, dispersing less readily than those associated with the stipe, and may have life cycles much longer than a year. Trophic modes also differ between the fauna of the two habitats, and are much more diverse among the holdfast communities. Stipe epiphytes support a few small mesograzers, principally gastropods and amphipods, and a minority of larger browsers such as juvenile sea hares, *Aplysia punctata* (Fig. 111), and sea urchins, *Echinus esculentus*, but the fauna otherwise consists of suspension feeders and small specialist predators, particularly sea slugs. The holdfast community includes suspension feeders, but also a wider range of small predators and many detritivores, feeding on the

FIG 111. The sea hare *Aplysia punctata* is one of the largest sea slugs found in British waters, and feeds on a wide range of seaweeds. (P. E. J. Dyrynda)

detrital load within the interstices of the holdfast. The relatively stable character of the holdfast habitat allows for maximum trophic diversity, encompassing many meiofaunal detritivores, especially harpacticoid copepods, ostracods, mites and nematodes, the taxonomic diversity and abundance of which have rarely been measured or estimated for northwestern European holdfast communities.

Primary production and trophic pathways

The contribution of *Laminaria* habitats to the energy budgets of northwest European coastal waters is enormous. Returning briefly to numerical aridity, it has been estimated that Norway's *L. hyperborea* fringe contributes around 1 kg of carbon per square metre of kelp bed annually, and that is around three times the measured production of phytoplankton for the same region (Fredriksen, 2003). Kelp-derived carbon is an important resource for kelp-associated animals, and this can be demonstrated by comparing ratios of naturally occurring stable isotopes of both carbon ($^{13}C/^{12}C$) and nitrogen ($^{15}N/^{14}N$) in kelps, their most important rhodophyte epiphytes and their associated faunas. (The delta (δ) notation expresses the ratio of the heavier to the lighter isotope, thus $\delta^{13}C$ and $\delta^{15}N$). The ratios change as $\delta^{13}C$ reduces by about 1% per trophic level while $\delta^{15}N$ increases by around 4% per trophic level; measures of carbon and nitrogen isotopes thus supply insights into food web structures in kelp communities. For example, $\delta^{13}C$ and $\delta^{15}N$ values measured in *Patella pellucida* were close to those for *L. hyperborea* tissue, indicating that it derived 100% of its carbon input through grazing kelp tissue. For *Lacuna vincta*, in contrast, only 60% of its carbon originated from *L. hyperborea*, the rest deriving from red algal epiphytes and a minority of animal material. The diet of *Aplysia punctata* was apparent from tissue carbon isotope values, which were close to those of epiphytic red algae, while *Echinus esculentus* was shown to derive 47% of its carbon input from kelp. Data for suspension feeders were equally interesting: for most sponges, hydroids, bivalves, bryozoans, brittlestars and ascidians, carbon isotope values were intermediate between those for phytoplankton and kelp, and the balance varied between species. The bivalve *Hiatella arctica* (Fig. 112) was found to derive as much as 55% of its carbon input from kelp, while herbivorous amphipods seemed to be feeding largely on phytoplankton. The kelp carbon signature (i.e. its $\delta^{13}C$) can be found throughout the food web based upon it: the Edible Crab, *Cancer pagurus*, derived 67% of its carbon input indirectly from *L. hyperborea*, via its prey. At the highest trophic levels, the small sucker fish *Liparis montagui*, contained 68% kelp-derived carbon and Cod 33%, while the diet of the fish-eating Cormorant, *Phalacrocorax carbo*, included 37% kelp carbon and that of the molluscivorous Eider Duck, *Somateria mollissima*, 49%. The trophic interrelationships of these top consumers

FIG 112. Shells of the crevice-dwelling bivalve *Hiatella arctica*. Scale bar: 1 cm.

are revealed by their carbon isotope values. Cod living within the kelp beds had a greater $\delta^{13}C$ content than those living and feeding in the open sea, while juvenile Saithe, *Pollachius virens*, which use the kelp bed as refuge, had nonetheless low $\delta^{13}C$ values, as they fed on kelp-associated amphipods that derived their nutriment from phytoplankton rather than from kelp.

More than 20 species of fish have been recorded from *L. hyperborea* (Norderhaug *et al.*, 2005). While some of these are of infrequent, almost accidental occurrence, such as Mackerel or Plaice, others seem more closely associated with the habitat. For some of these the kelp canopy is important as a refuge, or as a nursery, while for many species kelp beds are significant as feeding grounds. Fish feeding amongst kelp may be generalists, feeding on a broad range of prey at all levels within the bed, or specialists associated with frond, stipe or holdfast and displaying a limited choice of prey. Feeding behaviour, prey type and rate of consumption will often be conditioned by responses by the fish to the top predators, otters, seals and seabirds. Saithe are probably the most abundant and constant of kelp-associated fish species, and as noted previously these are often young fish for which refuge – shelter from predation – is the most important resource provided by the kelp canopy. This is reflected in their diet, for although juvenile Saithe have been found to feed on amphipods, isopods and the small gastropods associated with kelp fronds, their principal prey item is often the pelagic copepod *Calanus finmarchicus*. Cod, Ballan Wrasse, *Labrus bergylta*, and Bull Rout, *Myoxocephalus scorpius*, were the largest fish occurring commonly in

Norwegian *L. hyperborea* beds; all three are bottom feeders and were found to consume a wide variety of kelp-associated fauna, but each also showed a degree of specialisation. Cod and Bull Rout fed on crabs and small fish from among the holdfasts, while Ballan Wrasse consumed especially squat lobsters, *Galathea* (Fig. 113), and smaller crustaceans. The epiphyte-clad stipes attract a guild of smaller species of wrasse that feed principally on gastropods and amphipods, and particularly on *Rissoa parva* and *Lacuna vincta*.

The natural fertility of coastal waters thus creates the largest biological resource in northwestern European seas, which supports an enormous and diverse biomass of consumers. However, this resource is not easily exploited, because kelp tissue is extremely unpalatable. As with all other large, temperate-zone brown macroalgae, kelps are not unprotected from herbivores. The holdfast and stipe are structurally tough, and the apparently vulnerable lamina has antimicrobial surface properties that discourage prospecting invertebrate larvae from settling, and probably also deter microherbivores. All of these large seaweeds synthesise secondary metabolites – 'secondary' in the sense that they do not contribute to major metabolic processes within the plant, but instead consist of active chemical compounds that interact with receptors in competitors and herbivores, to suppress or deter them. Most important in fucoids and kelps are compounds termed phlorotannins that render the algal tissues distasteful (Toth &

FIG 113. The squat lobster *Galathea squamifera*. (P. E. J. Dyrynda)

Pavia, 2002b). Phlorotannin content of algal tissue is related in ways not presently understood to levels of nitrogen in the tissues. The ratio of carbon to nitrogen in kelp tissue is greater than 17 for much of the year; animals need a ratio of less than 17, and nitrogen limitation may thus result in the high primary production of kelp being unavailable to consumers; it is simply unsuitable as food (Norderhaug et al., 2003). Phlorotannin content of kelp tissue decreases slowly as it ages, but the C/N ratio does not change, and kelp production only finally becomes available to consumers following bacterial degradation of kelp tissues. Experiments with artificially induced bacterial degradation showed that the phlorotannin content of *L. hyperborea* tissue decreased from 2.5% to less than 1%, dry weight, after 44 days of bacterial activity, and the effect on the C/N ratio was even more dramatic. The C/N ratio of the kelp substrate and its bacterial community was halved in just three days of bacterial degradation and dropped to less than one-third by 44 days. Four amphipod herbivores, and the snail *Rissoa parva*, were tested with diets of either fresh or degraded kelp; the results were equivocal, in that *R. parva* and two of the amphipod species survived equally well on fresh or degraded kelp, but the other two amphipods survived, and showed growth, only when fed degraded kelp. A proportion of the kelp production was utilised by respiration and growth of the bacterial community, and it is possible that it was the bacteria that provided the primary nutrition for the two amphipod species that rejected fresh kelp. However, feeding experiments with two bivalve species, the oyster *Crassostrea gigas*, and the northwest Pacific mussel *Mytilus trossulus*, showed clearly that pellets of fresh kelp reduced feeding rate, and were actively rejected as food by both species, while pellets of artificially aged kelp tissue, with lowered phlorotannin content, were actively selected (Levinton et al., 2002).

SEAGRASSES

In contrast to kelp 'forests', seagrass meadows are reasonable analogues of their terrestrial eponyms: seagrasses are flowering plants, rooted in a soil that supplies nutrients required for growth and reproduction, and under favourable conditions seagrass meadows are both lush and verdant. There are 58 known species of seagrass, in 12 genera, accommodated within two angiosperm orders (Jernakoff et al., 1996). They can be found from high-latitude cool temperate seas to the tropics, and achieve their greatest diversity in the Australian region. The most extensive seagrass beds are those of Western Australia, where no fewer than 25 species, in ten genera, are recorded. The European seagrass flora is comparatively modest, with just six species, *Halophila decipiens*, *H. stipulacea*, *Cymodoce nodosa*, *Posidonia* oceanica, Zostera marina and Nanozostera noltii, and only the latter two species occur in north European seas. Tropical seagrasses belonging to the genus *Thalassia* are commonly termed turtle grass, while cool temperate species are often referred to as eelgrass, although botanists have frequently employed the descriptive, though confusing, 'grass wrack', denoting a grass-like seaweed. Species of *Zostera* and *Nanozostera* are classified in the same order as the freshwater pondweeds belonging to the genus *Potamogeton*, reflecting their terrestrial pedigree, and in low-salinity habitats in the northern Baltic Sea *Zostera marina* grows in mixed communities including species of *Potamogeton* and the Beaked Tassel Pondweed, *Ruppia maritima*.

Eelgrass flourishes in clear shallow waters, in sheltered embayments and within estuaries; *N. noltii* is largely intertidal in distribution, but spreading into the shallow subtidal in low salinity habitats, while *Z. marina* (Fig. 114) ranges from the low intertidal into the subtidal, with a maximum depth of around 10 m. Along the Atlantic coasts of Britain and Ireland the two species thus have discontinuous distributions; they range eastwards along the Channel coast as far as West Sussex, and from the Thames estuary to the Wash, and in the major estuaries of northeast England and eastern Scotland. The most extensive European seagrass meadows are found on the sheltered coasts of the southern North Sea and the western Baltic, and the coastal waters of Denmark, in particular, formerly supported enormous populations of the two species. The Danish fisheries scientist C. G. J. Petersen seems to have been one of the first ecologists to appreciate the ecological significance of eelgrass, early in the twentieth century; he calculated that *Zostera* meadows covered 6,860 km² of shallow sea floor around the coast of Denmark, with an annual production of 'eight million tons of dry matter' (Rasmussen, 1977). Petersen reasoned that such a huge primary production must be of profound importance to marine energy budgets, and concluded that eelgrass production formed the base of marine trophic webs. Plankton ecology was then in an early stage of development, and it was another decade or more before the importance of phytoplankton primary production in marine energy budgets was understood. This was emphasised from 1931 onwards, when North Atlantic *Zostera* meadows began to vanish before the spread

FIG 114. A patch of Common Eelgrass, *Zostera marina*, meadow. (J. S. Porter)

of an epidemic disease; the loss of the eelgrass had no measurable impact on marine energy budgets, and the fish populations that Petersen suggested had been dependent on eelgrass production were unaffected.

The perennial bulk of an eelgrass bed is an interwoven mat of elongate, branching rhizomes, up to 2 cm in diameter and bearing bundles of roots, which extend to a depth of around 20 cm within the sediment. The rhizomes produce slender green shoots that emerge from the sediment surface, each consisting of a cluster of narrow, strap-shaped leaves, enclosed basally by a sheath. Rhizomes and roots stabilise the sediment, which increases through the accumulation of leaf litter, algal fragments and fine detritus. Leaves grow continuously in subtidal *Zostera* populations and are shed frequently, each leaf having a short life of a week to a few months. Rhizomes may also have individually short life spans, but *Zostera* meadows spread, and are sustained, by vegetative propagation of rhizomes: populations may be partly or entirely clonal, and individual clones are thought to persist for decades. Flowering shoots are produced in the summer months, fertilisation is achieved through water-borne pollen, and both seeds and whole flower heads are dispersed by water currents. The structure of the eelgrass bed and the morphology of the plants reflect subtle interactions between the eelgrass and its physical and biological environment. Leaf density, leaf length and width, growth rates and production of both leaves and rhizomes, and reproductive output, all vary in response to variation in environmental parameters. Light, temperature, salinity and nutrient cycles, in relation to hydrodynamic factors, are especially important in seagrass ecology. However, seagrasses do not simply respond to variation in environmental parameters, but through their structure modify their immediate environment, mediate the effects of environmental factors, and create complex habitat space for other organisms.

Occurrence, distribution and densities

The Common Eelgrass, *Zostera marina*, was the most widespread and abundant species on north European coasts prior to the epidemic disease that decimated North Atlantic seagrass meadows in the 1930s. The Dwarf Eelgrass, *Nanozostera noltii*, appears to be less susceptible to the disease, and the two species are now probably equally common, although occupying different habitats. *Zostera marina* has a largely subtidal distribution; it is possible to paddle amongst *Z. marina* at extreme low water of spring tides (ELWS) on some coasts, but the plants do not flourish if subjected to frequent emersion. For most populations the optimum depth lies between 1 m and 4 m below chart datum (i.e. the level of the lowest astronomical tide); water clarity, with its consequences for photosynthesis, is one important determinant of depth limits, and in especially clear waters, such as those around the Isles of Scilly and southwest Ireland, *Z. marina* meadows flourish at depths as great as 10 m. *Zostera marina* is said to tolerate a considerable range of salinities, from undiluted seawater, at 35 psu, to brackish habitats in the northern Baltic, at 6.5 psu, and in southern England and on the Mediterranean coasts of France it can be found in lagoon habitats with low and fluctuating salinities. While a degree of euryhalinity – tolerance of fluctuating salinity – must be expected in coastal and estuarine organisms, the range of salinities experienced by *Z. marina* populations across the total geographical range of the species suggests, at the least, that it must exist as a series of physiological races, or ecotypes.

Zostera marina beds are established in sands and fine gravel; they accumulate a soil formed of eelgrass detritus, decayed drift algae and the detached and decaying remains of their associated fauna. They do not persist in turbid muddy conditions. The density of leaf shoots varies considerably; *Z. marina* meadows on the west coasts of Sweden grew at densities of 200–1,000 shoots, mean 245, per square metre (Jephson *et al.*, 2008), while in a north Baltic meadow, at a locality subject to a salinity of 6.5 psu, shoot density ranged from 50 to 500/m^2 (Boström & Bonsdorff, 2000). Around the coasts of Britain and Ireland shoot densities may be as low as 100/m^2 but probably display a range similar to that of the Swedish populations. Each shoot consists of 4–6 flat, elongate and parallel-sided leaves; they are typically up to 50 cm long, exceptionally reaching 1 m, up to 10 mm wide, and abruptly tapered at the tip. Shoot densities and leaf morphology vary seasonally, and in response to variation in environmental factors. Vegetative growth, of leaves and rhizomes, and sexual reproduction vary seasonally in relation to temperature. Growth is rapid at water temperatures of 10–15 °C; below 10 °C there is little active growth, while above 20 °C leaf growth ceases and leaves begin to be shed, although rhizome biomass continues to increase as the plants accumulate reserves for the spring resumption of growth. Vegetative growth peaks in June and July in north European habitats. Production of flower shoots occurs only within a narrow range of optimum temperatures; Baltic populations do not reproduce sexually because water temperatures are too low, and Mediterranean populations do not flower because temperatures are too high. Between these extremes flowering shoots achieve mean peak densities of 7.8/m^2 in the Ems estuary, east Wadden Sea, in September, declining to 2/m^2 by October (Erftemeijer *et al.*, 2008). Leaves are shed continuously, the rate of loss accelerating in late summer, but subtidal *Z. marina* meadows are never entirely unvegetated, although leaf morphology also varies seasonally: summer-grown leaves are longer than those present in the winter months.

Subtidal *Z. marina* meadows in brackish lagoonal habitats, and rare intertidal populations, display plant morphologies and ecological characteristics that distinguish them from *Z. marina* meadows in fully marine environments. The leaves are commonly only 15–30 cm long, with a maximum width of 3 mm, and notched at the tips; the proportion of rhizome biomass is generally low, and these beds appear to be sustained by sexual reproduction and seedling recruitment rather than by vegetative propagation. Leaves are shed with increasing frequency from September, and leaf loss may be total, with all above-surface vegetation lost by late December. Some specialists consider these characteristics to define a third north European species, *Z. angustifolia*, while others regard them as characterising an ecotype of *Z. marina*, and a consensus will probably only be achieved once the molecular genetics of *Z. marina* has been investigated among populations from across its entire geographical range.

Nanozostera noltii has a similar distribution to *Z. marina* on the western coasts of Britain and Ireland, but is much more abundant than the Common Eelgrass on the east coasts of England and Scotland, with especially expansive meadows in the greater Thames estuary, the Wash, the Humber estuary and the Firth of Forth. It is mostly intertidal in occurrence, with dense beds developed on sheltered muddy sands, between mean high and low water marks of neap tides. Individual shoots consist of 2–5 slender leaves, up to 20 cm long, 1.5 mm wide, and with a notched tip, and shoot densities may be extraordinarily high. *Nanozostera noltii* beds on intertidal sand flats in the Wadden Sea achieve densities of 1,000–23,000 shoots per square metre, with an average of around 12,000 (Polte *et al*., 2005). It has a quite different pattern of growth to *Z. marina*. The rhizome of *N. noltii* is much thinner than that of *Z. marina*, and the ratio of rhizome to shoot biomass is much smaller. *Nanozostera noltii* grows rapidly through spring and early summer; its growth is cued by seasonal increases in light and nutrients, and is not reliant on reserves stored in the rhizomes. The leaves last longer into the winter than those of *Z. marina*, and leaf loss may not begin until October, but through the winter grazing wildfowl may remove all remaining leaves, and the plants then survive only as rhizomes and roots. On marine coasts *N. noltii* is primarily an intertidal organism, but its distribution shifts downshore with decreasing salinity, and in enclosed brackish lagoons it becomes truly subtidal. Its very short, narrow leaves and its growth pattern distinguish it from *Z. marina* and *Z.* '*angustifolia*', but clearly the phenotypic and genotypic characteristics of *N. noltii* populations must be just as complex as those of *Z. marina*.

Physical environmental influences

Common Eelgrass beds may occur as continuous meadows, occupying hectares of sea floor, or as discrete patches ranging in area from just a few square metres to several hundred. Most often they form a mosaic of interlinked large and small patches, with intervening areas of bare sand, and some with ring formation, where a physical disturbance has broken the integrity of a patch and initiated erosion. *Zostera marina* beds are dynamic structures, persistent but frequently changing as bed morphology responds to local conditions. Hydrodynamic factors, in relation to water depth, are the primary forces determining bed morphology. Tidal and wave-driven water flow defines the limits of each patch and meadow, and flow velocity has important effects on bed structure, vegetation density and individual plant morphology. Above a critical velocity plants cannot produce sufficient rhizome and root mass to persist, and boundaries of beds are usually fairly sharply defined. At the lowest velocities accumulating silt and detritus, and consequent periods of anoxia and enhanced turbidity, are similarly inimical and the vegetation is sparse and diffuse. Tidal currents and wave-driven water flow have different effects. Eelgrass shoots respond to unidirectional tidal flow by flexing, and persistence and structure of the bed are functions of current strength and the ratio of leaf to rhizome and root biomass: slender, flexible leaves and an increased proportion of subsurface biomass indicate strong currents. The orbital movement of water resulting from wind-induced waves leads to a continuous back and forth flapping of the eelgrass leaves, which is potentially more damaging than a strong unidirectional current. Orbital flow velocity at the sediment surface, in relation to the duration of exposure to wave action, thus determines bed structure and individual plant morphology. On tidal coasts this is further modulated by the local tidal range and the depth of the bed.

Interactions between these factors have been demonstrated experimentally by transplanting *Z. marina* shoots at intervals along a depth gradient, from 0.15 m above mean sea level to 0.9 m below mean sea level, at three locations on the coast of the Dutch Wadden Sea, each subject to a different degree of wave exposure (van Katwijk & Hermus, 2000). At two of the locations transplants were successful at depths of 0.0–0.2 m below mean sea level, where relative exposure to wave action was less than 60% of a tidal cycle (Fig. 115). At greater or lesser depths the duration of wave exposure was too great and all the transplants were lost. The critical orbital water velocity was 0.165 metres per second (m/s); at the third, most exposed, site orbital water velocity was frequently greater than 0.6 m/s, at all depths, and no transplants survived. That hydrodynamic factors rather than depth-related light attenuation were the causative agents was demonstrated simply by providing transplanted shoots with perspex shields that sheltered them from wave exposure, with the result that transplants survived at all depths. These experiments were conducted along a depth gradient within the intertidal zone,

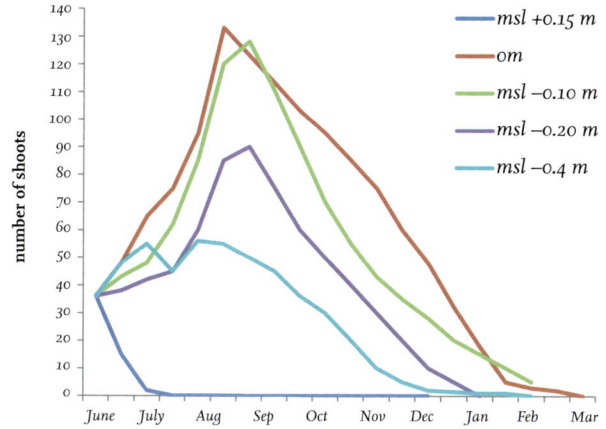

FIG 115. Effect of tidal depth on transplanted *Zostera marina* shoots at a site in the Dutch Wadden Sea. At the shallowest and deepest horizons the transplants failed. msl, mean sea level. (After van Katwijk & Hermus, 2000)

and showed that wave exposure and tidal range may determine the distribution and abundance of *Z. marina*, and that orbital water velocity and duration of wave exposure were critical factors. Wave activity, and the duration of wave exposure, decrease with increasing depth, and in response to the proximity of shelter provided, for example, by offshore banks and reefs, and then the vertical and horizontal distribution of eelgrass are dependent on light and nutrient supply, and to a lesser extent on hydrodynamics.

The effects of hydrodynamic factors on individual plant morphology are subtle, and variation in levels of hydrodynamic energy within eelgrass habitats results in responses which sometimes seem contradictory. For example, experimental culture of *N. noltii* plants in three flume tanks subjected to three different water flow velocities, 0.01, 0.1 and 0.35 m/s, showed that plant morphologies responded significantly within just four weeks (Peralta et al., 2006). With increasing flow velocity, leaf length decreased but leaf elongation rate increased, while root length increased and the ratio of above-surface to below-surface biomass decreased (Fig. 116a); all growth parameters increased with increasing current strength (Fig. 116b). Field measurements of *N. noltii* plants at sheltered and exposed locations in the Wadden Sea appeared to demonstrate the same relationship. At the sheltered site, with ambient flow velocity 0.08 m/s, plant density was 4,869/m^2, average shoot length was 16.18 cm and average leaf length was 10.46 cm. At the exposed site, with ambient velocity 0.26 m/s, density was only 1,987/m^2, and shoot and leaf length were both lower, at 10.72 and 2.02 cm respectively (Schanz & Asmus, 2003), but below-surface biomass was almost half that at the sheltered site. Shoot density, and shoot

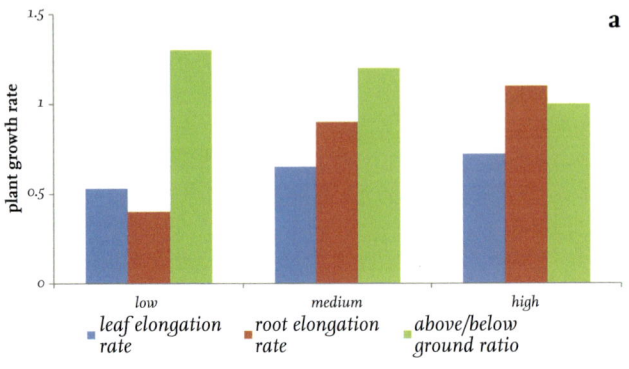

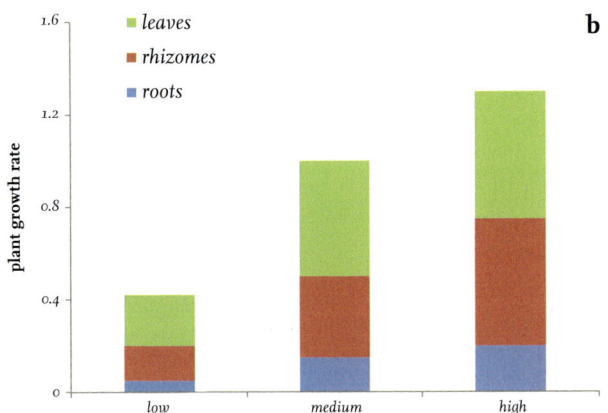

FIG 116. Characteristics of *Nanozostera noltii* plants cultured at three current velocities. Low: 0.01 m/s; medium: 0.10 m/s; high: 0.35 m/s. (a) Leaf elongation rate and root elongation rate, cm per plant per day, and ratio of above-surface to below-surface biomass. (b) Growth rate, mg dry weight per plant per day. (After Peralta *et al.*, 2006)

and leaf morphology, affect the growth efficiency of the seagrass bed, through variation in the ratio between photosynthesising and non-photosynthesising tissue, and thus physical environmental factors, have significant consequences for seagrass productivity. The apparent contradiction between the two examples discussed here simply reflects methodological differences. In the experimental investigation all factors could be controlled, and water flow was unidirectional. The two sites examined in the field study would have differed significantly, as the two different hydrodynamic regimes would have influenced sediment type and stability, and would have comprised varying proportions of orbital and unidirectional flow.

The eelgrass habitat

While the distribution and extent of eelgrass patches and meadows, shoot density and individual plant morphology are all determined primarily by hydrodynamic factors, these same biological attributes in turn have measurable effects on the hydrodynamic regime of the habitat, and consequently on its ecological characteristics. Within an optimum range of values, *Zostera marina* shows a positive response to increasing flow velocity, reflecting increasingly efficient molecular diffusion, of oxygen, carbon dioxide and nutrients, at the interface of water and leaf. However, eelgrass beds act as baffles, slowing currents and dampening wave-driven oscillation, and thus modifying their hydrodynamic environment (Peterson *et al.*, 2004). Within the eelgrass bed the plants create a low-energy environment, promoting deposition of fine sediment, reducing re-suspension and dispersion of material, and stabilising the sediment.

These effects are all attributable to physical processes within the boundary layer of the overlying water body; the boundary layer is that lower portion of the water body that is subject to frictional drag as it moves across the sea floor, and, depending on the degree of roughness of the sea floor and the flow velocity of the water body, it may be from centimetres to many metres thick. As the boundary layer thickens water flow is disrupted, breaking into eddies and forming a zone of turbulence, or turbulent flow. Eelgrass bends and sways in response to water flow, forming a canopy that may be continuous or intermittent, depending on the shoot density of the bed. Water is deflected above the canopy and around the edges of the eelgrass patch, accelerating relative to the ambient flow velocity, and decelerates abruptly as it passes beneath the canopy and between the shoots. The degree and rate of deceleration, established in both field and flume experiments, are striking. Initial velocities of 5–20 cm/s were reduced by up to 50% just 1 m into the bed, and the degree was density-dependent, with the greatest reduction occurring at the highest experimental shoot density, 1,200/m^2 (Gambi *et al.*, 1990). Field measurements in an eelgrass meadow subject to a moderate ambient flow velocity showed that flow velocity within the bed was 84% lower at high shoot density (600–1,000/m^2) than at low shoot density.

The most significant drop in flow velocity occurs within a transitional zone from the upstream edge of the eelgrass bed to a distance of about 1 m into it, and this has interesting ecological consequences. Most of the silt and detritus in suspension is abruptly deposited within this zone, and only the finest and lightest material is carried further into the bed. Passively dispersing invertebrate larvae, and drifting eggs of both invertebrates and fish, also drop out of the water column within this zone, and only strongly swimming larvae, and perhaps the most buoyant eggs, pass further into the bed. The layer of turbulent flow above and below the canopy in this transitional zone also promotes constant re-suspension of material, and a greater degree of turbidity than occurs beyond the zone and deeper into the bed. Patch size is an important ecological parameter in this context: small patches of eelgrass have a high perimeter-to-area ratio, and thus a greater proportion of transitional-zone habitat, characterised by an enhanced turbidity. The fauna associated with Z. marina may thus differ in composition and character in relation to patch size, reflecting hydrodynamic influences. It has been suggested that the hydrodynamic regime of the eelgrass bed is a more significant factor in the settlement and recruitment of juveniles of some bivalve species than predation.

Sediment stability is also related to flow velocity; at mean velocities greater than 0.2 cm/s fine material is re-suspended, and above 0.34 cm/s sand begins to ripple and bed erosion occurs (Widdows et al., 2008). Flow velocities may vary seasonally, increasing as a result of greater winter wave action. When orbital velocity is greater than ambient current speed, or when wave and current are aligned, a thicker boundary layer of turbulent flow results and bed erosion increases. Depth is thus important, and below wave depth critical eroding velocities may not be reached, and integrity of the seagrass meadow is maintained.

Production and trophic webs
Growth rates and production of eelgrass are determined primarily by the availability of light and the supply of nutrients, and the effects of both of these factors are mediated by depth and the hydrodynamic regime. Reduction of light, through water turbidity, or shading, or simply through gradual attenuation with increasing depth, has been shown to have a significant effect on growth and production in Z. marina. Photosynthesis by Z. marina needs a minimum ambient irradiance of 98 µE per square metre per second (E, an einstein, is a measure of photon flux), and this level of illumination must be exceeded for six hours daily in order for growth to occur (van Lent et al., 1995). Eelgrass meadows growing in clear, mesotrophic marine habitats in the Wadden Sea responded positively to experimental enrichment with nitrogen-based fertiliser, showing significant increases in growth and biomass, but the same parameters decreased significantly when shading screens reduced ambient illumination by 84%. The importance of light was clear: shaded portions of the meadow, provided with enhanced nitrogen supply, still showed a significant decline in growth rate and biomass. In a shallow-water eutrophic site shading had the same negative effect on growth rate

and biomass of the *Z. marina* population, but addition of fertiliser had no effect as the habitat was already rich in organic nitrates. Light reduction proved to be a further limitation on this population, which was unlikely ever to experience nitrogen limitation; mesotrophic habitats allow maximum light penetration, but eelgrass populations are likely to experience nitrogen limitation. Other facets of the life cycle of *Z. marina* are likely to vary in response to this ecological gradient. The frequency of sexual reproduction, the incidence of flowering shoots and the production of seed all appear to be influenced by nitrogen availability, perhaps in relation to salinity and temperature, but these relationships have yet to be explored by experiment.

The relation between light, depth and turbidity is further modulated by the effects of shading. This may be a consequence of greater density, and individual size, of eelgrass shoots in nitrogen-enriched habitats (van Lent et al., 1995), or of algal growth similarly boosted by increased nutrient levels. *Zostera* leaves provide surfaces for attachment and colonisation by many other organisms, and the longer and broader the leaves, and the greater the shoot density, then the greater the leaf area per unit area of sea floor. The epiphytes of eelgrass include sessile invertebrates, such as hydroids, bryozoans and ascidians, fungi, diatoms and other microalgae, encrusting coralline algae, and the large macroalgae, to which the term 'epiphytes' is most often applied. Macroalgal epiphytes include green, red and brown seaweeds, both annual and perennial species, with a wide range of sizes, and their patterns of growth, productivity and reproduction are subject to the same environmental factors that affect eelgrass ecology. Macroalgal growth, productivity and biomass are determined by light and nutrient availability, and mediated by hydrodynamic factors, and where macroalgal communities respond positively they begin to affect the eelgrass population negatively. Vigorous seaweed growth leads to increased shading, with the result that the photosynthetic efficiency of the eelgrass declines, growth rates decrease and plant morphology changes, and the structure of the eelgrass population also changes.

Eelgrass meadows contribute significantly to the primary production of coastal marine habitats, but the scale of the production is difficult to quantify. The total primary production of a seagrass bed represents the sum of the production by each of the different primary producers in the community. Apart from the eelgrass, primary producers in the meadow include phytoplankton, sessile microflora and the macroalgal flora. Production by individual *Zostera* shoots may be measured in experimental mesocosms, and has even been derived from field experiments in which bell jars were employed to isolate groups of plants. Production estimates may be similarly derived for individual plants of different macroalgal species, but the primary production of the phytoplankton and microflora components of the community is not so easily estimated. It has been suggested that as much as 30–50% of the primary production of seagrass meadows may be contributed by the associated macroalgal community, the latitude of this estimate reflecting variation in relation to the species of seagrass, and epiphytes, and the physical environmental characteristics of the habitat.

In Kiel Fjord, on the western Baltic coast, *Z. marina* meadows cover more than 20 hectares of sandy sea floor, between 1.5 and 6 m depth (Jaschinski *et al.*, 2008). Organic content of the sediment is low, at less than 1% by weight, and salinity fluctuates between 10 and 20 psu; at midsummer eelgrass comprised 91% of the plant biomass of the meadows, the balance consisting of red algae, especially *Delesseria sanguinea*, attached to pieces of hard substrata within the sediment of the bed. Epiphytic diatoms, sediment microflora and phytoplankton were also significant primary producers. Biomass values, as ash-free dry weight (AFDW) per square metre, showed a predictable proportion, 54 g for eelgrass, 4.9 g for red algae, and 1.3 g, 0.5 g and 0.3 g for sand microflora, epiphytic diatoms and phytoplankton. However, primary production of the eelgrass was equivalent to 57.5 mg of carbon per square metre per day, while for epiphytes and algae combined the figure was 89.9 mg. It was estimated that the summer maximum production rate for the total associated flora could be three times that of the eelgrass.

The primary production of *Z. marina* varies substantially across its geographical range, and in relation to physical environmental gradients, and the values derived for the Kiel Fjord meadows are modest in comparison to those obtained for some Danish coastal populations. For these, annual leaf production has been measured at 1,058–2,000 g dry weight per square metre (dwt/m^2), equivalent to 423.2–800 mg of carbon per square metre (C/m^2), and some West Atlantic eelgrass beds achieve annual primary production in excess of 1,000 mg C/m^2. In oligotrophic (i.e. nutrient-poor) northwest Mediterranean habitats mean annual leaf production of *Z. marina* may be as high as 1,425 g dwt/m^2, or 570 mg C/m^2 (Cebrián *et al.*, 1997), sustained by nutrient stores within the bed, while brackish-water populations in nutrient-rich areas of the Dutch Wadden Sea produce only 109 g dwt/m^2, 43.6 mg C, per year, reflecting other physiological constraints. *Nanozostera noltii* generally achieves lower production rates than *Z. marina*, with recorded values of 130–211 mg C/m^2 per year, probably also reflecting environmental limitations.

Eelgrass leaves are not very palatable: they contain a fibrous structural component, lignin, and being low in nitrogen are also not very nutritious. Most eelgrass species support very few species of herbivorous animals, and instead their production passes into marine trophic webs following bacterial degradation. The products of decomposition may support suspension- and deposit-feeding detritivores, as well as large bacterial populations, which in turn provide nutrition for microphagous organisms. The fate of eelgrass primary production, the trophic pathways it follows, and the rate of utilisation by primary consumers, all vary between the different seagrass species, and the two European species offer interesting contrasts. In the case of *Z. marina* about 90% of its annual production is lost to consumers, but only 5% is consumed by herbivores, while for populations of *N. noltii* up to 50% of the annual leaf production is grazed by herbivores (Cebrián *et al.*, 1997), in northern Europe particularly by Wigeon, *Anas penelope*, and Brent Geese, *Branta bernicla*. Easy access to *N. noltii* beds, exposed at each low tide, may be a factor in its exploitation by grazing wildfowl; edibility may be another. It is faster growing than *Z. marina*, with more rapid turnover of leaves, and the tissue contains proportionately less lignin, with a higher, more nutritious content of carbohydrates.

Although herbivory accounts for little direct loss of *Z. marina* foliage, it has indirect effects that are quite significant. Leaf surfaces are abraded by small grazers feeding on microorganisms, and the leaves themselves are chewed by tiny mesoherbivores, particularly isopods and amphipods (Robertson & Mann, 1980); the tissue is thereby exposed to bacterial attack, and decomposition is initiated. This is a slow process; it takes as long as 40 days for *Z. marina* leaves to lose 60% of their weight, and after six months 10% of their tissue still remains. A small proportion of the annual leaf production, perhaps as much as 10%, is buried within the soil of the bed, inaccessible to aerobic decomposition, and contributes to a nutrient sink. Leaf shedding accounts for the majority of the primary production of *Z. marina* that follows the decomposer route: some of this decomposes, and is consumed, within the bed but much of it is exported from the eelgrass patch or meadow, carried by tide and current, to be utilised in other habitats, and perhaps by very different communities than those associated with the living plants. Some of the large estuaries on the American gulf coast support very extensive seagrass meadows that shed hundreds of tonnes of leaf biomass annually, and much of this is carried far offshore, even to beyond the edge of the continental shelf, providing the energy input for deep benthic communities.

Eelgrass communities

The communities associated with eelgrasses may be as diverse as those associated with kelps, but they never achieve the same biomass per unit area. A square metre of kelp bed provides a far greater area of substratum for occupation than an equivalent area of seagrass, and a substantial part of it is perennial, ensuring persistence of the community. *Zostera* leaves have only a short life and their epiphytic communities are shed with them. However, *Z. marina* tolerates a wider range of environmental conditions than any species of *Laminaria*; the morphology and character of the eelgrass meadow change in response to physical environmental gradients, and the composition and structure of its associated assemblages also change accordingly. The epiphytes of *Z. marina* consist principally of filamentous, foliose and coralline macroalgae, and abundant diatoms. The species composition and density of the community will vary from one eelgrass patch or meadow to another, and will also display considerable seasonal variation reflecting the life cycles of both the eelgrass and its epiphytes. To a certain extent each *Zostera* bed is a unique habitat, defined by ambient environmental parameters, and the resulting morphology of the bed. Epiphyte abundance and diversity will thus be determined by the local hydrodynamic regime, temperature and salinity characteristics, nutrient resources, and the density and individual size of the eelgrass plants.

Seasonal variation in the composition and biomass of the community results from successional recruitment, and the increasing growth rate and turnover of leaves through the year. A Danish eelgrass bed supported no fewer than 24 species of filamentous algae, and an uncounted number of diatom species, which showed two peaks in biomass and production, with a succession of dominant species through the year (Borum *et al.*, 1984). Bacteria and diatoms were the first colonisers, settling on winter-grown leaves in January and developing dense mats that persisted until the leaves were shed. Filamentous green algae (chlorophytes) began to settle in January and were abundant by early April, the bulk of the biomass consisting of *Neotromatella monostromatica*, *Monostroma grevillei* and species of *Ulothrix*. Brown algae (phaeophytes) also appeared in January, but the main recruitment of phaeophytes did not begin until April and only achieved peak biomass in late May and early June. At least a dozen species of filamentous brown algae were common constituents of the epiphyte community: old eelgrass leaves were densely settled by *Ectocarpus siliculosa*, *Dictyosiphon foeniculaceus* and *Pilaiella littoralis*, while newly grown leaves were heavily colonised by diatoms, particularly *Cocconeis scutellum*. Epiphyte biomass peaked in May, at around 62 g dwt/m^2 of eelgrass bed, and accounting for 62% of the total above-ground macrophyte biomass. It dropped sharply in June, to a value equivalent to winter levels, but then increased again in August to a second, though very minor, peak value, less than 5% of the first. The August peak marked a significant change in the epiphyte community, when the first, winter-grown *Zostera* leaves were shed, at about 200 days old. New growth was rapid through the summer, shoots developing a new leaf every 10–14 days, but leaf turnover was also rapid, each lasting no more than 70 days, which was not long enough for the dominant brown algal species of early summer to re-establish, and through August and September about 90% of the biomass of epiphytes was lost as *Zostera* shed its leaves. The total annual primary production achieved by this Danish eelgrass habitat was equivalent to 70 g C/m^2, 80% of which accrued during the spring, when epiphyte productivity was maximal, at 2.1 g C/m^2 per day, while eelgrass production alone was measured at 1.3 g. The two production peaks recorded (Fig. 117) parallel the annual marine primary productivity cycle, the large spring bloom declining as nutrients are depleted, and a minor autumn peak fuelled by a recovery of nutrient levels driven by equinoctial turbulence.

The *Zostera* epiphyte community discussed above was ultimately limited by nutrient depletion, but in coastal habitats afflicted by anthropogenic nutrient enhancement, through agricultural runoff or waste-water outflow, epiphyte biomass may be sustained at high levels through the growing period, and may be augmented by dense mats of free-growing filamentous seaweeds. Eutrophication is often associated with a decline in eelgrass abundance, as shading created by the macroalgae and water turbidity created by phytoplankton blooms both reduce the photosynthetic efficiency of *Zostera* leaves, with the result that growth and productivity drop. However, the consequences of eutrophication are not always consistently negative, and its effects may be mitigated by interactions within the eelgrass communities. This was demonstrated by an ecological

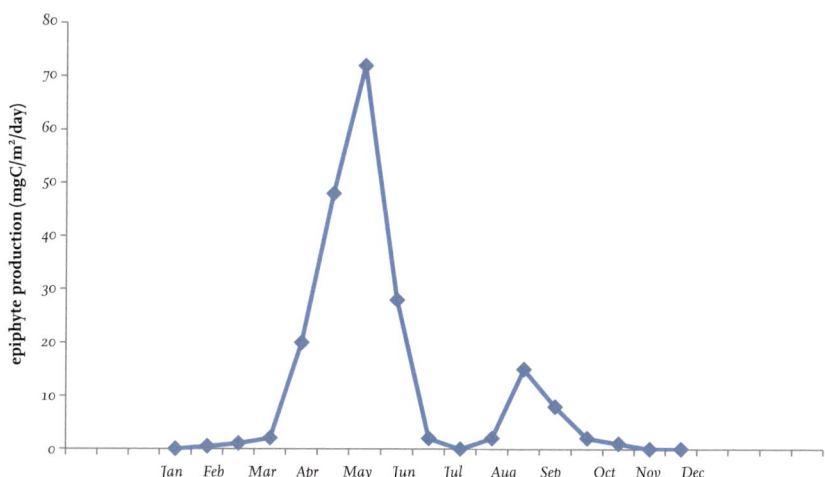

FIG 117. Epiphyte production per square metre of a Danish *Zostera* meadow. (After Borum et al., 1984)

study of *Zostera* habitats on the west, south and southeast coasts of Sweden, all subject to anthropogenic eutrophication, but which showed striking differences between the associated communities (Jephson *et al.*, 2008). The direct effect of eutrophication is referred to as a 'bottom-up' process: enhanced nutrients stimulate primary production, and lead to an increasing biomass of the primary producers, phytoplankton and macroalgae, and increasing numbers of consumers. These primary consumers may be limited by predation, which thus reduces grazing pressure on the primary producers, and this effect is referred to as 'top-down' control. In the Swedish study, west-coast eelgrass meadows supported a significantly greater biomass of filamentous algal epiphytes than those on the south and southeast coasts (Fig. 118), and their total cover of *Zostera marina* had declined by more than 50% in the preceding two decades. About 40% of this biomass consisted of species of the phaeophyte *Ectocarpus*, 34% was species of chlorophyte belonging to *Ulva* and *Cladophora*, and 27% was rhodophytes of the genera *Polysiphonia* and *Ceramium*. These dense algal assemblages provided habitat for tube-building amphipods, mostly suspension feeders, and also for small mesograzers, amphipods, isopods and gastropods that fed on the algae rather than the eelgrass. Two species of grass shrimp, *Palaemon elegans* and *P. adspersus*, and small fish, especially gobies, pipefish and sticklebacks, were the principal predators of the mesograzers. The south and southeast coast sites, though equally exposed to eutrophication, supported a much more sparse epiphyte community, with a biomass as low as 5%, and never more than 20%, of that of the west coast habitats. The same algal species were present, but the rhodophytes were dominant. The same suites of mesograzers and predators were also found, although as salinity decreased from the west to the Baltic southeast coast, brackish-water gastropods became more predominant. The small fish predators declined in abundance, but large individuals of the amphipod *Gammarus locusta* and the isopod *Idotea balthica* were common.

While there were qualitative differences between eelgrass communities on the south and southeast coasts of Sweden, the most striking, and significant, differences were between these communities and those on the west coast, and these differences were attributable to the effects of predation. Overfishing along the western coast of Sweden had depleted populations of Cod, and probably of other large bottom-feeding fish. The smaller predators that had formed a substantial part of the diet of these larger fish – crabs, shrimps and gobies – flourished to the extent that they limited the populations of mesograzers, and with grazing pressure thereby lowered, the epiphytic macroalgal community was unchecked. Fishing pressure was much less severe on the south and southeast coasts, and a stable Cod population exploited, and limited, the populations

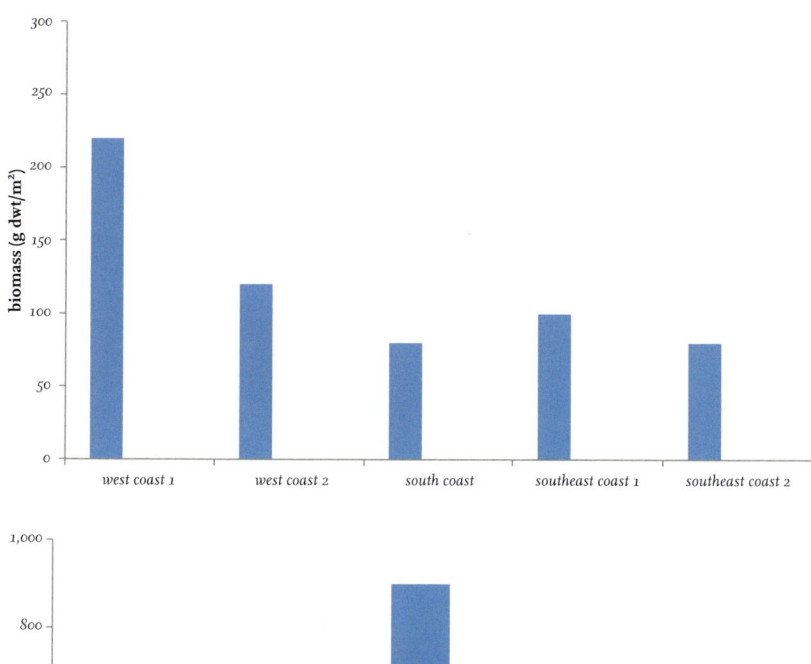
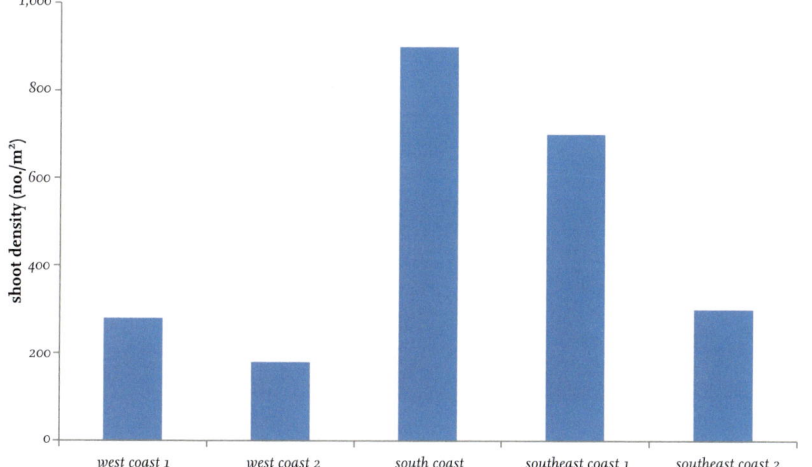

FIG 118. Biomass and shoot density of *Zostera marina* at five eutrophic sites on the Swedish coast. (After Jephson *et al.*, 2008)

of small predators. Consequently, mesograzer populations were unchecked, and limited the biomass of filamentous epiphytic algae. In comparatively low-diversity ecosystems marine food webs may become simplified to just three trophic levels, and when strong trophic interactions result in alternating abundancies ecologists recognise a trophic cascade, in which predators benefit producers by limiting herbivores. On the west coasts of Sweden removal of large benthic predators led to a significant increase in numbers of small predators; these maintained mesograzer populations at levels too low to have a limiting effect on algal populations (Fig. 119), and three, alternating high, low, high, abundances constituted a trophic cascade. The southern and southeastern *Zostera* meadows were within the foraging range of a stable Cod population that limited the populations of small predators, with the result that algal biomass was controlled by a greater biomass of mesograzers. With less predation pressure *Gammarus locusta* and *Idotea balthica*, the most significant crustacean mesograzers, were much larger than those in west coast habitats, and thus had a size refuge from predation by the omnivorous grass shrimps, *Palaemon elegans* and *P. adspersus*, which switched to herbivory, grazing seaweeds and further limiting epiphyte biomass. So predation did not benefit primary producer populations, and there was no trophic cascade effect.

Interactions between the successive trophic levels of an eelgrass community vary in intensity along environmental gradients, and in relation to the biodiversity of each level. Where nutrient levels are minimal epiphyte growth is immediately consumed by modest mesograzer populations, and the eelgrass increases the proportion of below-surface biomass through additional root growth; predation has no effect on algal biomass, and the eelgrass may be grazed by a different suite of mesograzers, especially small gastropods. Taxonomic diversity within a trophic level will weaken top-down effects because highly diverse assemblages will include a proportion of predator-resistant species, and will also promote greater niche diversity, and fewer strong interactions; eelgrass habitats are structurally complex, and complexity is enhanced by

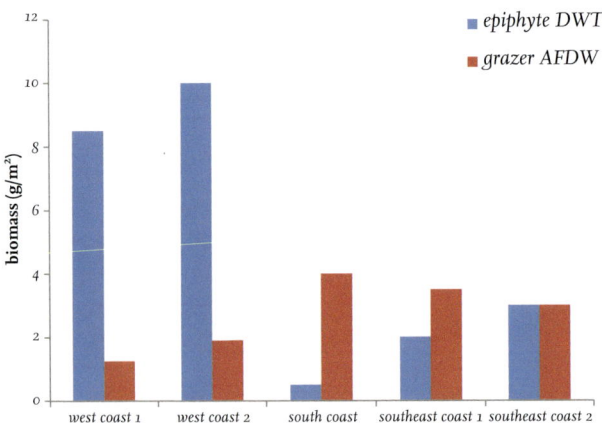

FIG 119. Biomass of epiphytes and mesograzers on *Zostera marina* at five sites on the Swedish coast. AFDW, ash-free dry weight; DWT, dry weight. (After Jephson et al., 2008)

epiphyte growth, providing maximum refuge from predation for mesograzers. However, these mitigating effects may still be negated by especially effective keystone predators which may strongly affect the structure and function of the assemblage. For example, pipefish are visual predators adapted for foraging in complex, three-dimensional habitats, and it has been demonstrated in Mexican *Z. marina* beds that two or three adult pipefish per square metre of eelgrass habitat provide effective control of mesograzer populations (Jorgensen *et al.*, 2007). Hydrodynamic factors may also drive trophic cascades within seagrass beds. In brackish habitats the small mud snail *Peringia ulvae* is an ecologically significant mesograzer, and *Nanozostera noltii* beds in the leeward shelter of Scolt Head, on the north coast of Norfolk, supported a mean of 32 invertebrate species, and 6,568 individuals per 0.1 m^2; total faunal density scaled up to 51,300 per square metre, of which 43,000 were *P. ulvae* (Barnes & Farnon Ellwood, 2011). At such densities, in low-energy environments, *P. ulvae* will effectively control epiphyte biomass, but as current strength increases, in high-energy habitats the snails are swept away, epiphyte growth is unchecked (Fig. 120), and seagrass growth is inhibited (Schanz *et al.*, 2002). The primary producers benefiting from trophic cascades within eelgrass beds are, of course, the epiphytic macroalgae, while the eelgrass is actually disadvantaged through competitive pressure. There is an interesting contrast here with kelp habitats: kelps are negatively affected

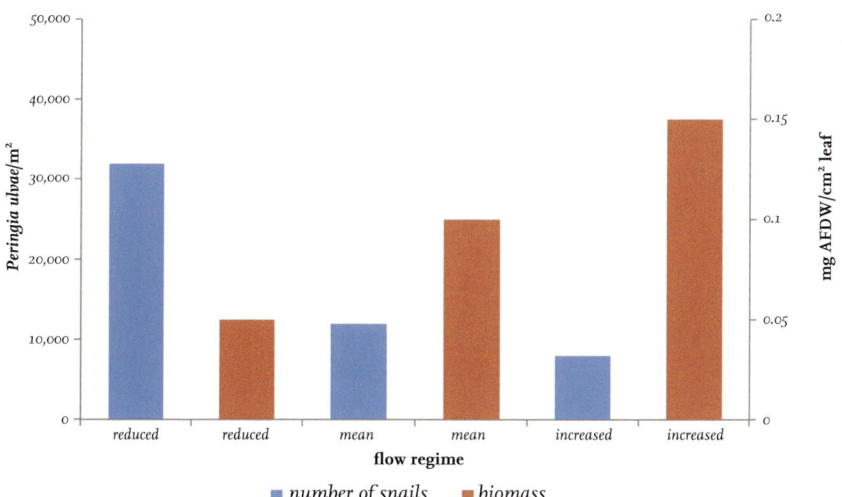

FIG 120. Abundance of mud snails and biomass of epiphytes on *Nanozostera noltii* subjected to three experimental flow regimes. AFDW, ash-free dry weight. (After Schanz *et al.*, 2002)

by herbivory, and herbivore pressure is contained through predation of grazing echinoderms by lobsters and fish. Eelgrass has a positive mutualistic association with mesoherbivores at one end of a nutrient-enrichment gradient, where epiphytes are checked by grazing, and a negative relationship at the other, where the mesograzer community consumes eelgrass tissue in mesotrophic systems.

The density and composition of the eelgrass epiphyte community change along environmental gradients, but also seem to show quite abrupt discontinuities between eelgrass patches over very short distances. A survey of three *Zostera* beds off Plymouth, comparable in area and depth range, showed that while diversity and density of epiphyte communities changed little within each bed, there were sharp and significant differences between the epiphyte assemblages in beds just 1 m apart (Saunders *et al.*, 2003). The epiphyte assemblage of each *Zostera* patch seems to be unique, reflecting the colonisation, growth and succession of the community, in relation to the ecological characterisitcs of the particular patch. Additionally, epiphyte assemblages must be expected to show considerable variation over the greater scales of latitude and longitude. The Plymouth epiphyte communities were quite different from those of the Swedish eelgrass habitats. Mean epiphyte densities ranged from 1.21 to 7.96 individual plants per centimetre of *Zostera* leaf; 99% of the epiphyte species recorded were either filamentous or encrusting coralline morphologies, the former more abundant, with a mean 2.03 plants per centimetre of leaf, the latter with a mean of 1.44. Although epiphytes provide the most significant source of food for mesoherbivores, and are responsible for the larger proportion of energy cycling within an eelgrass habitat, in the Plymouth study there seemed to be no relationship between the density of epiphytes and the abundance of gastropod grazers, either between eelgrass beds or between sampling points within the beds. There was a strong positive link between epiphyte density and amphipod abundance, but this was related to habitat complexity: the amphipods were principally suspension-feeding species that construct permanent tubes, and they benefited from the increased surface and shelter provided by the epiphytic algae.

Amphipods and isopods are characteristic of all temperate eelgrass habitats, the taxonomic composition and density of each assemblage, as expected, varying in response to local environmental factors. No species is entirely restricted to *Zostera*, all can be found wherever suitable substrata and shelter occur, but most are much more abundant amongst eelgrass than in adjacent, unvegetated areas, and some species achieve very large biomass and density. In the Swedish study (Jephson *et al.*, 2008), at the western sites with high epiphyte densities, the tubicolous amphipods *Erichthonius difformis* (Fig. 121a), *Monocorophium insidiosum* and *Corophium volutator*

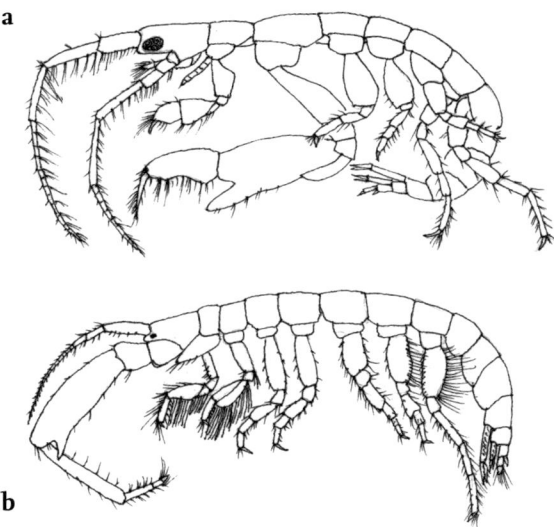

FIG 121. Two tube-building amphipods. (a) *Erichthonius difformis*; (b) *Corophium volutator*. Not to scale.

(Fig. 121b) were recorded at combined densities of more than 25,000 individuals per square metre of eelgrass bed, with a mean biomass of 0.8 g AFDW/m²; the predatory *Microdeutopus gryllotalpa* occurred at more than 400/m², and *Gammarus locusta* at more than 200/m². Small grazing snails, mainly species of *Rissoa* and *Hydrobia*, occurred at densities of 400–1,700/m² at the western sites, and in similar numbers on the south and southeast coasts, although the principal species in the southeast were the brackish-water *Ecrobia ventrosa*, *Theodoxus fluviatilis* and *Radix balthica*. The most abundant isopod mesograzers at the south and southeast sites were three species of *Idotea*, *I. balthica*, *I. chelipes* and *I. granulosa*, but numbers were modest, with a maximum density of 128 per square metre for *I. balthica* at one site.

In the Gullmarsfjord, 58° N on the Swedish west coast, *Zostera marina* occurred at 0.7–2.5 m depth, intergrown with the seaweed *Fucus serratus* (Pihl Baden & Pihl, 1984; Pihl Baden, 1990). The total area of the meadow studied was 30,000 m², but patches of *F. serratus*, 20–30 m² in extent, and isolated plants of *F. vesiculosus*, comprised about 40% of the meadow's area. Seven species of amphipod, and the isopod *I. balthica*, were the numerically most abundant component of the invertebrate community, and for most samples were the largest proportion of the biomass as well. *Monocorophium insidiosum* and *Erichthonius difformis* were overwhelmingly abundant, and during the late summer maximum their combined densities exceeded 80,000 per square metre, and their tubes were packed together at 2–3 per square centimetre of eelgrass leaf. Density and biomass of crustacean mesograzers fluctuated seasonally, reflecting the different periods of reproduction and recruitment for each species, and in October total amphipod biomass peaked at 0.7 g AFDW/m^2. Juvenile bivalves were a significant proportion of the community during June and July, but their numbers decreased sharply after August. Six species were represented. The most abundant was the Blue Mussel, *Mytilus edulis*, which settled at mean densities of 6,423 per square metre in June, rising to 11,032 in July, but declining abruptly to just a few hundred by September. This drop in numbers may have been partly a result of predation pressure, but it also reflected the post-settlement migration of juvenile mussels to adult habitats. Juvenile Sand Gaper clams, *Mya arenaria*, reached peak numbers, 3,726 per square metre, in July, declining even more sharply than *M. edulis* in August. Predation might have been a factor in this decline also, but, as for *Cerastoderma edule*, *Abra alba* and *Macoma balthica*, small numbers of which were recorded among the epifauna of the eelgrass bed, most of the juvenile gaper clams would have recruited to the infauna soon after settlement, and thereafter would not have been recorded in epifaunal samples.

The epibenthos of eelgrass habitats consists largely of small predatory crustaceans and fish. Most are migratory to a greater or lesser extent: some are seasonal residents, present only during the spring, summer and early autumn, for others the eelgrass provides a habitat for only a part of the species' life cycle, and in temperate Atlantic regions no species is strictly limited to *Zostera marina*. The Green Crab, *Carcinus maenas* (Fig. 122), is practically ubiquitous in northwest European coastal habitats, including eelgrass, amongst which populations may be very high. Green Crabs are significant primary consumers in eelgrass communities, preying especially upon bivalve spat, Brown Shrimps, *Crangon crangon*, polychaete worms and the larger amphipod species. However, these are invariably juvenile crabs, newly recruited to the benthos, or belonging to 1-group or 2-group cohorts (i.e. in first or second year classes), and the most important resource provided by the eelgrass is refuge from larger predators. As they grow larger, Green Crabs become less frequent amongst *Zostera* beds, as their foraging range and optimum prey size both increase.

The Brown Shrimp (Fig. 73), and the grass shrimps *Palaemon elegans* and *P. adspersus*, are also important predators, feeding selectively on amphipods. Brown Shrimps were found to consume a greater proportion of available prey in Swedish eelgrass beds than Green Crabs (Möller *et al.*, 1985), while *P. elegans* fed on *Gammarus locusta* at such a rate that, in the absence of larger predators, a trophic cascade ensued that resulted in a fivefold increase in the biomass of filamentous green algae, and a concomitant decline in *Zostera* biomass (Persson *et al.*, 2008). The two species of *Palaemon* are probably more closely associated with

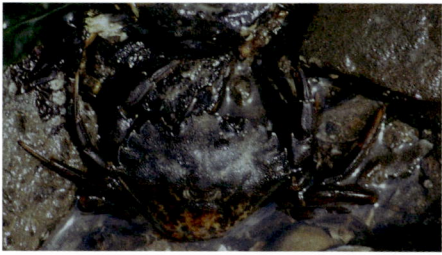

FIG 122. Juvenile 'Green' Crabs, *Carcinus maenas*, are typically mottled in shades of cream, brown and olive.

vegetated habitats than *C. crangon*, which is abundantly present in all sheltered, fine-sediment coastal environments, but all three move offshore in order to reproduce, and into deeper waters during the winter, and so their populations show often sharp fluctuations through the year.

Gobies are prominent among the small fish consumers, especially *Pomatoschistus microps*, *P. minutus* and *P. pictus*, three very similar little fish, all often confused as 'sand gobies' (Fig. 72). These also show seasonal migrations, and eelgrasses constitute only part of their inshore range. Sticklebacks (*Gasterosteus aculeatus*, *Pungitius pungitius* and *Spinachia spinachia*) are less peripatetic, and numbers probably only dip in the shallowest eelgrass patches, during the coldest months. Pipefish (Fig. 123) seem to be characteristic of seagrass habitats worldwide; of the two species common in the western Baltic and eastern North Sea, *Syngnathus typhle*, the Deep-nosed Pipefish, seems more closely bound to eelgrass habitats, while *Nerophis ophidion*, the Straight-nosed Pipefish, is perhaps equally common amongst rocky-shore seaweeds. Ten species of small fish, the three shrimp species and the Green Crab form the majority of the epibenthic predators in Swedish and Danish eelgrass assemblages. Further west and south this tally increases; around southwest England small wrasses, and additional species of goby (Fig. 124) are often common among *Zostera*, and the complement of shrimps may include several species of *Hippolyte*, the humpback shrimps.

Most research into the ecology of eelgrass faunas has been concerned with the total fauna, both epifauna and infauna, or with just the epifauna, while only few studies have focused particularly on the infauna. Also, in all cases the meiofauna has been almost completely ignored, although its potential significance is occasionally hinted at. In the Gullmarsfjord, Norway, detrital-feeding harpacticoid copepods were consistently numerically more abundant than amphipods and isopods, with a mean monthly abundance of 30,499 per

FIG 123. The Greater Pipefish, *Syngnathus typhle*. (J. S. Porter)

FIG 124. The Two Spot Goby, *Gobiusculus flavescens*. (J. S. Porter)

square metre, compared to 23,387 per square metre for amphipods, but while the latter represented a maximum biomass of 0.7 g AFDW/m², the figure for the harpacticoids was only 0.08 g (Pihl Baden, 1990).

The sedimentary environment of a *Zostera marina* bed provides further niches within the eelgrass habitat, with further potential for exploitation. The hydrodynamic regime within the bed results in the accumulation of the finest material, < 63 μm diameter, and the organic content of the soil is augmented by the decay of eelgrass leaves and algal epiphytes. Below the surface of the sediment the rhizomes and roots of the plants provide a structurally heterogeneous environment. It has been assumed that all of these conditions promote density and diversity of the infaunal community, which will be comparatively greater than in adjacent soft-sediment habitats. The size of the eelgrass bed is seen to be a significant factor affecting abundance and diversity of infaunal communities. Large patches of Z. *marina* accumulate larger proportions of fine sediment than small patches, and sediment characteristics are important determinants of infaunal community composition. Small patches have a higher edge-to-area ratio. The edge of a seagrass patch is a transitional zone subject to greater disturbance than its centre, and perhaps also to greater predation pressure. The central areas of large patches, it follows, must be more stable than peripheral areas, and are thus colonised by long-lived, K-selected species. Conversely, disturbance and predation around the edge of the bed result in a community of short-lived, opportunistic, r-selected species. However, there is a contrary view of this apparently neat paradigm. It has been noted that many of the macroinfaunal taxa recorded at high density in *Zostera marina* beds are found in equal, or greater, abundance in adjacent unvegetated habitats, and that the below-surface mat of roots and rhizomes may be responsible for a decreased abundance of some taxa. Species that occur more abundantly around the edges of an eelgrass patch than within it may be occupying a transitional zone marking a sharp decrease in population density from an adjacent, homogeneous soft-sediment habitat to the heterogeneous infaunal environment of the *Zostera* bed. A survey of Z. *marina* beds at three localities around the Isles of Scilly, with comparable depth range and plant density but each subject to a different degree of wave exposure, recorded 89 infaunal macroinvertebrate taxa (Bowden et al., 2001). The figure is higher than might be expected for macroinfaunal soft-sediment communities, at comparable depth and with similar hydrodynamic regimes, but the density of individuals recorded, from around 7,000 to above 26,000 per square metre, is modest in comparison to densities recorded for many soft-sediment habitats. There have been no comparative studies of the infauna of eelgrass beds and adjacent unvegetated soft sediments, and it is not clear whether eelgrass accumulates species from other habitats or is a source of emigration. The enhanced structural complexity of eelgrass habitats certainly seems to have a positive effect on taxonomic diversity of the associated fauna, but may not necessarily have a similarly positive effect on biomass and abundance.

In analysing the diets of the primary consumers in eelgrass communities it is apparent that, with a few exceptions, fresh *Zostera* leaves feature only marginally in trophic pathways. One significant exception is the isopod *Idotea balthica*, an omnivore feeding on a wide variety of foods, including microalgae and epiphytic macroalgae, eelgrass leaves and small invertebrates. Where *Z. marina* is its major food source, as in some parts of the Wadden Sea, it plays an important role in shredding fresh leaves, promoting the onset of bacterial decay, and thus contributing indirectly to the detrital food web.

It is difficult to construct food webs and measure energy flow for eelgrass communities. Eelgrass meadows may be foci of high primary production, but the total primary production for a given bed is compounded from the contributions by epiphytic macroalgae and sessile microalgae, including sediment-associated microflora, as well as that of the eelgrass, and the relative contributions by each source will vary along environmental gradients. Allochthonous (foreign, originating from elsewhere) carbon sources may also contribute to energy flow through the food web, in the form of phytoplankton, algal detritus, and perhaps even material of terrestrial origin, all carried into the bed by coastal currents. It is relatively straightforward to estimate production by each of the primary producers in an eelgrass community, but it is difficult to estimate the proportion of each utilised by each primary consumer. Stable carbon isotope ratios can be employed, but it is often not possible to distinguish between values derived from primary producers within the habitat, and those from carbon sources outside of the habitat. However, some precision may be achieved through analysis of fatty acid content of particular consumers. Certain fatty acids are specific for particular groups of producers, and the relative proportions of different fatty acids in a consumer – its 'fatty acid signature' – will reflect the relative proportions of the different producers in its diet. Further, while carbon isotope ratios decrease with each trophic level, fatty acid transfer is more conservative, making it possible to trace trophic pathways from primary producers to at least secondary consumers.

The Kiel Fjord eelgrass meadows supported 30 main consumers: two polychaete worms, four molluscs, 13 crustaceans, the starfish *Asterias rubens*, nine small predatory fish and the Cod as top predator (Jaschinski *et al.*, 2008). The food web was largely based on epiphytic diatoms and benthic microflora; phytoplankton and the red alga *Delesseria sanguinea* contributed a smaller proportion of the energy flow, while *Z. marina* leaves were of negligible significance. Fatty acids characteristic of *Z. marina* were present in the fatty acid signatures of all consumers but at levels of only around 1%, while all contained high levels of fatty acids derived from diatoms. Two gastropods, *Rissoa membranacea* and *Littorina littorea*, constituted the bulk of the grazer biomass; the fatty acid signature of the former indicated a diet with a high diatom content, while *L. littorea* fed more on red algae. A small population of *Mytilus edulis* fed predominatly on phytoplankton, and its fatty acid signature was evident in the predatory *A. rubens*. *Idotea balthica* and the amphipod *Ampithoe rubricata* appeared to have mixed diets, embracing all of the categories of primary producers, and perhaps a proportion of the smallest primary consumers, such as bivalve spat, but another amphipod, *Gammarus oceanicus*, fed preferentially on red algae. All species of isopod and amphipod were seen to constitute critical links between primary producers and higher trophic levels, as they were the main food sources for small epibenthic predators, *Carcinus maenas*, *Crangon crangon*, *Palaemon adspersus* and the six small fish species.

Total consumption by small epibenthic predators in a shallow eelgrass bed in the Gullmarsfjord, west Sweden, was equivalent to 30 g AFDW/m^2 per year (Möller et al., 1985). The principal predator was the shrimp *Palaemon adspersus*; 50% of its prey consisted of amphipods, isopods and other small herbivores, and another 15% of infaunal species, including such primary consumers as nereid polychaetes. Most of the primary production at this site was generated within the bed, phytoplankton input being negligible, but energy passing to the upper trophic levels was ultimately exported, as all of the epibenthic predators migrated to deeper water with the onset of winter. Up to 90% of epibenthic production thereby passed into offshore benthic food webs.

Zostera marina is a resilient species. It is subject to often sharp fluctuations in physical environmental pressures related to quite natural climatic fluctuations, and eelgrass landscapes consequently undergo natural expansion and contraction, fragmentation and consolidation. Populations display continuous fluctuation in density and extent, reflecting seasonal, annual and longer-term cycles, and a certain degree of physical disturbance may be a requisite for a healthy, productive population. An eelgrass meadow damaged by an exceptional storm, with large areas uprooted and dispersed, or buried in loose sediment, will decline rapidly, losing 60% of its area in just a few seasons. However, recovery may be just as rapid, with the bed restored to its original density and extent in around five years. Recovery depends partly on successful recruitment of new seedlings and partly on conditions for growth. Populations occupying optimum shallow water sites display greatest resilience, while deep-water populations, at the limit of the species' vertical range, recover only slowly, and may disappear completely if disturbance continues.

Anthropogenic pressures impose further stresses on eelgrass habitats, but to a great extent their impact can perhaps be countered by restoring and sustaining the natural biodiversity of eelgrass beds. Physically disturbed areas can be reseeded with *Zostera* plants or seedlings, and once established new meadows will spread naturally. Eutrophication could be limited by passing waste waters through coastal reed beds and salt marshes. In the western North Atlantic seagrass meadows provide habitat for a number of large filter-feeding bivalves, and there is experimental evidence that these have a positive effect on their immediate environment. *Zostera marina* was formerly abundant in the barrier island estuaries of Long Island, New York, with an associated community of large bivalves – Bay Scallops, *Argopecten irradians*, Eastern Oyster, *Crassostrea virginica*, Hard Clams, *Mercenaria mercenaria*, and Blue Mussels. Settlement, recruitment and survivorship of bivalve spat are greatest amongst those attaching to eelgrass leaves, and adult densities are positively correlated with density of eelgrass shoots. The bivalves benefit from reduced predation pressure within the eelgrass canopy, and also from the influx of fine suspended material trapped by the meadow. The faeces and pseudofaeces produced by the bivalves contribute to the organic content of the sediment within the bed, and ultimately to eelgrass production.

Long Island eelgrass meadows have contracted dramatically in recent decades, in some areas now occupying less than 14% of the estuarine benthic environment, bivalve populations have been severely reduced, and the estuaries have become subject to frequent and prolonged phytoplankton blooms. All of these events may be related to increasing anthropogenic eutrophication, but through indirect pathways. Water clarity is critically essential to the growth and productivity of eelgrass meadows. With increasing eutrophication estuaries and coastal lagoons, optimum environments for *Zostera*, degrade to low-diversity environments dominated by phytoplankton and microbial communities. A healthy estuary will have a benthic community characterised by a high biomass and a high productivity which, in association with Z. *marina*, forms a diverse and stable community that resists changes driven by eutrophication. The first step in the degradation of Long Island eelgrass meadows appears to have been the destruction, through overfishing, of the bulk of the populations of suspension-feeding bivalves. This probably occurred in the historical past, long before any industrialisation of the region, and with little observable effect, but once coastal eutrophication began the impoverished *Zostera* habitats, lacking resistance, declined. *Mercenaria mercenaria* was recorded at mean densities of 0.8 per square metre in a survey of Long Island's south shore estuaries, but was most abundant, at 5.2 per square metre, in or adjacent to Z. *marina* beds (Carroll *et al.*, 2008). Eelgrass productivity was positively associated with water clarity but negatively associated with chlorophyll *a* abundance: as phytoplankton density rose and water clarity decreased, so eelgrass productivity declined as light levels decreased. Field experiments showed that Z. *marina* would produce more leaves per shoot, and greater leaf area, in response to nutrient enrichment, but only if water clarity was maintained. Experimental plots, 0.0625 m^2, shaded to reduce ambient light by 40%, showed a significant decrease in *Zostera* productivity, while the addition of fertiliser or Hard Clams (one per plot), or both, significantly increased productivity, even in the shaded plots. The influence of light was paramount, but the presence of the clam was essential in maintaining water clarity. It was suggested that, in the area investigated, a density of 16 clams per square metre would increase light availability and enhance nutrient levels sufficiently to significantly increase growth of the Z. marina population.

There is thus good evidence of a mutualistic association between *Zostera marina* and its bivalve guests (Fig. 125), although the strength of that association must be related to the population structure of the bivalves. Maintaining water clarity demands that a clam population clear the water column above it at a

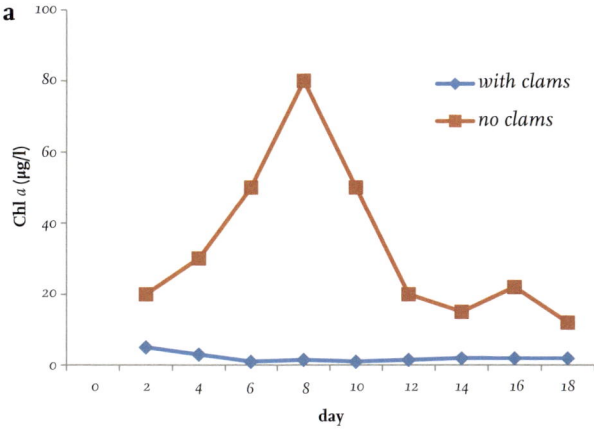

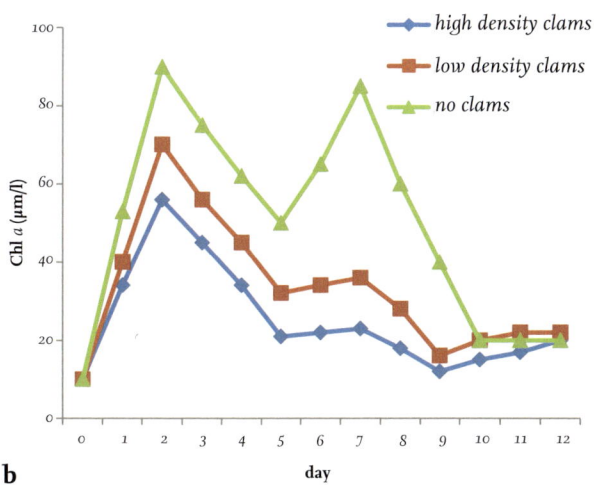

FIG 125. Temporal changes in chlorophyll *a* concentrations. (a) Experimental results with historical Hard Clam, *Mercenaria mercenaria*, density of 57 individuals per m2 and control without clams. Estimated turnover time 1.1 days. (b) Experimental results with high-density (29 individuals per m2) and low-density (14 individuals per m2) clams, and control without clams. Estimated turnover times 2.2 days (high) and 4.5 days (low). (After Wall *et al.*, 2008)

particular rate, and the time taken to achieve clearance, the 'turnover time', will depend upon the mean individual filtration rate and mean individual size, as well as the mean density of the population (Wall et al., 2008). Hard Clams with a mean length of 52.2 mm and an individual filtration rate of 0.41 litres per hour, per gram AFDW, had a turnover time of 4.5 days, at a density of $14/m^2$, while a density of $29/m^2$ cleared the same volume in 2.2 days. The significance of these data lies in the development times for phytoplankton blooms: any turnover time greater than 3.7 days was unable to stall a bloom and maintain water clarity. Individual, weight-specific filtration rates are also critical; *Crassostrea virginica*, with a mean length of 84.5 mm, had a weight-specific filtration rate of 1.91 litres per hour, per gram AFDW, and at a density of just $4/m^2$ had a turnover time of 2.5 days. Blue Mussels had the lowest weight-specific filtration rate, 0.289 litres per hour, per gram AFDW, and the lowest mean length, 39.0 mm, and even at a density of $229/m^2$ turnover time was 3.6 days, too low to avert a bloom. Population structure of the bivalve community must be an important factor, with a critical minimum proportion of large, old individuals necessary to enhance stability of the *Z. marina* population. There is an important positive feedback in which populations of associated species – bivalves – by improving the habitat, through long-term persistence, enable the foundation species, *Z. marina*, to extend its range and broaden its niche, and strengthen the resistance of the habitat to physical environmental stress. Stable, diverse and stress-resistant communities are not susceptible to trophic cascades; epiphytic macroalgae support, and are checked by, a diversity of mesoherbivores, which in turn support numerous small predators and ultimately stable stocks of large predators. Maximum diversity at each succeeding trophic level promotes community stability.

CHAPTER 6

More Key Habitats: Maerl Beds and Biogenic Reefs

Maerl beds develop from the growth and accumulation of branching coralline algae, and consist of a thin, living, upper layer above an interlocked gravel of dead corallines. 'Biogenic reefs' are habitats created by invertebrate foundation species, and include beds of Horse Mussels, *Modiolus modiolus*, banks of the deep-water Stone Coral, *Lophelia pertusa*, and certain polychaetes and bryozoans which may form aggregated populations. This is perhaps an unfortunate extension of the meaning of reef, which formerly denoted an emergent submarine hazard for Vikings. Extending it to embrace rocky communities of tropical corals is logical, but the biogenic reefs of northwest Europe are mostly flat, rippled or mounded, and never emergent. Both maerl beds and biogenic reefs, as with kelps and seagrasses, are 'key' habitats, in creating significant habitat heterogeneity. However, while kelps and seagrasses contribute importantly to the primary production of the communities they engineer, it is of minor significance in maerl communities, and practically none at all in the case of invertebrate reefs. However, maerl beds and biogenic reefs support dense, species-rich assemblages which are responsible for enormous secondary production.

MAERL

The habitat provided by maerl results from the responses of a slow-growing, living substratum to subtle biological and environmental influences. It is a Breton word for a knobbly, pink calcareous gravel, formerly used as a soil fertiliser. The biological term maerl is generic, encompassing a range of growth

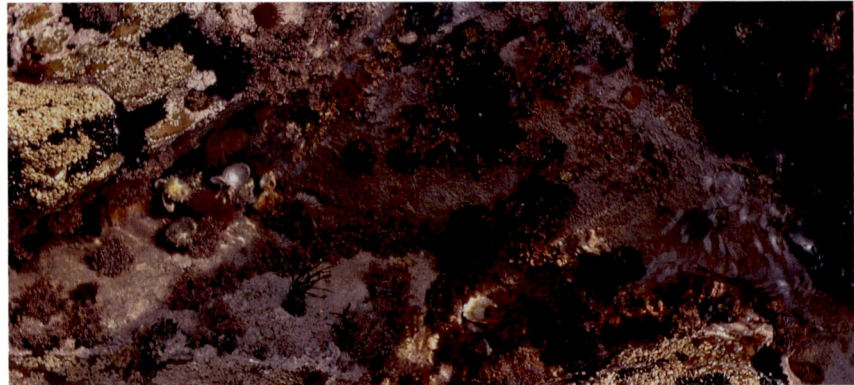

FIG 126. Sheet-encrusting coralline algae in an exposed shore rock pool.

forms displayed by a number of species of coralline algae under particular ecological conditions. Usually forming encrusting sheets on littoral and sublittoral hard substrata (Fig. 126), these pinkish, chalky corallines may form unattached, nodular, cylindrical or branching structures that accumulate as deep layers – maerl beds – of latticed structures providing space and surfaces for occupation by a striking diversity of free-living and encrusting organisms. Branching may be sparse or dense, in two or three dimensions, and branches may be elongate or short and stubby. The branches, or thalli, may be 2 or 3 cm long and up to 5 mm diameter, and their form varies continuously in response to local current regimes and degree of disturbance. The slow-growing, interlocking mesh of the maerl bed (Fig. 127) may take several hundred years, or perhaps thousands of years, to achieve the extent seen in north European shallow coastal environments. The depth of the bed depends upon local hydrodynamic factors, but may be in the range of 10–20 cm; only the upper layer, a few centimetres, consists of live, growing coralline algae, the rest is dead and bleached, but encrusted with live sessile animals, and bedded in a layer of silt. Thus, although it is not attached to the bottom, the close-packed structure of the maerl bed provides a degree of stability, and long-term persistence.

Maerl habitats can be found along the western coasts of Europe, from Iberia to Brittany, southwest England and western Ireland, and on the west and north coasts of Scotland, Orkney, Shetland and northern Norway, reaching a latitudinal limit at around 80° N. To the south, maerl occurs at suitable localities in the Mediterranean, and on the Atlantic coast of Morocco. There are three especially significant genera of maerl-forming corallines, encompassing a still-debatable number of species (Irvine & Chamberlain, 1994). *Phymatolithon*

FIG 127. Unattached branches – thalli – of free-growing coralline algae forming the upper layer of a maerl bed. (J. S. Porter)

calcareum and *Lithophyllum racemus* are important constituents of maerl beds in the Mediterranean, together with at least three species of *Lithothamnion*. *Phymatolithon calcareum* seems to be a particularly eurytopic, or environmentally tolerant, species, ranging from the Mediterranean northwards to southern Norway, and is one of the most important maerl-forming species along the west coasts of the British Isles and Ireland. *Lithothamnion corallioides* is present in the Mediterranean, and ranges northwards to Brittany, southwest England and western Ireland, but north of Galway Bay and the northern Irish Sea it is replaced by *Lithothamnion glaciale*, which ranges to the northern limit of maerl, where it co-occurs with *L. lophiforme*.

Carbonate sediments are all biological in origin: deep-sea calcareous oozes represent the skeletal remains of thousands of generations of microscopic planktonic organisms, coccolithophores, while shelf-sea carbonates are the accumulated shells, tests and skeletons of almost every type of marine benthic invertebrate. Coral reefs are the single most significant carbonate producers in the marine realm as a whole, and it was formerly considered that in cold northern and southern latitudes production of calcium carbonate could never equal that achieved by carbonate producers of warm temperate and tropical environments. It was also thought that cold waters were deficient in dissolved calcium carbonate, and that carbonate structures were swiftly degraded and dissolved after the death of the organism. The latter assertions are immediately refuted by the observation that shell gravels, of varying grade according to their composition, occur commonly in north European shelf environments. Northern seas are actually supersaturated with calcium carbonate, and while no carbonate producer displays rates of production at all similar to that achieved by coral reefs, some, including maerl, may fix calcium carbonate at a rate equivalent to some tropical, reef-associated organisms (Henrich et al., 1995). Maerl-forming corallines, like all algae, require light as an essential component of photosynthesis and growth, and light is thus a primary factor determining the distribution of maerl communities. To the north, cloudy skies and the seasonally varying angle of incidence of the sun's rays further limit the depth of penetration of sunlight. In clear Mediterranean waters, maerl is able to maintain photosynthesis to depths as great as 100 m. Brittany communities flourish as deep as 30 m, and western Irish beds at around 20 m. This polewards shallowing, or polar emergence, of maerl habitats, achieves a zenith in northern Norway, where maerl beds are limited to a narrow range of depths, 6–15 m.

Maerl requires clear water, allowing maximum penetration of sunlight, but water clarity may be affected by sediment load and turbidity. Water turbidity is a second significant factor affecting the ecology of maerl beds, and its effects are complexly interconnected with those arising from current speed and wave exposure. The accumulated maerl thalli do not form a single rigid structure. The living upper layer may be partially fused by continuously growing corallines, and the dead lower layers are sporadically cemented together by other encrusting organisms, but the bed as a whole is unattached to the sea floor and is vulnerable to dispersal by severe water movement. The genesis and development of a maerl bed must be a long and slow process; it has never been recorded and documented, but it is generally thought that while it may be sustained to some extent by growth within the bed, the foundation of a maerl community, and to a degree its persistence, depend upon the growth of branching coralline algae on adjacent hard substrata, and their dislodgement and accumulation by hydrodynamic factors (Freiwald et al., 1991). Bottom currents tend to entrain detached maerl thalli, and they become concentrated in areas of optimum current regimes, too slow to disperse the tumbled thalli further, but strong enough to concentrate them continuously. However, wave action may disrupt the integrity of the bed, overturning and scattering it, and thus maerl communities tend to develop in areas of moderate currents, but sheltered from the direct effects of wave action, especially seasonally variable wave action. Regular winter storm waves would readily disperse summer accumulations of maerl, and it is probable that some decades of relatively constant hydrodynamic conditions are required for a maerl community to become established. Continuous water movement, provided by a moderate bottom current, is necessary, however, for the maerl to flourish.

The three-dimensional structure of the dead lower layers serves to trap and accumulate silt and detritus, providing habitat and food for small detritivores, tube dwellers and burrowers. Conversely, the living upper surface must be kept clear of silt, as this would shade the plants, reducing the rate of photosynthesis, and hence growth. So the optimum maerl environment is delicately balanced within a narrow zone of current-swept sea floors sheltered from the direct and damaging effects of waves. Such environments are few. Sheltered estuaries are not favoured by corallines; apart from lowered salinities, fluctuating seasonally in response to freshwater influx, the outflow of terrigenous material increases water turbidity and, again, depresses photosynthesis, and hence growth, of the maerl. Rias – deeply drowned river valleys with low riverine input – provide ideal conditions for maerl, as do the ice-deepened sea lochs of western Scotland and the dissected coastlines of the western and northern isles of Scotland. In these environments shelter is maximal, water movement optimal, and the water is clear, with little dilution or turbidity from river runoff. Maerl communities have been described and detailed from 16 sites around southwest England and southwest Wales, with that in the seaward reach of the River Fal ria, Cornwall, most notable, but they are known from almost 400 sites around the western and northern coastlines of Scotland (BIOMAERL Team, 1999).

Maerl thalli grow slowly, expanding peripherally through the extension and division of chains of cells, termed filaments, with the terminal cell of each chain continually dividing to form new cells as it advances. Vertical growth is essentially the same: the thallus thickens as the cells of the surface layer continue to divide, forming new layers beneath the actively growing frontal layer (Irvine & Chamberlain, 1994). Growth of maerl corallines is light- and temperature-dependent, and is thus related to water depth. Rates of growth vary in relation to depth and latitude, and almost certainly vary between the different species of coralline algae constituting the bulk of each bed, but it is very difficult to measure and quantify growth rates in maerl corallines. There have been some estimates of average annual elongation of maerl thalli: around Britain unspecified maerl corallines have been stated to increase in length by less than 1 mm per year, although annual elongation of *L. incrustans*, from an unspecified locality, has been estimated at 6–7 mm. However, it is difficult to understand how these figures have been derived for very slow-growing plants that increase by annual thickening over the whole living surface of the thallus. This thickening can be measured by microscopic examination of sections of the plants. The frontal growth of corallines tends to be episodic, reflecting light, temperature and nutrient levels, and growth is thus shown in bands: summer growth bands tend to be broad and light-coloured, while winter bands are thin and dark. Annual growth of each maerl species is thus recorded in the sequential light and dark bands revealed by microscopic sections. Collecting such data is tedious, and it is perhaps not surprising that there are few published estimates of annual growth rates of maerl corallines.

Another measure of maerl growth is the annual production of calcium carbonate ($CaCO_3$), usually expressed as grams per unit area per year, and there are many accounts of calcium carbonate production by different species of maerl, at particular temperatures and depths, and at particular latitudes. Growth and carbonate production of a Brittany maerl bed, consisting largely of *Lithothamnion corallioides*, have been documented in detail (Henrich et al., 1995). Water depth was 4–8 m; the mean winter seawater temperature was 8–9 °C, while summer temperatures averaged 16 °C. The winter growth bands of *L. corallioides* had a maximum thickness of 0.12 mm, and summer bands a maximum of 0.38 mm; the annual thickening of the maerl thalli ranged from 0.3 to 0.5 mm, and the calculated calcium carbonate production ranged from 100 to 500 g per square metre per year. At Tromsø, Norway, around 70° N, maerl beds are found at depths of 7–20 m; water temperature averages 3–4 °C in winter and 14 °C in summer, and light is here a significant factor in maerl growth. From late November to mid-January total darkness pertains, but from late May to mid-July the midnight sun provides almost continuous illumination. The maerl coralline forming the bulk of the bed, probably *L. glaciale*, showed winter growth increments totalling just 0.32 mm and summer bands of 1.20 mm thickness, giving an annual thickening of 1.0–1.5 mm. Notwithstanding this modest annual growth, the Tromsø maerl beds accumulate 900–1,400 g of $CaCO_3$ per square metre, at 7 m depth, and 400–600 g at 18 m depth, both values exceeding the production rate recorded for the Brittany beds. Nutrient levels have a further influence on maerl growth. Off northwest Africa these are low but constant, while northwards nutrient levels become increasingly seasonal. Nutrients required for all growth, and reproduction, through the year by the Tromsø population must be absorbed and stored through the period of the spring bloom, and in sufficient quantity to sustain the algae through the sunless weeks of mid-winter.

While watching maerl grow is a rather dull business, considering its site and manner of growth reveals some intriguing puzzles. At depths of 40 m around Malta maerl corallines grow in knobbly spherical forms, termed rhodoliths (BIOMAERL Team, 1999). They are enabled to develop this distinctive morphology because gentle bottom currents, or wave action, periodically overturn the rhodoliths, allowing continuous growth over the whole of the surface, while neither damaging nor dispersing them. The key to rhodolith formation in Mediterranean maerl communities is a low level of disturbance, and it is apparent that the rhodoliths are formed in situ. Each probably commences as a single microscopic spore settled upon a minute piece of substratum, such as a small shell fragment. Considering the slow rate of frontal thickening of maerl corallines, rhodoliths of 20 cm or more diameter must represent many years of continuous growth. On northeast coasts of Canada rhodolith beds occur offshore, and whether or not they are actively growing, it seems that each rhodolith originates as an erect, three-dimensional growth from an encrusting base. These are then loosened and eventually detached by bioerosion, and then carried offshore to accumulate as rhodolith banks. In the Baie de Morlaix, Brittany, maerl beds seem to originate in essentially the same manner as the Canadian rhodolith banks, and there is a distinct ecological gradient from sheet-encrusting coralline communities, to communities of attached, branching forms, and then offshore to unattached maerl beds (Henrich et al., 1995). The primary species here are *Lithothamnion corallioides* and *Phymatolithon calcareum*. These encrust rocky shoals, at around 10–12 m depth, and produce the first branching growths, which are detached and transported, and accumulate in the lee of the shoals. There is a clear seaward gradient reflecting hydrodynamic effects: closest to the origin, maerl fragments are admixed with gravel and shell fragments, but these mixed deposits give way to a thick layer of live, branching and interlocked maerl thalli. Further seawards the maerl is increasingly disturbed by water movement; the bed is wave-rippled, and individual thalli are smaller and more sparsely branched. Eventually, maerl bed peters out into an unconsolidated algal gravel of worn, dead corallines. On Norwegian coasts *Lithothamnion glaciale* communities display the same gradation, from attached, encrusting, to unattached, branching growth, to dead algal gravel, reflecting again local hydrodynamic effects (Freiwald et al., 1991). However, this zonation has not been described for maerl habitats around the British Isles, and while the main maerl-forming corallines – *L. corallioides, L. glaciale* and *P. calcareum* – are stated to be erect forms of otherwise sheet-encrusting species, the encrusting forms appear not to have been recorded from British coasts. It has to be assumed that these maerl beds are truly autochthonous (i.e. originating where they are now found), but the genesis of a new bed has not been observed, and probably never will be, given the continual high level of disturbance of sea-bottom habitats that now afflicts all shallow shelf environments.

   The three-dimensional structure of a maerl bed provides habitat complexity on otherwise featureless sea floors, supporting dense and diverse communities of organisms (BIOMAERL Team, 1999). The living reticulate layer experiences constant water flow through its lattices, and the interstices are thus well oxygenated, well lit and clear of silt. Sessile diatoms and macroalgae flourish under such conditions, and are cropped by a guild of small, grazing, prosobranch snails. The accumulation of silt and detritus in the dead lower layers, together with the faecal output of the grazers of the upper layers, supplies nutriment to detritivores, and both of these trophic groups are preyed upon by numerous small predators. Large macroinfaunal invertebrates contribute to the complexity and integrity of the maerl habitat. A degree of stability is imparted by the long tubes of the polychaetes *Owenia fusiformis* and *Chaetopterus variopedatus*, and the tube anemone *Cerianthus lloydii* (Fig. 41), which pass through the structure of the bed and deep into the underlying sediment, acting as stanchions pinning the layer of maerl to the sea floor. Large, deep-burrowing clams, such as *Mya truncata*, inhabit vertical burrows in the substratum beneath the maerl bed; sea cucumbers, *Neopentadactyla mixta*, heart urchins, *Echinocardium pennatifida*, and most characteristically the chunky mud shrimps *Upogebia deltaura* and *U. stellata* (Hall-Spencer & Atkinson, 1999), maintain permanent chambers spreading deep below the maerl bed, with channels opening at the surface, through which respiratory and feeding currents are drawn. The mounds of sediment and faecal pellets ejected by these busy engineers contribute further to the heterogeneity of the habitat, and the constant inflow of water aerates the lower layers of the maerl bed, and the underlying sediment, to a considerable depth, as well as drawing in and distributing fresh plankton production and fine detritus. Flushed with food and oxygen the extensive burrows and galleries of the deep-burrowing macrofauna provide living space for numerous smaller organisms, filter feeders and deposit feeders, and a range of small predators and scavengers. As the diversity of the community increases it perhaps achieves increasing stability and persistence. However, the dynamics of maerl communities have scarcely been investigated. While many large-sized, long-lived organisms such as Dog Cockles (Fig. 16), scallops and gaper clams are common in maerl beds, and perhaps characteristic, maerl also shelters large numbers of individually small, short-lived species, which probably vary in abundance both seasonally and annually, and the populations of which may be quite transient.

The taxonomic diversity of maerl communities has only begun to be computed, and there is no estimate of the total number of species that might be expected to occur in pristine maerl habitats. Surveys of maerl beds at five locations – Malta, Alicante, Galicia, Brittany and the Clyde Sea – revealed the total species inventory of seaweeds, invertebrates and fish to range from 400 at Malta to 611 in the Clyde Sea (BIOMAERL Team, 1999). These surveys were detailed, but necessarily labour-intensive, and effort was thus directed to the identification of macrofauna and macroflora, with no surplus to devote to the meiofauna, microfauna and microflora. Although thus unrecorded, it was noted that had it been possible to sample these categories also they would have added significantly to the total biodiversity at each site. Maerl-associated organisms encompass all major taxonomic groups of marine invertebrates, as well as fish and all categories of macroalgae. In any particular maerl bed the groups occur in differing proportions, with each representing a different proportion of the total local fauna of that group. In Clyde Sea maerl beds, molluscs, annelid worms and crustaceans were the most abundant groups, in terms of numbers of species. Most diverse were the molluscs, although the 137 species recorded represented a little less than one-third of the known molluscan fauna for the area. A total 118 species of annelid was recorded, representing more than one-half of the total Clyde Sea fauna, but the 96 crustaceans constituted only about one-quarter of the species known for the region. However, of more than 1,600 species of seaweed, invertebrates and fish recorded from the Clyde Sea region, 566 were found on just two 50 m-radius maerl beds.

The structural complexity of a maerl bed allows for the development of multi-species communities, displaying wide trophic diversity. Regarding molluscs, for example, the top few centimetres are occupied by large, suspension-feeding, epifaunal bivalves, especially the scallops *Pecten maximus* (Fig. 128), *Aequipecten opercularis*, *Chlamys varia* and *Palliolum tigerinum* (Fig. 129). The first two named are vagile, and active, and densities and distribution of these within the bed probably vary continually. Infaunal bivalves tend to be more sedentary, and more than 30 species of these occur in Clyde Sea maerl beds. Thick-shelled, shallow-burrowing species such as *Glycymeris glycymeris*, *Venus casina* (Fig. 154), *Tapes rhomboides* (Fig. 130) and *Clausinella fasciata* live on or within the top layers of living maerl, while the vertical burrows of the deep-burrowing *Lutraria angustior*, *Mya truncata* and *Ensis arcuatus* extend into the sediment below the dead lower layers of the bed. All of these are suspension feeders, filtering food from self-generated respiratory currents, but at intermediate depths, amongst the muddy lowest layers of dead corallines, a minority of deposit-feeding clams can be found, such as the primitive nut shell *Nucula nucleus*,

FIG 128. The Great Scallop, *Pecten maximus*. (a) Upper surface, with a thatch of small red seaweeds. (b) Lower surface of the same individual. (J. S. Porter)

FIG 129. Left and right shell valves of the small scallop *Palliolum tigerinum*. Scale bar: 2 cm.

FIG 130. The carpet shell *Tapes rhomboides* in life position in the top layers of a maerl bed. (J. S. Porter)

and the thin-shelled *Gari tellinella* and *Abra alba*. Numerous small, active snails are abundant amongst the pink living layers of maerl, especially top shells, including perhaps four species within the genus *Gibbula*, and many tiny species of *Rissoa*, *Alvania* and *Onoba*. Together with two species of tortoiseshell limpet, *Tectura virginea* and *Testudinalia testudinalis*, and as many as six species of chiton, these small micrograzers feed on the microflora and fine detritus that gathers on the upper surfaces of the maerl bed, performing an important function in maintaining the photosynthesising layers clear of fouling and thus able to function at optimum rate.

Dense and diverse communities naturally promote many mutualistic associations, especially exemplified by *Kurtiella bidentata*. This small, suspension-feeding bivalve is found in often immense numbers in unconsolidated substrata, and always associated with large macroinfaunal species, especially brittlestars (page 119). It lives within the burrows of its host, feeding on fresh phytoplankton and other material drawn in by the host's feeding currents. Amongst communities of the brittlestar *Amphiura filiformis* in muddy sediments, *Kurtiella bidentata* may be counted in thousands per square metre of sea floor. It has been recorded at densities of up to 200 per square metre in Clyde Sea maerl beds, where it was associated in particular with the burrow-building mud shrimp *Upogebia deltaura*. Another family of molluscs, the enigmatic Pyramidellidae, are tiny snails with slender, spired shells mostly less than 5 mm in length; they are all micropredators or ectoparasites of other molluscs, especially bivalves, and of large, tube-dwelling polychaetes, and a substantial proportion of the 40 or so species known from British seas occur in maerl communities. Molluscan predators are especially represented by an array of small sea slugs, such as colourful species of *Doto* (Fig. 99), *Cuthona*, *Eubranchus* (Fig. 131) and *Facelina*, each adapted to a narrow range of hydroid prey species. The internally shelled cephalaspidean sea slugs, *Retusa truncatula* and species of *Philine*, are predators of polychaetes and other small molluscs, while the encrusting communities of bryozoans and hydroids are preyed upon by nudibranch sea slugs, such as stiff-bodied *Onchidoris bilamellata* and *O. muricata*, and the flamboyant *Limacia clavigera* and *Polycera quadrilineata*. Small predatory snails include the cowries *Trivia monacha* and *T. arctica* (Fig. 150), specialist predators of colonial sea squirts, and the sponge-feeding *Marshallora adversa* and *Bittium reticulatum*. The largest molluscan carnivores also abound, in particular the necklace shells *Euspira*

FIG 131. A predatory sea slug, *Eubranchus tricolor*. (J. S. Porter)

montagui and *E. nitida*, which feed mostly on small bivalves, and the whelks *Buccinum undatum*, *Neptunea antiqua* (Fig. 55) and *Colus gracilis* (Fig. 56), which function as both non-selective carnivores and scavengers. The diversity of fish associated with maerl beds has not been examined in detail, and it is unlikely that any species is especially associated with the maerl community. However, it is evident that maerl beds are important nurseries, providing food and shelter for juvenile gadoids, specifically Cod, Pollack, *Pollachius pollachius*, and Saithe (Kamenos *et al.*, 2004).

None of the species recorded from maerl habitats are obligate members of the community, although many may be more abundant, or at least as abundant, on maerl as on other benthic substrata; a few seem to be prominent components of every maerl bed, and some species fulfil important ecological functions (BIOMAERL Team, 1999). The sea anemone *Edwardsia claparedii* is a highly modified burrower, with a slender, worm-like body up to 7 cm in length, and bearing just 16 short tentacles. It has a patchy distribution around Britain, in fine muddy sands, and is nowhere common, yet it was found burrowed into the muddy lower layers of Clyde Sea maerl beds at densities of up to 50 per square metre. The most abundant bivalve in Clyde Sea maerl was *Dosinia exoleta* (Fig. 132), at 6–30 individuals per square metre, while the most frequently recorded tubeworm was *Owenia fusiformis*, which contributed up to 3.3% of the total wet weight of the biomass in some samples, equivalent to 9–18 g wet weight of biomass for each square metre of maerl.

The brittlestar *Ophiocomina nigra* (Fig. 74) lives in dense, layered, lenticular aggregations, the distribution of which is determined by the same hydrodynamic

FIG 132. Shell valves of the bivalve *Dosinia exoleta*. Scale bar: 5 cm.

factors that define the distribution of maerl. Their co-occurrence in the Clyde Sea survey, where in some samples *O. nigra* constituted 46.7% of the total wet weight of biomass, was perhaps coincidental, and indeed the dense, smothering mass of brittlestars would not seem to be conducive to a healthy environment for live maerl. Conversely, two sea urchin species that are also facultative members of the maerl community, which also often occur at high densities, appear to have important ecological roles within the community. In the Clyde Sea the Green Sea Urchin, *Psammechinus miliaris* (Fig. 133), was common in the surface layers of living maerl, with as many as 14 individuals present in 0.1 m² grab samples. In Galway Bay, maerl beds supported populations of the Purple Sea Urchin, *Paracentrotus lividus*, at densities of up to 1,600 per square metre. Both species are common in the Bay of Brest, Brittany, together with the large and spiny urchin *Sphaerechinus granularis*, and all three are significant consumers of large brown macroalgae (Guillou *et al.*, 2002). On healthy maerl beds in the south of the bay *Ps. miliaris* and *P. lividus* were present at modest densities of 10–12 individuals per square metre, and *S. granularis* was not recorded. To the north, the maerl community was contaminated by sewage-laden effluent from Brest, and sea urchin populations were profoundly affected. Large, old individuals of *S. granularis* were the only permanent residents in these northern beds, and at densities of fewer than one individual per square metre; small individuals of *P. lividus* were recorded in June, only, of each year monitored, while *Ps. miliaris* was absent. Close to Brest, seaweed growth was vigorous, and the biomass of macroalgae achieved a summer maximum equivalent to 72.7 g dry weight per square metre of maerl. Nutrient enrichment, provided by the urban effluent, undoubtedly promoted enhanced seaweed growth, but the contamination of the habitat also suppressed sea urchin populations, and thus removed the most important control of macroalgal

FIG 133. The Green Sea Urchin, *Psammechinus miliaris*. (P. E. J. Dyrynda)

biomass. At the clean southern site, with constant sea urchin populations, seaweed biomass achieved a summer maximum of only 2.2 g dry weight per square metre, as a direct consequence of sea urchin grazing, combined with reduced nutrient input. Here, the populations of *Ps. miliaris* and *P. lividus* showed a healthy distribution of individual sizes, indicative of successful breeding, and successful settlement and recruitment of juveniles. Sea urchins were thus found to be very important in limiting the annual growth of macroalgae, which, unchecked, would otherwise have smothered the slow-growing maerl corallines. It is not known, however, to what extent maerl production is directly affected by seaweed shading.

The tiny heart urchin *Echinocyamus pusillus*, which grows no longer than 15 mm, was abundant in Clyde Sea maerl communities, with up to 10 individuals per 0.1 m² grab sample. It is found across the whole of the northwest European continental shelf, and in the Mediterranean, burying itself in coarse sands and gravels. In such habitats, and being so small, it is unlikely to maintain a permanent burrow, as do its larger relatives, such as species of *Echinocardium*, and perhaps is best regarded as an interstitial species. Perhaps the relative stability of the maerl environment particularly suits it. *Echinocyamus pusillus* is regarded as a 'high-fidelity' species, to be expected wherever maerl occurs (BIOMAERL Team, 1999), but the fact that it may be equally abundant in pristine maerl beds and in those damaged by anthropogenic impacts suggests that it is the physical nature of its immediate environment that is the principal factor determining its distribution and abundance.

The mud shrimp *Upogebia deltaura* is a bulky animal, often more than 10 cm in length, and it is an energetic engineer of its environment. *Upogebia deltaura* is especially associated with muddy gravels, and the maerl habitat, with its additional advantage of enhanced structural stability, seems to be particularly attractive to it. Its association with maerl was not always apparent (Hall-Spencer & Atkinson, 1999); the burrows of *U. deltaura* extend as deep as 60 cm below the sediment surface, and standard benthic grabs simply destroy evidence of the burrows while failing to catch the deeply buried mud shrimps. However, a combination of remotely operated vehicles and scuba demonstrated how common *U. deltaura* may be amongst maerl communities. Burrow openings are inconspicuous, flush with the sediment surface and less than 1 cm diameter, but may be recognised by a greyish aureole; burrows are round-sectioned, about 2 cm diameter, with smooth walls of fine, closely packed, mucus-bound sediment. Burrow openings are spaced as close as 2–3 cm apart; each individual burrow system may have two or three separate openings, but accommodates usually just one or two mud shrimps, and they do not interconnect. Each system is a branching, three-dimensional structure, and densities of *U. deltaura* may be as high as five individuals per square metre of maerl bed surface. The hard-packed layer of fine sediment lining the burrows provides a smooth, low-friction surface, allowing a constant laminar flow of water driven by the beating pleopods of the mud shrimps. Although it has been reported to prey upon other small invertebrates, *U. deltaura* is primarily a suspension feeder, filtering and sorting fine particles from the incurrent water flow. Rates of flow have been measured at 4.2–34.1 ml of water per minute, representing a considerable volume in a 2 cm diameter tunnel, and while providing food for the mud shrimp, it also creates niches for other species. Most importantly the burrows are aerated: oxygen-rich and food-laden currents are drawn deep into what would otherwise be organically rich but anoxic sediments. Other species are thus able to live and feed deeply within the sediment. Some perhaps feed upon suspended material, like the mud shrimp; others may feed upon organic material within the sediment, but the walls of the chambers provide surfaces for the growth of bacteria and other microorganisms, which may be a further source of nutrition for both the mud shrimp and its co-habitants.

Maerl beds are characterised by high benthic biodiversity and biomass, and their importance for the maintenance of marine biodiversity is emphasised in the European Union Habitats Directive of 1992 (92/43/EEC), in which both the habitat and the two most significant coralline species, *Phymatolithon calcareum* and *Lithothamnion corallioides*, are afforded protected status (Hall-Spencer, 1998). The vulnerability of the maerl habitat is now quite evident. Maerl is a source of high-grade agricultural lime, easily and cheaply extracted, and usable without costly processing. Overexploitation of live maerl is readily resolved, simply by limiting and licensing, or banning, extraction, but the scale of other threats has only recently become apparent, and these are less susceptible to resolution. Maerl is especially vulnerable to bottom fishing. Not only does it suffer the same physical damage as other benthic habitats, but its recovery depends upon its intrinsic biological properties: it can only recover through growth. Maerl-forming corallines are extremely slow-growing, and their growth rate is further depressed, or even halted, by the reduced water clarity and increased turbidity caused by dredging. Both the Great Scallop (Fig. 128) and the Queen Scallop, *Aequipecten opercularis* (Fig. 140), live at moderate densities in maerl habitats. It is probable that both species were formerly abundant or perhaps more abundant, in unconsolidated bottom habitats, but around most of the British Isles both have been fished extensively, and in some areas to exhaustion. Maerl was generally avoided by scallop dredgers as it impeded or damaged the fishing gear, but the development of more powerful boats and heavier fishing gear in the latter half of the twentieth century opened it up to exploitation, and it ceased to serve as a refuge, or reservoir, for all commercially significant fish, shellfish and crustaceans.

FIG 134. The Gaping File Shell, *Limaria hians*, showing its fringe of pallial tentacles. (J. S. Porter)

Of the many species of mollusc recorded from maerl beds, the most singular is the Gaping File Shell, *Limaria hians* (Fig. 134). This small clam, up to 40 mm long, has elongate oval, convex shell valves which are in contact only at the short hinge line and at the opposing ventral edge; around the rest of their periphery the two valves gape widely. The finely ridged shells are a dirty white colour, but a coiling mass of brilliant orange and red tentacles extends from the gape. *Limaria hians* occupies the interstices of kelp holdfasts, spaces within Horse Mussel beds, or cavities and crevices within any stable hard substrata, but it was formerly abundant amongst unconsolidated gravels as a significant habitat engineer, binding and stabilising the sediment, and entrapping silt, and its own faecal material, in a byssal mesh (Hall-Spencer and Moore, 2000a). File shells of this species are free-living, but they build substantial domiciles, or 'nests', employing tough and elastic byssus threads secreted by the foot to glue pieces of shell, gravel and maerl. These are impressive circular structures, reportedly up to a metre in diameter and protruding as much as 20 cm above the surrounding surface. Each nest shelters a number of file shells, of different sizes, and is typically colonised by a sessile fauna of hydroids, sponges and other filter-feeding organisms, further enhancing the heterogeneity and diversity of the habitat.

Limaria hians is able to swim, but it makes slow and stately progress, quite unlike the swift, looping movement of scallops. An initial expulsion of water from the mantle cavity propels the file shell off the bottom, and it then swims by a deliberate rowing action of its extended tentacles, the rowing movement proceeding in waves from the leading ventral edge of the shell to the dorsal hinge line. Also unlike scallops, which swim with the shells horizontal, the file shell swims with its paired valves orientated perpendicularly to the substratum. The slow, deliberate swimming cannot be considered to be an escape response; the bright and conspicuous tentacles secrete an acid-rich mucus that deters predators, and any that are nonetheless lost are regrown rapidly. The file shell belongs to a group of free-living limids that do not use their byssal secretions simply as anchors; it shuffles into loose shell and gravel, constructing an internal scaffolding of byssal threads that secure and stabilise the surrounding sediment. While its 'nest' probably provides an additional protection from predators, its wider biological significance remains unknown; it is perhaps important as a refuge from physical disturbance, such as current and storm surge, and the balletic swimming may simply enable the file shell to relocate in the event of it being dislodged.

For much of the Clyde Sea region, and perhaps for gravel habitats around western Britain in general, *Limaria hians* assemblages represent a lost biotope. It persists in some Scottish sea lochs relatively undisturbed by ground fishing, forming continuous beds over hectares of sea floor. In such a community in Loch Fyne, Gaping File Shells were present at densities above 700 per square metre, in complex, galleried nests, each accommodating as many as 100 individuals, including groups of the smallest juveniles and the largest, solitary, adults (Hall-Spencer and Moore, 2000a). The diversity of the community – 284 species of algae and invertebrates recorded from six nests, on one sampling occasion – seems at least as great as that recorded from northwest European maerl habitats. The extraordinary behavioural complexity of *Limaria hians*, and the ecological intricacies of the community it creates and sustains, are largely uninvestigated, and seemed likely to remain unknown, destroyed by just a few decades of dredging. The first quantitative study of *Limaria hians* communities were conducted only in 2006–2007 at two further sites on the Argyll coast (Trigg *et al.*, 2011). Ten cores, each 10 cm diameter and 15 cm deep, were collected at each of two sites, in Loch Fyne and Loch Creran, and yielded 282 taxa for a total sampling area of 0.16 m². This included 40 species of algae, 10 species of sponge and 17 bryozoans, none of which were counted; the 215 species individually enumerated comprised 7,275 individuals. Annelids, molluscs and crustaceans were the most diverse groups, and together accounted for 66.9% of all species identified.

Organic enrichment of inshore benthic habitats through human activity also results in profound ecological changes that, again, are especially damaging to maerl communities. Marine fish farming is particularly significant in this respect. The sea lochs of western Scotland provide sheltered sites for salmon farming, and production of farmed salmon by Scottish fish farms increased by more than 30-fold in the two decades from 1980, with predictable consequences for the benthic environment in the areas around the farms (Hall-Spencer *et al.*, 2006). Unconsumed protein-rich food pellets, and fish faeces, accumulated beneath the holding cages. This additional organic input into the benthic environment provided nutriment for bottom-living fish, and attracted the usual array of scavengers, crabs, whelks and starfish, but was detrimental to much of the resident epifauna and infauna. As the waste accumulated, the organic content of the sediment increased, the biological oxygen demand (BOD) rose, with resulting anoxia, the biomass of anoxia-resistant species increased, and faunal diversity decreased dramatically.

Moving the cages out of the sheltered environment of the sea lochs to higher-energy environments seemed to be the solution to the problem: more vigorous water movement would disperse organic waste more widely, effectively diluting it, and it would be readily incorporated into benthic food webs without adverse effects on the communities. However, the most suitable alternative sites, characterised by substantial water movement but sheltered from direct wave exposure, are precisely those favoured by maerl, and it was soon apparent that the waste effluent from the rearing cages, rather than being dispersed, began to accumulate within the maerl beds. The consequences of this were revealed in a study of the effects of fish farm wastes on maerl communities at three sites, in South Uist, Orkney and Shetland, where the ecological states of contaminated maerl beds were compared with those of uncontaminated reference sites (Hall-Spencer et al., 2006). The differences between impacted and unimpacted maerl habitats were sharply evident. An initial effect of marine aquaculture was increased shading of the beds by the cages and, despite bottom currents of 0.4–0.7 m per second, an enhanced turbidity, which resulted in a coating of silt on the maerl thalli. Experiments on tropical maerl communities showed that the thinnest layer of silt reduces the production rate of maerl corallines by as much as 70%, not simply through a decrease in photosynthetic rate, as a consequence of shading, but also as a result of the inability of the algae to absorb nutrients and exchange gases across fouled surfaces. Further, and again in spite of substantial bottom water flow, the organic waste was not dispersed, but instead was trapped and retained within the maerl lattice, which became packed with black, muddy sediment. Immediately below rearing cages maerl was killed by the mounting anoxic waste; its associated fauna, of encrusting and sessile organisms, grazers and small predators, disappeared. Large scavengers were abundant and these, decapods, whelks and starfish, disturbed the integrity of the maerl bed, creating pits and depressions which collected and retained more of the rain of organic waste.

Comparison of K-dominance curves (page 58) for contaminated maerl communities in the vicinity of rearing cages with clean reference sites showed striking differences (Fig. 135). At 100 m from the contamination, curves were steep, indicating species-rich communities with high evenness values, but closer to the cages the curves became progressively flatter as species diversity decreased, and directly beneath the cages the community became dominated by few, abundant opportunistic species, with as much as 80% of the total represented by just one species. It is evident that maerl communities cannot resist, or adapt to, the slightest degree of organic contamination, and that rotating cage use to allow 'fallow' periods, as suggested by some researchers, does not obviate the problem. Once damaged, a maerl bed may take decades, or perhaps centuries, to recover, simply because of the extremely slow rate of growth of the corallines, which, further, can only be sustained in clear, well-lit and silt-free water.

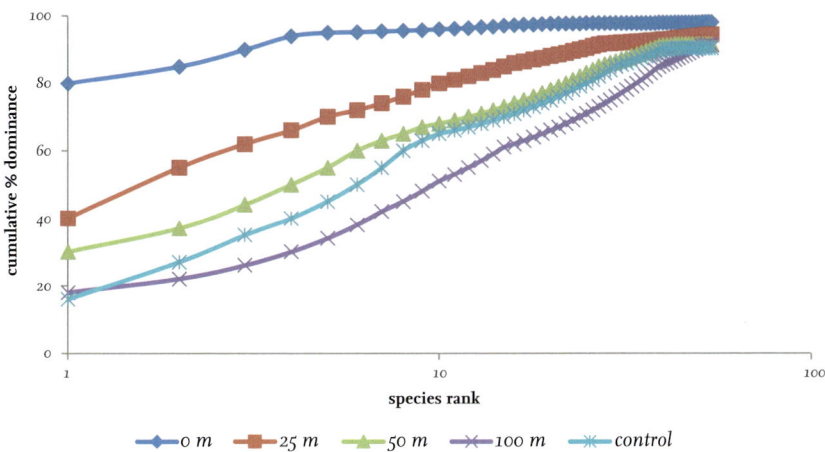

FIG 135. K-dominance curves for maerl communities at successive distance from salmon-rearing cages, and from uncontaminated reference sites. (After Hall-Spencer *et al.*, 2006)

HYDROID, BRYOZOAN AND POLYCHAETE ASSOCIATES

Many epibenthic marine invertebrate species live in aggregations sufficiently dense to modify their immediate environment, creating stable species-rich habitats, as diverse and extensive as those formed by *Zostera* meadows and maerl beds. Such habitat engineers, naturally, are sedentary or sessile in habit and are necessarily long-lived. They may found their communities on rocky reefs and outcrops, but most of the more expansive communities occur on mixed grounds, muddy sands and gravels, and fulfil important ecological roles in stabilising the sediment, creating habitat complexity and promoting faunal diversity. The infaunal assemblages of the underlying sediment increase in diversity, and total diversity is further augmented by the settlement and establishment of additional sessile and sedentary species, crevice dwellers and small epifauna. The variety of marine invertebrate habitat engineers is interestingly broad and includes many clonal, modular groups, such as hydroids and bryozoans. Clumped populations of the larger hydroid species (Fig. 136), such as *Nemertesia antennina*, *N. ramosa*, *Sertularia cupressina*, *Kirchenpaueria pinnata* and species of *Halecium*, provide shelter within their tangled hydrorhizal stolons for minor crustaceans and polychaetes, and support populations of small predatory sea slugs, in particular brightly coloured species of *Doto* (Fig. 99), but otherwise their structural complexity is low and their stinging cnidocytes discourage other potential associates. Two bryozoan species, the stiff, frondose *Flustra foliacea* (Fig. 47), common in inshore

FIG 136.
(a) Colonies of the large hydroid *Nemertesia antennina*. (b) Base of a *Nemertesia* clump, thickly overgrown with Jewel Anemones, *Corynactis viridis*, (below) and encrusting and erect bryozoans. (J. S. Porter)

waters on most British coasts, growing patchily on reefs and vertical rock faces, and as extensive beds on coarse sandy gravels, and the more substantial, coral-like *Pentapora foliacea* (Fig. 86), common on rocky western coasts of Britain and Ireland, are long-lived and persistent components of their benthic communities, and support rich associated faunas, with species diversity often counted in tens and numbers of individuals in thousands (pages 160–4).

Tube-building polychaetes are another group of habitat engineers. *Owenia fusiformis* and *Lanice conchilega* construct long, thin tubes of cemented sand grains and shell fragments, deeply anchored in fine sands in the low intertidal and shallow subtidal. Both sometimes occur at densities sufficient to modify water flow within their aggregations, causing finer sediment to accumulate and creating niches for a small associated fauna, especially amphipods. However, neither of these polychaetes is a prominent part of the benthic fauna; their distribution is governed by wave and tidal current, and their communities seem not to show long-term persistence. *Serpula vermicularis* is a small tubeworm that secretes a sinuous white calcareous tube; it occurs commonly around the western, northern and northeastern coasts of the British Isles, and around Ireland, and is a frequent component of the hard-substratum encrusting epifauna. Under very special conditions, on muddy sands in very sheltered, semi-enclosed sea lochs, *S. vermicularis* forms gregarious populations. This appears to be a consequence of a limited supply of hard substrata, post-larval individuals having to settle on and attach to adult tubes, and perhaps also of limited larval dispersal. These aggregations may form the bases of small communities, but they are subject to catastrophic physical collapse following natural or anthropogenic perturbation, and perhaps also as a result of changing age structure of the worm population.

In the sheltered environment of Loch Creran, western Scotland, *Serpula vermicularis* aggregations form lenticular mounds, 0.5 m high and long, covering about 25% of the loch floor from low water to 15 m depth. A quantitative study of the fauna associated with *S. vermicularis* mounds, based on a sample of ten, with areas 0.0014–0.1767 m^2 and volumes 101–53,034 cm^3, revealed a total of 278 invertebrate taxa (Chapman *et al.*, 2012). The number of species was positively correlated with the size of the aggregation, rising steeply from 34 in the smallest sample, but reaching an asymptote at 185 species; a mound with area 0.1 m^2 had a community of 163 species, totalling 12,756 individuals. The most species-rich groups were polychaetes, represented in the combined samples by 94 species, molluscs, mostly gastropods, with 70 species, and small crustaceans, especially amphipods and isopods, with 45; these three groups together comprised 75% of all species identified. The most abundantly represented group was the crustaceans, which comprised 48.5% of all individuals; polychaetes made up a further 32.5%, but the taxonomically diverse molluscs amounted to only 8.1% of the total of individuals. *Serpula vermicularis* mounds contributed substantial habitat heterogeneity to the shallow, muddy sand environment of Loch Creran. The erect calcareous tubes provided abundant surface area for colonisation by sessile organisms, while the interstices between them provided living space for crevice dwellers, and accumulated deposits of fine sediment that could be exploited by small infaunal animals. The larger mounds, with the greater biomass of associated fauna, would have provided food and shelter for communities of motile species, particularly fish and decapods, which would not have been represented in the samples collected, and the actual diversity of the assemblages would have been higher. However, none of the species recorded is limited to this particular habitat, and about two-thirds were equally abundant in Horse Mussel beds elsewhere in the loch.

The most conspicuous and persistent of the community-building polychaetes are the honeycomb worms *Sabellaria alveolata* and *S. spinulosa*, in both of which gregarious larval settlement occurs, larvae being attracted by the mucus cement employed by the worms in creating their tough, gritty, sand-bound tubes. They are stout, short-bodied worms, up to 30 mm long with a maximum width of around 4 mm across the crown of stiff chaetae that seal the tube entrance. The tubes are aligned along their long axes, the slightly flared rims closely packed together and very suggestive of the combs built by bees. *Sabellaria alveolata* (Fig. 137) is a southern, warm-water species distributed from Morocco to the west coasts of Britain, and common throughout the Mediterranean. It reaches its northern limit in the Outer Hebrides, and in Britain is mostly distributed on southern and western coasts, where it is locally abundant on moderately exposed rocky sea shores. *Sabellaria spinulosa* occurs off all northwestern European coasts, except for the shores of the Baltic, and is strictly subtidal in distribution; it is often found as solitary individuals, or in small groups, among sessile epilithic communities, but under certain conditions develops extensive colonies on rocky outcrops, or on coarse, mixed substrata.

The longevity of *Sabellaria spinulosa* is not precisely recorded, but individual worms may live for five years, perhaps longer; their tubes survive them and the colony grows as succeeding generations settle upon and reinforce the constructions of their predecessors. The occurrence and distribution of *S. spinulosa* colonies depend largely upon hydrodynamic factors. The worms have a protracted reproductive cycle; the larvae are planktonic, returning to the benthos after a free-swimming larval period of several weeks, and while they are able to select their preferred substratum, adult tubes, on which to settle and metamorphose, they are dependent on wind-driven surface currents and

FIG 137. Portion of a reef formed by the shallow-water honeycomb worm *Sabellaria alveolata*.

bottom circulation to concentrate them in areas of suitable habitat. Larvae may be dispersed along coasts or circulate around coastal embayments, and recruits to each population are usually allochthonous (originating from elsewhere). Recruitment may show great annual variation, partly resulting from environmental effects on larval growth and development, and partly as a consequence of the effects of unseasonal weather on larval dispersal. Bottom currents have additional roles in supplying sand grains of a size suitable for tube construction, and in carrying away fine particles that might otherwise smother the worms.

Sabellaria spinulosa colonies may cover large areas of sea floor as discontinuous crusts and mounds, many hectares in extent, and up to 0.5 m thick. There seem to be cycles of growth, expansion and decay in these populations, the initially thin crusts increasing to thicker sheets and mounds, which then degrade and decline, perhaps as a consequence of senescence and fluctuating recruitment. That such cycles are natural is suggested by research on the huge colonies of *S. alveolata* in the Baie Saint-Michel, France, which found that in degraded areas of *Sabellaria* crust the worm population consisted mostly of small individuals, representing recent recruits to the damaged areas. Tube extension in another population of *S. alveolata* was found to average 0.7 mm per day, a rate sufficient to maintain and repair tube rims, but in parts of the colony that were experimentally manipulated tube growth averaged 4.4 mm per day, suggesting that the worms recovered swiftly from quite significant damage (Vorberg, 2000).

Both species of *Sabellaria* may develop dense populations and provide the foundation for very diverse communities. The French *S. alveolata* populations had mean densities of 35,000 per square metre, with an associated fauna of 63 invertebrate species, some of which achieved densities as great as that of their host. In the current-swept central area of the Bristol Channel *S. spinulosa* has been recorded at densities above 4,000 per square metre, with an associated community of 70 macrofaunal invertebrates (George & Warwick, 1985). The majority of the species recorded in association with the two species of *Sabellaria* are other species of polychaetes, including many small tube-builders and crevice dwellers, but tube-building amphipods, such as species of *Corophium*, and the crevice-dwelling bivalves *Hiatella arctica* (Fig. 112) and *Venerupis senegalensis* may also be abundant, while the often elusive sipunculan *Golfingia vulgaris* occurred at densities of 1,492 per square metre among *S. alveolata*.

HORSE MUSSELS

The Horse Mussel, *Modiolus modiolus*, is an especially significant reef builder in northwest European shelf environments. The epithet 'horse' implies a large, coarse-fleshed mussel, not as esculent as its littoral counterpart the Blue Mussel, *Mytilus edulis*. It is certainly large, slow-growing beyond the first four years of life, and very long-lived, so it is unsurprising that it has never been favoured by epicureans. The Horse Mussel has a northerly distribution; it is widespread in the cold temperate, eastern North Atlantic, ranging beyond the Arctic Circle to northern Norway and the White Sea, and reaching a southern limit in the Bay of Biscay. It is present off all British coasts, but is most abundant to the north and west. Sparse populations, typically of large, old individuals, are found rarely in rock pools on exposed rocky coasts, but it is essentially a subtidal species distributed below 5 m and to at least 100 m deep. *Modiolus modiolus* does not tolerate salinities much lower than 30 psu, but colonises the widest range of substrata, from bedrock to unconsolidated muds and sands. On vertical rock walls, boulders and rocky reefs Horse Mussels occur as solitary individuals, small groups or larger patches, but from the Irish Sea and the Yorkshire coast northwards they form very extensive, dense, layered and undulating beds with a richly diverse associated fauna and flora (Fig. 138). These beds are persistent features of the sea floor: one large bed south of the Isle of Man has a recorded history of more than 150 years (Lindenbaum et al., 2008).

The Horse Mussel is the bulkiest of the bivalve species found in the British sea area. It grows to beyond 25 cm in length, and the largest individuals will have

MORE KEY HABITATS: MAERL BEDS AND BIOGENIC REEFS · 257

FIG 138. Close-up view of part of a Horse Mussel, *Modiolus modiolus*, bed, with associated brittlestars, *Ophiothrix fragilis*, showing the orange siphons of the mussels protruding through accumulated shell debris and biodeposits. (J. S. Porter)

a girth of almost 10 cm; maximum longevity is unknown but is probably above 50 years, and in well-established populations the modal age is around 20 years. Both growth and reproduction are sensitive to environmental influences, particularly fluctuations in temperature and food supply, and show great variation across the considerable geographical and bathymetrical range of the species. Sexual maturity is reached at 3–4 years of age in southern populations, when the animals are 35–40 mm long, but at 6–8 years in subarctic populations. Reproductive cycles are very variable; in some populations eggs and sperm mature and are released over a protracted period, while in others maturation shows clear seasonal peaks. Spawning is thought to occur within a narrow temperature range, 7–10 °C, and again shows distinct summer peaks in northern populations, but is slow and continuous among populations to the south. Reproductive cycles have been documented for a few *Modiolus* populations but on the whole are poorly known, although it is clear that they are strongly affected by environmental factors, and that there is no overall pattern. Some populations may breed every year, while others breed only at intervals of two or three years; individuals in isolated deep-water beds perhaps never breed at all, and are sustained only by sporadic influx and recruitment of allochthonous larvae. In all populations, settlement and recruitment often show remarkable seasonal and annual variation. The hydrography of enclosed water bodies, such as Strangford Lough in Northern Ireland, often serves to retain rather than disperse larvae, and the adult population is maintained by its own reproductive output, but elsewhere larvae are dispersed widely and new recruits to the adult population generally originate from distant beds. Hydrodynamic factors are important: in shallow, high-energy environments settlement and recruitment tend to be high, with little interannual variation, but both decline with increasing water depth and decreasing water movement. Dense and successful settlement, however, does not invariably result in high levels of recruitment, and predatory crabs, whelks and starfish inflict a high degree of post-settlement mortality. The life cycle of the Horse Mussel is perhaps an adaptation to this huge mortality; the new recruits grow rapidly through their first 4–6 years of life, allocating all surplus resources to somatic growth rather than to reproduction. At around 45 mm length it achieves a size refuge from its predators, mortality decreases sharply, and individuals attain sexual maturity. Growth slows as an increasing proportion of resources is devoted to gamete production, and continues at a slowly declining rate through the rest of the animal's life span. This is evident in size frequency distributions of M. *modiolus* in optimum habitats: they tend to be bimodal, with a small peak of newly recruited juveniles and a second, much larger peak of reproductively active adults.

Horse Mussels form stable beds at 5–50 m depth along the northwestern and northern coasts of the British Isles. Despite decades of damage by bottom-fishing gear they are still very extensive, some beds covering hundreds of hectares of sea floor, and perhaps formerly covering hundreds of square kilometres. The morphology of the bed is determined by local hydrodynamic regimes; off moderately current-swept coasts *Modiolus* beds may be a metre or more thick, with an undulating surface of troughs and ridges, creating wave forms, with crests up to 18 m apart, orientated across the current (Lindenbaum *et al.*, 2008). The bed is bound together by a tough mesh of byssal threads secreted by the mussels, in increasing quantity in response to water turbulence. Adult mussels are closely packed and anchored amongst the accumulated shells of previous generations; juveniles are wedged in amongst the adults, with the smallest amongst the tangle of byssal threads, and the mussel biomass is multiplied by an enormous associated fauna.

It is difficult to estimate density of *Modiolus modiolus* beds. Dredge sampling leads to underestimated population densities: the largest individuals are so tightly bound to the bed that the dredge fails to dislodge them, while the smallest are detached from the fabric of the bed and lost. Counts from photographic samples, or within fixed quadrats by divers, grossly underestimate the number of smaller Horse Mussels hidden within the byssal mesh of the adults. Such methods have suggested that densities might range from 20 to 40 individuals per square metre; visual estimates by divers of mussel densities off the coast of north Wales were around 100 per square metre (Sanderson et al., 2008), but quadrat sampling of the same bed, using a diver-operated hydraulic suction sampler, demonstrated the degree of underestimation, yielding densities of 1,400 per square metre (Rees et al., 2008).

The structural base of the *Modiolus* community consists of Horse Mussels, together with a variety of other sedentary bivalves and the accumulated dead shells of all species present, stabilised by the tangle of byssus threads, while its biotic environment derives from the biological activities of the live mussels. Large, suspension-feeding bivalves such as *M. modiolus* filter impressive volumes of water, extracting equally impressive quantities of particulate matter. The filtration, or clearance, rate of actively feeding individuals in a Newfoundland population during spring and early summer ranged from 0.84 to 1.95 litres per hour for small animals, averaging 1.5 g tissue dry weight, to 1.03–3.61 litres per hour in large animals, averaging 6.68 g dry weight, and for the whole population most values were in the range 2–3 litres per hour over periods of 10–12 hours' continuous feeding (Navarro & Thompson, 1996). Assuming the minimum density of 100 large Horse Mussels per square metre, recorded by visual estimate, the volume of water filtered per unit area is sufficient to reduce significantly the density of suspended material in the overlying water column. Clearance rates for the Newfoundland Horse Mussels showed considerable seasonal fluctuation; in particular, feeding was intermittent during the winter months and clearance rates ranged from 0 to 2.59 litres per hour for medium-sized animals. However, this variability was related to food quality and quantity rather than to temperature, or other environmental influences, and even in deep winter an enhancement of food supply might stimulate active feeding and rising clearance rates. Particles carried into the mussel's mantle cavity by the inhalant water stream are filtered and sorted on its broad, folded gills; selection of food particles depends partly on quality, non-nutritious material being rejected, and partly on size, as many otherwise nutritious diatoms are too large to pass between the gill lamellae toward the labial palps and the animal's mouth. While feeding, *M. modiolus* sheds a constant trickle of small, dark-coloured faecal pellets that pass out of the mantle cavity with the exhalant water stream. The rejected material gathers below the folds of the gills, close to the inhalant siphon as larger, looser and light-coloured packages, termed pseudofaeces, which are abruptly ejected by periodic reversal of the inhalant flow. Faecal pellets and pseudofaeces, collectively biodeposits, sink within the framework of the Horse Mussel bed, providing the most important source of energy for the community. *Modiolus modiolus*, as the dominant primary consumer in the assemblage, is a critical trophic link between the primary producers in the overlying water column and the deposit-feeding benthos.

The amount of material recycled by Horse Mussels varies in relation to phytoplankton cycles. The Newfoundland population produced pseudofaeces only during the spring phytoplankton bloom, between April and May, when clearance rates were maximal (Navarro & Thompson, 1997). The total production of biodeposits increased as the bloom progressed, reaching a maximum mean value of 40.9 mg dry weight per day for a medium-sized mussel, but dropped to a mean 4.3 mg per day as the bloom ended. Faecal production exceeded pseudofaecal production fivefold throughout the whole period of the bloom. Both contributed equally to chlorophyll *a* recycling at the beginning of the bloom, but at its peak pseudofaeces were the most significant source of chlorophyll *a* for benthic consumers. With the onset of the bloom faecal production rose to 27.3 mg dry weight per day, including 7.6 µg of chlorophyll *a*, while a pseudofaceal production of 4.8 mg contained 5.8 µg of chlorophyll *a*; at the peak 33.1 mg of faeces contained 14.4 µg of chlorophyll *a*, but 7.8 mg of pseudofaeces contained 41.9 µg. For a brief spring period pseudofaecal production was the most important source of organic material for the *Modiolus* community: 28–54%, by weight, of pseudofaeces consisted of organic matter, while 76–91% of faecal production consisted of inorganic material. However, although pseudofaeces always had the higher content of chlorophyll *a*, faecal production was still important in recycling inorganic nutrients to the benthos and in providing substrates for the growth of microorganisms, and thus ultimately contributing to energy flow within the community. Suspension feeding by Horse Mussels drives energy transfer between marine trophic levels, concentrating and packaging particulate material, including fresh phytoplankton, for consumption by deposit-feeding benthos. Scaling up the daily biodeposition recorded for the Newfoundland populations, for a minimum density of 100 individuals per square metre, suggests that it might be measured in kilograms per square metre per year, sustaining the high productivity and biodiversity characteristic of the habitat.

There is no defined *Modiolus modiolus* community; instead, the mussel bed forms the framework for a fauna that varies in composition, diversity and density in response to local physical and biological factors. However, current classification (Connor et al., 1997) recognises several marine biotopes characterised by *Modiolus modiolus* that appear to have restricted distributions. In the shelter of Strangford Lough Horse Mussels and the two small scallops *Chlamys varia* (Fig. 139) and *Aequipecten opercularis* (Fig. 140) are co-dominants in a rich assemblage of suspension feeders, including the brittlestars *Ophiothrix fragilis* (Fig. 138) and *Ophiocomina nigra* (Fig. 74), many species of sponge, and

FIG 139. Left and right valves of the small scallop *Chlamys varia*. Scale bar: 3 cm.

FIG 140. Shells of the Queen Scallop, *Aequipecten opercularis*. Scale bar: 5 cm.

large cnidarians such as *Alcyonium digitatum*. In some Scottish sea lochs *A. opercularis* and the two species of brittlestar occur frequently as part of the *Modiolus* community, but large solitary ascidians, such as *Ciona intestinalis* and *Ascidiella aspersa* (Fig. 89), are particularly characteristic. On more exposed and current-swept coasts more sparsely developed beds are present on rocky ridges and coarse, mixed sediments, often in association with *Sabellaria spinulosa* and *Alcyonium digitatum*, and with a blanket of brittlestars.

As many as 300 species of macrofaunal invertebrates and algae have been recorded from a single survey of a Horse Mussel bed, although total taxonomic diversity has not been computed for any *Modiolus* assemblage. Motile species, particularly widely ranging predators, have almost certainly been under-recorded, encrusting modular organisms such as sponges, bryozoans (Fig. 141) and ascidians are often overlooked, and while the meiofauna is likely to be as diverse as the macrofauna it has never been systematically sampled. Off the north Wales coast *Modiolus* beds had an associated fauna of 213 macroinvertebrate species, excluding encrusting, modular organisms (Rees *et al.*, 2008). Mean total density ranged from 8,800 individuals per square metre where the predominant substratum was dead shell, to almost 23,000 per square metre amongst live mussels. Epifaunal animals will include many predatory and scavenging species. Presence and density of the ambulatory starfish *Asterias rubens* probably depend upon the age structure of the *Modiolus* population, as only the smallest mussels are vulnerable to its attentions, but decapods may occur in striking variety and considerable numbers. Besides the omnipresent Green Crabs, spider crabs (Fig. 146), nut crabs, *Ebalia tuberosa* (Fig. 147), Round Crabs, *Atelecyclus rotundatus* (Fig. 142) and several species of swimming crab, *Liocarcinus*, are often the most frequent among perhaps 20 or more decapod omnivores, and the suspension-feeding porcelain crab *Pisidia longicornis* may occur at densities above 100/m². The brittlestars *Ophiothrix fragilis* and *Ophiocomina nigra* may comprise a significant proportion of the biomass of suspension feeders, but the bulk will consist of sessile organisms, especially

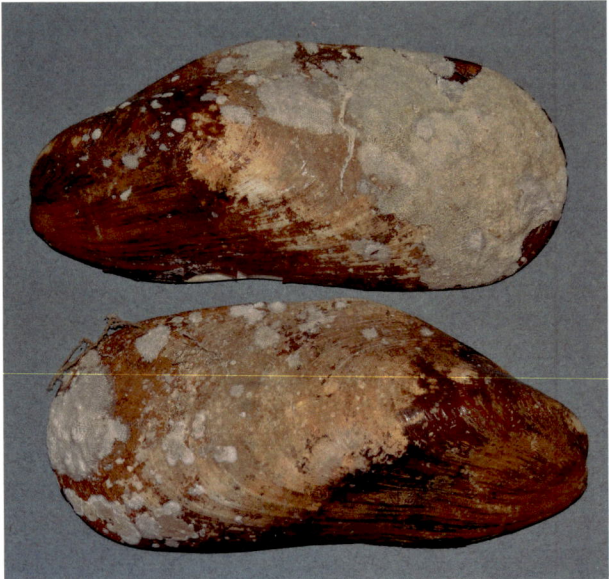

FIG 141. Shells of the Horse Mussel, *Modiolus modiolus*, encrusted with a dense community of bryozoans.

FIG 142. The Round Crab, *Atelecyclus rotundatus*. (J. E. Lancaster)

sponges, cnidarians, often large numbers of balanomorph barnacles, and numerous species of ascidian. The soft coral *Alcyonium digitatum* (Fig. 82) is a frequent associate of Horse Mussel beds. It grows as fleshy, pinkish white or orange lobes, often 20 cm or more high, and being a very long-lived species it may contribute significantly to the permanent fabric of the bed. Soft corals are particulate suspension feeders and probably benefit from the combined respiratory current generated by the mussels. Individual large sponges are also likely to persist for many years once established, but populations of some other substantial filter feeders may be more transient; the sea squirt *Ciona intestinalis* (Fig. 84) forms dense clumps that monopolise surfaces and prevent settlement of other sessile species, as well as larval *Ciona*, but with only a brief life span the population is only transitory. The continually accumulating biodeposits provide food and habitat for many species of burrowers and tube-builders within the structure of the Horse Mussel bed. Many small species of polychaete and amphipod thrive within this organically rich habitat, which may even support soft-sediment infaunal bivalves such as *Kurtiella undata* and *Abra alba*. Some of them occur in such huge numbers that the combined densities of the two bivalves, for example, may represent as much as 50% of the total density of the community.

Horse Mussel beds are regarded as highly productive; some ecologists suggest that they are the most important primary consumers in shallow shelf habitats, and especially significant in the transfer of pelagic production to the benthos. However, there is a paucity of data relating to production rates of *M. modiolus* and its associated fauna, and little information on trophic interactions and energy flow within the community. Some information is available from a study of rather atypical *M. modiolus* assemblages in the Bristol Channel, where the species is close to the southern limit of bed formation (George & Warwick, 1985). Here, the assemblage develops on rocky ridges and adjacent gravel grounds. The Horse Mussel population was sparse, with a mean density of only 245 per square metre, and it was not the dominant species in the community. Individual mussels were small, mostly less than 13.5 mm long, with a mean biomass of 141 g dry weight per square metre (dwt/m^2), comprising just 0.57% of the mean total biomass of the community. *Ophiothrix fragilis* was present at a mean density of 404/m^2 and a mean biomass of 17.3 g dwt/m^2, representing 70.8% of the total biomass, while *Sabellaria spinulosa*, at a density of 4,039/m^2 and a biomass of 4.55 g dwt/m^2, accounted for another 18.6%. Total diversity at the sample station exceeded 220 species, and estimates of total biomass and production for the community were based on analyses of the 21 top-ranked species, all but four of which were suspension feeders. The mean annual biomass was calculated at 24.43 g dwt/m^2, and the mean production at 34.1 g dwt/m^2 per year; this gave a production/biomass (P/B) ratio of 1.4, more than twice that for Bristol Channel *Venus* communities (0.56) and just higher than that for the notably productive *Abra alba* community (1.3). The Horse Mussels were small and slow-growing, and had an estimated P/B of just 0.5, but the majority contribution to total community production was that of *Ophiothrix fragilis*, which had a P/B value of 1.8. Suspension feeders were responsible for 96% of the measured production; 92% was attributable to *O. fragilis* alone, and it was suggested that the dominance of this species depleted food available to other suspension feeders, and that resource limitation might have accounted for their individual small size and slow growth.

STONE CORAL

The Stone Coral, *Lophelia pertusa*, must be the most spectacular benthic foundation species in northwest European shelf environments. It occurs as huge banks, often several thousand square metres in extent and tens of metres in height, but it is distributed mostly deeper than 200 m, and the spectacle can only be imagined. Sonar recordings may plot height, thickness and area of Stone Coral banks, remotely operated videos may picture a few gloomy square metres, and dredging may yield battered fragments of the coral (Fig. 143), but the overall architecture of the habitat it creates is not easily appreciated. *Lophelia pertusa* belongs to the Caryophylliidae, a geographically widespread family of scleractinian corals comprising several hundred species, both solitary and colonial, including the familiar European cup corals *Caryophyllia smithii* and *Balanophyllia regia*. Temperate species, and those living below the euphotic zone, lack the symbiotic dinoflagellates (zooxanthellae) that promote enhanced calcification, sustained growth and reef formation in shallow-water tropical species, but some achieve impressive size despite slow growth.

The Stone Coral is a modular, or colonial, species; each colony originates as a single polyp, developed from a sexually produced larva, which secretes a calcified external skeleton, the corallite. The colony spreads through continuous growth and division of polyps and corallites, developing a three-dimensional lattice

FIG 143. A fragment of the Stone Coral, *Lophelia pertusa*, dredged from the western edge of the shelf. Scale bar: 5 cm.

of cylindrical branches, anastomosed and fused at irregular intervals to form a rigid structure. Living polyps are confined to the outer layer of the colony; the common skeleton, or coenosteum, forming the inner framework is essentially a dead calcareous substratum that is soon colonised by sessile organisms, and penetrated by boring invertebrates, especially sponges. This burden weakens the framework of the colony, which eventually collapses, but it does not die and it is not dispersed. Living portions of the colony are able to re-attach to larger, stable fragments and continue to grow, while the exposed inner skeletal fragments supply new surfaces for settlement by invertebrate larvae, including Stone Coral planula larvae. Through continuous growth, collapse and regeneration the originating colony spreads upwards and outwards (Wilson, 1979), and through decades and centuries evolves into the substantial mounds revealed by sonar. The growth is clonal, but occasional recruitment of new larvae probably ensures the aggregation of numerous clones within each mound.

Lophelia pertusa has a worldwide distribution between 71° N and 55° S, but it is most abundant, or has been recorded most frequently, in the North Atlantic (Rogers, 1999). To the east it has a known range from North Cape, Norway, to the Strait of Gibraltar, and in the west it appears to reach a southern limit off the edge of the Florida continental shelf. In the eastern North Atlantic Stone Coral banks are most common at 200–400 m depth, along the edge of the continental shelf, but the Florida banks are best developed on the continental slope, at 500–600 m. In some west Norwegian fjords Stone Coral banks are found as deep as 400 m, but also occur as shallow as 39 m, and the mean depth of occurrence for these fjord populations is only 186 m (Mortensen et al., 2001).

Investigating the ecology of deep benthic habitats is a very expensive undertaking, beset by obvious practical difficulties, and it is not surprising that almost nothing was known of the ecology of L. pertusa until very recently. Offshore fishermen charted occurrences of the coral, as areas to avoid, but with the development of heavy ground-fishing gear such precautions became unnecessary, and rough grounds were found to be rich sources of demersal fish. As concern arose over the effects of offshore dredging and trawling on Stone Coral banks, Norwegian biologists began systematic exploration of their distribution and occurrence over an area of the west Norwegian continental shelf between latitudes of approximately 62° and 65° N (Mortensen et al., 2001; Fosså et al., 2002). The coral was found to be most abundant between 200 and 400 m depth, with bottom water temperatures of 4–8 °C, and salinity no lower than 34 psu. The banks were mapped using multibeam sonar: height and shape distinguished them from mounds of sponges or glacial moraines, and ROVs were deployed sporadically for ground truthing, verifying their identification. Within the recorded optimum depth range Stone Coral banks occurred in a 5 km-wide zone along the edge of the continental shelf. Individual banks covered areas of 2,500 to almost 40,000 m^2, with a mean of 5,628 m^2, and it was estimated that between the two latitudes as much as 35 km^2 of sea floor was covered with L. pertusa banks.

Substratum is perhaps the least important physical factor determining the distribution of Stone Coral; it will grow attached to stable rock substrata, but over much of the continental shelf of western Norway unconsolidated glacial deposits predominate and a Lophelia colony will utilise pebbles as multiple anchoring points. Also, continuous episodes of collapse provide continuous resources for settlement in the form of coral fragments. Temperature, conversely, is an important factor influencing L. pertusa, and changing bathymetric distributions across its geographical range reflect its narrow temperature tolerance. Most simply, it will be found at greatest depths at the warm southern end of its range, and at increasingly shallower depths towards the cooler north, a phenomenon termed polar emergence. However, bathymetric ranges that change over a smaller scale reflect the distribution of water masses with temperature ranges that are optimum for coral survival and growth. Stone Coral banks of huge extent are found in some west Norwegian fjords, at depths far shallower than those it lives at on the outer continental shelf. Temperature is certainly one factor of significance in the distribution of fjord populations: fjords are highly stratified through much of the year, the coral will tolerate an upper limit of only 10 °C, and both temperature and salinity of the bottom layers are stable in comparison to the water layers above the thermocline. Shelter may also be a factor influencing bathymetric distribution of shallow-water fjordic coral banks, as the upper depth limit of the coral may be determined by wave action. Hydrodynamic factors appear to affect the distribution of the Stone Coral in several respects. Wave and tidal velocity determine the persistence and stability of coral banks, and a threshold maximum velocity of 1 m/s has been suggested to limit their distribution on the continental shelf to depths below 100 m. However, there must also be a threshold minimum velocity, below which increasing sedimentation limits the development of suspension-feeding Stone Coral communities. Hydrodynamic characteristics of the environment are also interrelated with sea-floor topography and water-mass characteristics, with further consequences for the distribution of coral banks, which seem to be most densely distributed, and most extensive, along ridges and across slopes. These sea-floor features deflect water currents, creating turbulence and generating internal waves within the water mass. An initial benefit to the coral community may be the dispersal of sediment, but enhanced water movement may also facilitate vertical mixing and consequent downward flux of primary production. The salinity and temperature tolerances of the Stone Coral restrict it to water masses with appropriate salinity and temperature characteristics, and where the water body forms an interface with water of different characteristics, and thus with a different density, internal waves may again be generated, resulting in the same potential benefits to the coral. Both the density and individual size of coral banks might thus increase towards the boundaries of preferred water masses.

 The biological profile of *Lophelia pertusa* is sketchy; detail accrues only slowly and data related to feeding, growth and reproduction are sparse. Growth rates have been crudely estimated from accumulations of coral rubble of known age, and suggest linear extension of corallites in the range of 3.5–7.5 mm per year, varying according to depth and latitude. In contrast to shallow-water corals, the Stone Coral inhabits an environment characterised by minimal seasonal fluctuation in physical factors, and skeletal growth shows no corresponding banding that might be employed to calculate age and growth rate. Colonies collected from a Norwegian fjord population and maintained in aquaria grew at a rate equivalent to a corallite extension of 9.4 mm per year (Mortensen, 2001), while others from a Mediterranean population grew at a rate of 15–17 mm per year (Orejas et al., 2008).

Stone Corals colonising oil-drilling rigs in the northern North Sea provided marine biologists with a unique opportunity to monitor growth rates in natural populations. The reproductive biology of *Lophelia pertusa* is still unknown, and it was the discovery of these new colonists that proved for the first time that the coral produced planula larvae, and that new populations might be established by dispersing larvae. They occurred within a narrow depth zone, 59–132 m, corresponding to the cold Atlantic water mass below the thermocline (Gass & Roberts, 2006). The mean extension rate for 15 young colonies monitored in the field was 26 mm per year. This is substantially higher than the figures for the Norwegian and Mediterranean populations, and perhaps reflects increased levels of suspended particulate material, originating from the sessile assemblages encrusting the oil rigs.

Stone Corals grow by apical extension and division of the colony branches, each branch dividing to produce two or more new corallites per year. The factors initiating corallite division and determining growth rates are unknown, although nutrition seems likely to be important. The greatest rate of growth, as measured by rate of calcification, and hence linear extension, is achieved by the apical corallites of each branch, which seem to display swift, episodic growth (Maier et al., 2009). Calcification rates of apical polyps may be equivalent to a maximum increase of 1% of their skeletal mass per day. Calcification by older, subapical polyps occurs at a much lower rate, barely one-fifth of that of the apical polyps; linear extension is equally smaller, and corallites greater than 7 mm diameter show no measurable increase at all.

The distribution, density and size of Stone Coral banks correlate with sea-floor topography, and reflect benthic hydrodynamic regimes. These parameters determine food supply for the coral and its associated fauna, and food is probably the most important factor limiting the community. However, it is quite unclear what the diet of *L. pertusa* consists of. It has been assumed that particulate organic matter comprises the bulk: decayed phytoplankton, zooplankton faeces and casts, but perhaps also fresh phytoplankton and the lavish reproductive output of plankton communities. Analysis of stable isotopes of carbon and nitrogen in *Lophelia pertusa* tissue and in phytoplankton cells revealed differences between the values for $\delta^{15}N$ at a scale indicative of more than one trophic level, demonstrating that the coral could not have been feeding directly on phytoplankton (Duineveld et al., 2004). Stone Corals have been observed to ingest live copepods and cumaceans, and it may be that such small crustaceans provide the trophic link between phytoplankton and coral. Another intriguing possibility is that one important source of energy for L. pertusa may be microbial populations associated with the living tissue of the coral colony (Neulinger et al., 2008). Live polyps support mixed assemblages of bacteria, the composition of which differs between different populations of the coral, probably reflecting differing environmental factors, so no bacterial species, or assemblage, can be regarded as specific to L. pertusa. However, at any particular location the bacterial community associated with the coral is always significantly different in composition from bacterial populations present in the water column, or in the sediment accumulated in and around the coral bank. All coral communities, both temperate and tropical, seem to support characteristic bacterial assemblages. Some are known to be pathogenic, such as those causing black band disease in tropical reef corals, while others may be harmless symbionts utilising coral mucus and waste products as sources of nutrition. Some tropical corals are known to ingest their own mucus, together with its adhering detritus and the bacterial flora it supports, and this may be a significant source of energy: the corals are, effectively, cultivating and harvesting a microbial garden. This phenomenon has not been observed in L. pertusa. Yet, of the numerous bacterial taxa commonly associated with the Stone Coral, some are known to break down cellulose, others are able to metabolise chitin, and many derive energy from the dimethylsulphoniopropionate (DMSP) released by algae and phytoplankton. The microbial communities associated with L. pertusa may thus prove to be another important trophic link between the Stone Coral and pelagic production.

A review of the biology of L. pertusa listed more than 800 macrofaunal species recorded from coral banks in the northeast Atlantic region (Rogers, 1999). Within ten years this total had almost doubled (Roberts, 2005), and the inventory continues to mount as the diversity of the habitat is more thoroughly investigated. Not all animal groups are equally represented. Sedentary and sessile species are easily collected, together with their substratum, but vagile species can only be sampled by nets or box corers, difficult and expensive to deploy at depth, and the smallest crustaceans and polychaetes will always be under-represented in such samples. The composition and density of the coral-associated fauna varies in relation to latitude, depth and bottom hydrodynamics, and its nature thus changes from locality to locality, as does the structure of the Stone Coral population. An early conclusion was that Lophelia pertusa did not support a distinct macrofaunal community, which instead was simply determined by local physical and biological environmental characteristics. Further, it was mooted that no macrofaunal species could be regarded as specifically associated with the coral. Considering that practically nothing was then known about the ecology of the Stone Coral or its attendant fauna, these were rather premature opinions, as subsequent research has begun to demonstrate. A frequent associate of *L. pertusa* banks is another cold-water coral, *Madrepora oculata*, the geographical distribution of which more or less coincides with that of the Stone Coral. *Madrepora oculata* develops lamellar or shrubby colonies, with alternating corallites forming distinctive zigzag branches. It is far more fragile than *L. pertusa*, achieving a height of no more than 50 cm, and does not form persistent banks. However, in association with the robust growth of *L. pertusa*, its colonies are more densely distributed and longer lasting, and it contributes substantially to the skeletal framework of the coral bank. At some deeper localities, for example in the Rockall Trough, *M. oculata* may constitute the dominant component of coral carbonate mounds.

Sponges are the most abundant and diverse of the sessile, suspension-feeding animals associated with Stone Coral. The total number of species recorded in the northeast Atlantic region probably exceeds 150, and as many as 70 have been identified at single sites. These encompass practically the entire range of morphologies described from temperate-zone sponges. Carbonate-boring taxa, exemplified by numerous species of *Cliona*, are constantly present and, as noted previously, are significant factors in the cycles of growth, collapse and regeneration that characterise coral banks. A majority of the sponge fauna associated with *Lophelia* consists of encrusting species developing thin, thick or knobbly sheets, and others forming small nodules, mounds or digitate growths. However, there is a small group of species, principally within the family Geodiidae, that form very large, lobed, nodular, mound-like or cup-shaped colonies, commonly achieving diameters of 50 cm, and perhaps growing to more than 1 m in height, with weights well in excess of 25 kg. These huge sponges seem to occupy a narrow and clearly defined depth zone that may overlap that occupied by the Stone Coral (Klitgaard & Tendal, 2001). Where the sponges and the coral occur together, they may be co-dominant in terms of space occupied, but the sponges will be dominant in terms of biomass. Interactions between Stone Coral and the larger sponge species have still to be investigated, but it is probable that biological interactions determine their differing distributions; sponges most often win in spatial competition with other sessile taxa. *Geodia barretti*, a very large sponge, is known to support microbial communities on its surface, and to be a source of bioactive compounds that deter settlement of invertebrate larvae on it. These characteristics may together comprise a biochemical armoury that enables it to displace other sessile suspension feeders, including *Lophelia pertusa*. Although these large sponges were long known to sponge taxonomists, their abundance and diversity only became apparent during extended benthic surveys around the Faroe Islands in the last decade of the twentieth century. Fisherman had been well aware of them, and the grounds they occupied had a local name, *ostur*, or cheese bottoms, referring to the consistency of the sponges, particularly *Geodia* species, and the strong odour they released when landed. Cheese bottoms were areas to be avoided, as trawls were liable to be filled with tonnes of sponge colonies, creating a severe and dangerous threat of instability to the fishing boat. It is now evident that sponge banks occur commonly throughout the northeast Atlantic, as the principal component of a distinct biotope, supporting a rich associated fauna of vagile invertebrates and fish (Klitgaard, 1995).

Cnidarians are the next most abundant group of sessile suspension feeders associated with Stone Coral banks, and the recorded diversity of the phylum as a whole probably rivals that of the sponges. It includes around 70 species of Hydrozoa; many of these will be small species of hydroid that contribute only minimally to the biomass of the community, but there will also be a few species of heavily calcified hydrocorals that add to the framework of the habitat. Soft corals include species of *Alcyonium*, which form substantial fleshy colonies but are probably only sparsely present at the depths favoured by *Lophelia*, and a number of species characterised by small, creeping colonies, such as *Sarcodictyon roseum* (Fig. 144). Sea fans, especially *Paragorgia arborea* and *Paramuricea placomus*, are often abundant in *Lophelia* communities, their erect, shrubby colonies projecting well above the coral rubble substratum. *Madrepora oculata* is the largest of the Hexacorallia found in *Lophelia* banks. Several others characterised by small

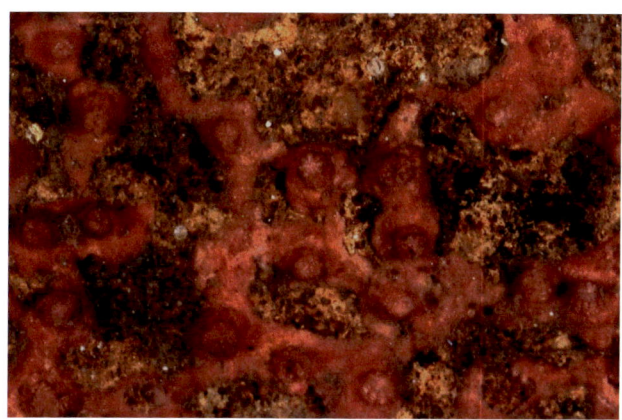

FIG 144. A creeping colony of the stoloniferan soft coral *Sarcodictyon roseum*. (J. S. Ryland)

colony size have also been recorded, such as species of *Epizoanthus* (Fig. 145), but the majority are non-modular, though probably clonal, species, including many species of sea anemone.

Bryozoans are another phylum of sessile, colonial suspension feeders that achieve remarkable taxonomic diversity within *Lophelia* communities. More than 100 species of Bryozoa have been recorded encrusting Stone Coral rubble; the total rises with every new survey, and is likely to include species entirely new to science. Most of these are rather inconspicuous, occurring as small patches, sheets or nodules, competitively inferior to sponges and cnidarians, and occupying spatial refuges deep within the coral rubble. However, a few robust species, such as *Porella compressa*, *Tessaradoma boreale*, *Hornera lichenoides* and *Stegohornera violacea* form erect, heavily calcified, and apparently long-lived colonies that probably comprise a not insignificant proportion of the carbonate matrix of their respective niches.

The four groups of non-modular invertebrates that contribute most to the taxonomic diversity of the *L. pertusa* community, the polychaetes, arthropods, molluscs and echinoderms, also provide the ecological and trophic complexity of the habitat. Among the polychaetes are small, sessile suspension feeders building carbonate tubes, such as *Hydroides norvegica* and *Serpula vermicularis*, and many others inhabiting more or less permanent tubes formed from proteinaceous

FIG 145. Colonies of the zoanthid coral *Epizoanthus incrustans*, in characteristic association with the hermit crab *Anapagurus hyndmanni* (J. R. Ellis (CEFAS))

secretions, or of mucus-bound silt or coarser material, within the interstices of the coral rubble or burrowed into the sediment accumulated around it. Trophic modes among these are various; many will be suspension feeders, such as the Parchment Worm, *Chaetopterus variopedatus*, and large-bodied species of Sabellidae and Terebellidae, but others are deposit feeders, including species of *Pectinaria* and large bamboo worms such as *Euclymene robusta*, and yet others, in the family Cirratulidae, are essentially scavengers. Many of the free-roaming polychaetes, including numerous species of paddle worm in the genera *Eulalia* and *Eumida*, scale worm (*Harmothoe* species) and ragworm (*Nereis* and *Neanthes*) seem to occupy that part of the trophic spectrum between deposit feeding and scavenging, and others, such as the robust Sea Mouse, *Aphrodita aculeata* (Fig. 43), are partial predators. However, there are also many, often large-bodied, obligative predators, particularly among the Nephtyidae and Glyceridae. Eunicids are large-bodied worms that occupy stiff tubes with a parchment-like consistency from which they partly emerge to forage over living coral polyps. They are frequently recorded from cold-water coral habitats, and their tubes are usually intimately intergrown with the coral colony, but it is only very recently that one species, *Eunice norvegica*, has been shown to form a mutualistic association with *Lophelia pertusa*. The worm is provided with secure accommodation, and the coral with continually increasing substratum: as the worm lengthens its tube, the coral tissue secretes an enveloping carbonate sheet, from which new corallites eventually grow. The worm appears to feed upon detritus accumulating within the colony framework, and perhaps competes with the coral polyps for food resources; however, it also removes material gathered in empty corallites and debris fouling the surface of the live colony, in both cases to the benefit of the coral. Finally, a scientist at the Dunstaffnage Marine Science Laboratories, Oban, observing live coral and its eunicid symbiont in an aquarium, demonstrated how extraordinary their association was (Roberts, 2005). Worms were recorded emerging from tubes within *L. pertusa*, grasping detached pieces of the colony and dragging them back to the main colony fragment. The worm was then able to fuse its tube to empty tubes within the smaller colony piece, and to extend its foraging range, but in doing so it provided the coral with the opportunity to fuse the two pieces of the colony. The most significant consequence of this association would seem to lie in the potential of the eunicid population to increase stability and expansion of its host coral colony.

Crustaceans comprise the majority of the species of arthropod recorded from Stone Coral banks, and many of the peracaridan crustaceans, predominantly isopods, amphipods and tanaids, are small animals inhabiting tubes of fine silt constructed amongst the coral debris. They are primarily suspension feeders,

FIG 146. The large spider crab *Hyas coarctatus*. (J. S. Porter)

FIG 147. The nut crab *Ebalia tuberosa*. (J. S. Ryland)

or captorial detritivores. More than 40 species of peracarids are known to occur amongst coral rubble, a total that almost certainly underestimates their diversity, as sampling is difficult and experienced taxonomic specialists are few. The eucaridan crustaceans, lobsters, crabs and prawns, are more easily collected and less difficult to identify, and their recorded diversity is probably close to reality. This includes several species of squat lobster, in the genera *Galathea* and *Munida*, the huge, long-legged Stone Crab, *Lithodes maja*, the large-bodied spider crabs *Hyas araneus* and *H. coarctatus* (Fig. 146) and more delicate, truly spidery species of *Inachus*, as well as nut crabs (*Ebalia* species) (Fig. 147) and hermit crabs, Paguridae, and more than a dozen species of prawn. Most of these larger species of crustacean are fairly omnivorous, as scavengers or opportunistic predators, but the smaller species of squat lobster probably function also as suspension feeders.

The molluscan associates of *L. pertusa* include a number of chiton species, all microphagous grazers, and a small variety of sea slugs, which are almost all predators of sponges or cnidarians, but the majority divides more or less equally between the bivalves and the prosobranch gastropods. The diversity of bivalves encompasses infaunal species associated with muds and muddy sands, such as species of *Abra* and *Ensis*, and more robust species of cockle and venus shell that burrow shallowly in coarse sands and gravels, as well as a vagile epifauna of numerous small scallops. Some are known to be crevice dwellers, such as *Hiatella arctica* (Fig. 112) and the ark shell *Bathyarca pectunculoides*, while several species of file shell, *Limaria loscombi*, *Limatula subauriculata* and *Limaropsis aurita*, appear to be nest builders, perhaps ecological equivalents of *Limaria hians* (Fig. 134) in maerl rubble. Finally, sessile bivalve species include those that attach permanently to their substratum, such as the saddle oyster *Heteranomia squamula* and the scallop *Talochlamys pusio*,

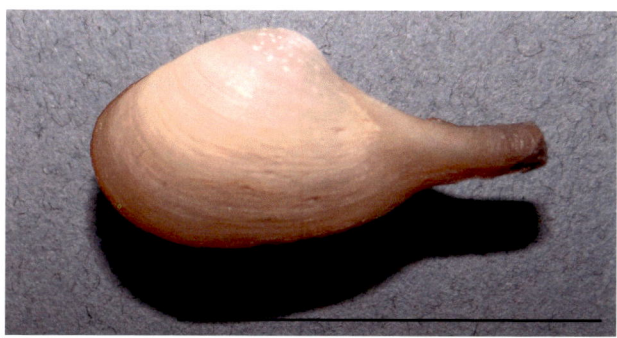

FIG 148. The deposit-feeding bivalve *Cuspidaria rostrata*. Scale bar: 2 cm.

and those with byssal attachments, such as mussels, Mytilidae, that retain the potential to move.

Trophic diversity of bivalve molluscs is as high as that of the polychaetes. A majority will be suspension feeders, but there are some facultative deposit feeders, such as *Abra alba* and *A. nitida*, and some interesting obligate deposit feeders, the nut shells, species of *Nucula* and *Nuculana*, which feed directly on the sediment. The most extraordinary adaptation is seen in *Cuspidaria rostrata* (Fig. 148), which employs a pumping respiratory current that collects particles of organic detritus and entraps small polychaetes and crustaceans, and thus functions as a facultative predator. Predators and carrion feeders are especially prevalent among the gastropods associated with Stone Coral. These include an impressive variety of large whelks, no fewer than three species of *Colus* as well as the Common Whelk, *Buccinum undatum* (Fig. 149), and numerous small species in the genera *Nassarius*, *Oenopota* and *Trophon*. Most of these are ominivorous, consuming a variety of prey, but specialists include the cowrie *Trivia arctica* (Fig. 150), the velvet shell *Velutina velutina*, and two species of *Lamellaria*, all of which feed on ascidians. Slit limpets, *Emarginula* species, and top shells, *Calliostoma* species, are microphagous detritivores or grazers, while the pelican's foot *Aporrhais serresianus* is an impressively modified suspension feeder, trapping particulate material in a mucus net, but deposit feeding is not seen widely

FIG 149. Shell of the Common Whelk, *Buccinum undatum*. Scale bar: 5 cm.

FIG 150. Shells of the carnivorous cowrie *Trivia arctica*. Scale bar: 2 cm.

amongst gastropods. Many of the smallest gastropod species associated with Stone Coral are probably feeding upon microbial communities, foraminiferans and perhaps *Lophelia* mucus.

Ninety-four species of echinoderm have been recorded from the shelf seas bounding the British Isles, while 111 were listed as occurring among northeast Atlantic Stone Coral banks (Rogers, 1999). Some sea urchin species, and some starfish, are individually large animals, and some brittlestars occur at very large population densities, so at some locations the echinoderms may comprise a significant proportion of the biomass of the fauna associated with the Stone Coral. The echinoids recorded are a good representation of the sea urchin fauna of the British sea area, including the infaunal heart urchin *Echinocardium flavescens*, the gravel-associated *Echinocyamus pusillus* and three species of the epifaunal *Echinus*. The list of asteroids also includes species familiar from British coastal waters, such as *Porania pulvillus*, *Hippasteria phryngiana*, *Crossaster papposus* (Fig. 151), *Asterias rubens* and two species of *Henricia*. However, the largest predators, *Luidia ciliaris*, *L. sarsi* and *Marthasterias glacialis*, are missing, and it is probable that *A. rubens* is only sparsely present in Stone Coral habitats. More than half of the echinoderm species reported from Stone Coral banks, 57 in total, are brittlestars, and fewer than ten of these have been collected from the British sea area. This diversity reflects both the latitudinal range and the bathymetric range of *L. pertusa* in the northeast Atlantic. The few familiar species include *Amphiura chiajei*, found on offshore muddy sands, especially around the northern half of the British Isles, *Ophiothrix fragilis*, common in shallow inshore habitats around the whole of Britain and Ireland, and several species of *Ophiura*, which

FIG 151. The Common Sun Star, *Crossaster papposus*. (J. S. Porter)

occur principally as part of the epifauna on fine to coarse sands. Sea cucumbers (holothurians) show a reverse pattern, with a low diversity associated with Stone Coral, just nine species. All but one of these have been recorded from other habitats around the British Isles, while another 25 species in the British fauna have been reported from both inshore and offshore localities but not associated with *L. pertusa*. Some of these may simply have been overlooked in surveys of Stone Coral communities, although the 11 infaunal species of Synaptidae, delicate, worm-like holothurians with often small individual body size, are most often found in soft muddy sands, and perhaps will not tolerate the coarse environment of coral rubble.

The total number of species associated with Stone Coral habitats increases with each additional survey, but it remains difficult to measure the taxonomic diversity of individual coral colonies or banks, and the relative abundance of the associated species. Similarly, while the trophic modes of the many different species recorded from *Lophelia* communities may be known, constructing food webs and estimating energy flow for a specific community are elusive objectives. Quantitative studies of *L. pertusa* communities are sparse, and limited in scope. A sample of 18.5 kg from a Faroe Islands survey yielded an assemblage of 4,626 individuals, representing 256 associated species (Jensen & Frederiksen, 1992); an unknown number of individuals and species of mobile organisms would have been lost from the dredges, but the figures nonetheless provide an indication of the biodiversity of the habitat. Another indication was provided by a remotely operated video survey of two patches of Stone Coral, at 82–87 m depth, in a Swedish fjord (Jonsson et al., 2004). The larger patch, 200 m^2, comprised 150 discrete coral colonies and the smaller, 100 m^2, just 50. The ROV transect tracks covered 443 m, recording assemblages of large epifaunal organisms over an area of 265.8 m^2. A total of 47 species could be identified and counted in the video records, and expressed as individuals per square metre of sea floor. The most abundant were crinoids of two species, *Antedon petasus* and *Hathrometra sarsi*, which could not be distinguished in situ, and which occurred at joint densities of up to 119 per square metre on the large coral patch and at a maximum 120 per square metre on the smaller. Three species of unitary ascidian, *Ciona intestinalis*, *Polycarpa pomaria* and *Ascidiella aspersa*, were present at combined densities of 100–330 per square metre on the large patch, and 4.62–101.6 on the smaller. Four large decapods could be identified, *Cancer pagurus*, *Lithodes maja*, *Munidopsis serricornis* and *Munida rugosa*, the latter at densities of up to 1.83 per square metre. Small prawns were frequent, but could not be separately identified from the video records, although it was known that three species of *Pandalus* had been recorded previously from the area. The most abundant cnidarian was a small anemone, *Protanthea simplex*, which achieved a maximum density of 10.51 per square metre. Only two species of fish were recorded, both at very low densities, and this unexpected result was attributed to sustained overfishing, previous surveys having recorded ten species and considerable densities. Conversely, densities of ascidians and crinoids were unusually high in comparison to Stone Coral communities in Norwegian fjords. Of the 47-species total, for all groups, 20 were vagile and 27 sessile; 18 species were filter feeders, two were deposit feeders and 10 were omnivores, while 17 species were considered to be predators. That interactions within this community might be quite complex was suggested by variation in the proportion of each trophic type in the community along the video transects. Few species were present at constant densities across the whole of the two coral patches. Predators and filter feeders were most abundant at the centre of each patch, where live coral was predominant, and declined through successive zones of coral rubble, and coral and boulders. Deposit feeders occurred at low densities over the whole of the large coral patch, but in the small patch were significantly more abundant around the periphery than at its centre.

CHAPTER 7

Time and Change

The marine benthic communities revealed through two centuries of exploration of the northwest European shelf are of comparatively recent origin, with a maximum age of less than 18,000 years. Relics of older communities can be seen in shelly sand deposits in Belgium and the Netherlands, and in East Anglia, where they are referred to as 'crag' deposits. The oldest of these, the Coralline Crag, preserves the remains of marine biota at least 3.4 million years old, while the Red Crag and the overlying Norwich Crag were laid down between 2.6 and 1.8 million years ago. Fossils found in these sands include many extinct species, such as the large brachiopod *Terebratula maxima* from the Coralline Crag (Fig. 152), and others now no longer to be found

FIG 152. The Pliocene brachiopod *Terebratula maxima*, from the Coralline Crag at Butley Creek, south of Gedgrave, Suffolk. Scale bar: 5 cm.

FIG 153. The whelk *Neptunea contraria*, from the Pleistocene Red Crag at Walton-on-the-Naze, Essex. Scale bar: 5 cm.

FIG 154. Shell valves of the clam *Venus casina*. Scale bar: 5 cm.

in the region, such as the left-handed whelk *Neptunea contraria* from the Red Crag (Fig. 153). Other crag fossils, particularly among the bivalves, are still extant, including *Glycymeris glycymeris* (Fig. 16), *Aequipecten opercularis* (Fig. 140), *Venus casina* (Fig. 154) and *Arctica islandica* (Fig. 155), with modern geographical ranges that extend to the north of the British Isles, or from south western Britain to the Mediterranean, and even to West Africa.

The present marine fauna of the region thus represents an intermingling of northern, cold-water, or boreal, species and southern, warm-water, or Lusitanian, species; northwest European shelf habitats occupy a biogeographical boundary zone, partly attributable to present-day hydrographical regimes, but also partly to environmental factors that have fluctuated over the past few million years. Environmental change has been driven by climatic cycles, with short-term periodicities measured in decades, and by others with longer-term multidecadal periodicities, and the longest extending over millennia. Through the past thousand years the expanding human populations of northwest Europe have

FIG 155. The molluscan Methusalah: in some habitats *Arctica islandica* has a potential life span in excess of four centuries.

had an increasing impact on coastal, and latterly offshore, shelf ecosystems, resulting in changes in benthic habitats that in some cases have only recently been appreciated. Coastal development, environmental contamination and rising fishing intensity have led to increasing habitat homogeneity, decreasing taxonomic diversity and the collapse of populations of many ecosystem engineers. A further agent of environmental change that is likely to have an increasing effect through the present century is global warming of the oceans, originating in large part from human activity.

THE LONGER TERM

During the early Pliocene, around four million years ago, tectonic activity led to the formation of the Isthmus of Panama, closing the seaway between the Atlantic and Pacific oceans, and resulting in profound changes in ocean circulation. The northern hemisphere began to cool as a consequence, and polar ice began to accumulate. The cooling trend was reversed for some quarter of a million years during the middle Pliocene, but accelerated again around 2.4 million years ago, when ice sheets expanded and the Red Crag began to be deposited. The Coralline Crag possibly dates from prior to the middle Pliocene warm period; oxygen isotope analysis of some fossils indicates a warm-water environment, but shells of the Queen Scallop, *Aequipecten opercularis*, yielded isotope ratios suggesting a water temperature range from 6 to 16 °C, the same as that which determines the geographical range of living populations of the species (Williams et al., 2009). This anomaly has not been explained: warm-water and cold-water elements of the Coralline Crag fauna might represent different stratigraphical horizons, and thus different environments, perhaps indicating that the palaeoenvironments of the deposit may have been transitional, marking the commencement of the middle Pliocene warm period (Williams et al., 2009). Climatic cooling continued into the Pleistocene, and through the latter half of that epoch, over a span of about 700,000 years, the northern hemisphere experienced cycles of glaciation, with a duration of around 100,000 years, alternating with short, warm, interglacial periods (Hewitt, 1996). The last glacial cycle began around 135,000 years before the present, and the last ice age peaked 20,000 years ago, when the Scandinavian ice sheet achieved its maximum extent, at 52° N, mean sea level was 100 m lower than today, and mean global temperatures were 4–5 °C cooler. Most of the shelf lying within the present-day 50 m isobaths would have been low-lying tundra, and much of that now between 50 and 100 m deep would have been ice-bound: the northwest European shelf was either glacier or tundra.

Many marine benthic species became extinct as the ice sheet spread and the sea retreated, but many others responded to the cooling climate by changing their latitudinal range. For fish this would have been achieved through migration, particularly through modification of seasonal migration patterns in response to changes in sea temperatures and food resources. For benthic invertebrates new latitudinal ranges would have become established as populations declined, and the species contracted, at one end of the original range, but flourished at the other end as populations expanded in response to ameliorating conditions. No species is equally abundant across the whole of its geographical range. Populations tend to be most dense, and biomass greatest, under optimum conditions for the species; individual populations may adapt to specific parameters to the north and south of the optimum, but density declines until eventually the species reaches the limit of ecological tolerances.

This last multi-millennial cycle may explain some of the modern-day geographical distribution patterns seen among the benthic biota of northwest Europe. A hypothetical response to environmental change, in particular temperature, is shown in Fig. 156. As the glaciers grew, the latitudinal ranges of many shelf species shifted southwards into the western Mediterranean, while those of many warm-water species, then at their northern geographical limits, contracted further southwards towards the coast of West Africa. Late Pliocene and Pleistocene fossil deposits of Italy record a warm-water, or Senegalian, fauna declining through

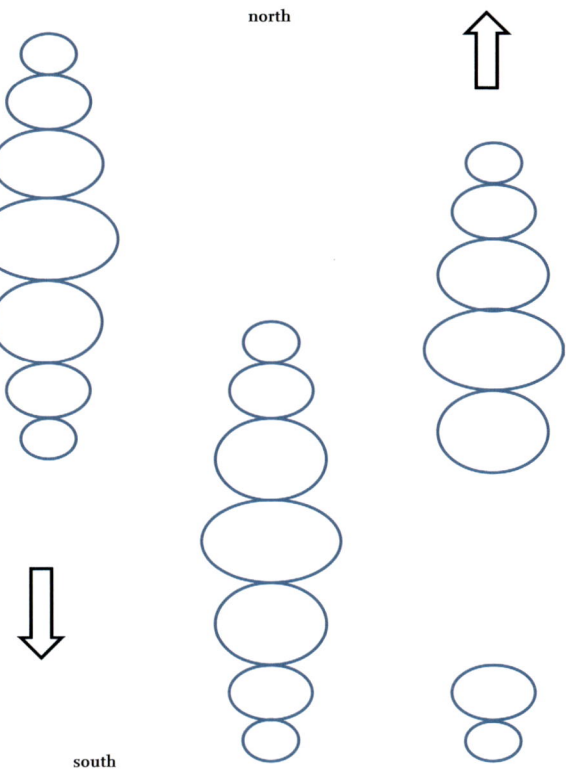

FIG 156. As the latitudinal range of a species changes in response to environmental change, populations may become disjunct, with relicts at its northern and southern limits.

time, and successive immigrations of northern temperate species (Malatesta & Zarlenga, 1986; Ragaini et al., 2007). Italy's Pleistocene fossils include shells of bivalves, gastropods and tusk shells (scaphopods), bryozoan colonies and the tests of numerous species of microfossil, particularly foraminiferans, some species of which still live today in north European seas. They are regarded as boreal climate markers, and are often referred to as 'northern guests' (*nordische Gäste*) or 'Pleistocene guests'. The total number of Pleistocene visitors in Mediterranean sediments is difficult to enumerate, but at least 60 species of mollusc have been recorded with certainty from various Italian Pleistocene horizons, some of them very familiar in modern-day northwest shelf faunas.

The two mussels, *Mytilus edulis* and *Modiolus modiolus*, are common in northwest shelf habitats; the former ranges into the subarctic Barents Sea, the latter to northern Norway. Both are uncommon south of Brittany, and their distributions do not extend beyond southern Biscay, but they have been recorded from Italian fossil horizons. Fossil records of *M. edulis* are sparse, but

M. modiolus seems to have been present in the Mediterranean through much of the Pleistocene. The large cockle *Acanthocardia echinata*, shells of which are a common sight on sandy beaches around most of Britain, has a modern-day distribution from northern Norway to the Bay of Biscay, has been reported from a few deep-water localities off the Atlantic coasts of Spain and Morocco, and is also known from several Pleistocene sites in Italy. The small, ribbed scallop *Pseudamussium peslutrae* (Fig. 157) has a more southerly distribution than that of *A. echinata*, from western Norway to Morocco, with perhaps a few outlying populations in the western Mediterranean, where it seems to have been common in the early Pleistocene. *Spisula solida* and *S. elliptica*, infaunal bivalves common in north European shelf habitats, also show contrasting modern-day distributions, the former from northern Norway to Morocco, the latter from the Barents Sea to north Brittany, and sporadically in the Bay of Biscay. *Spisula solida* seems to have been a later migrant to the Mediterranean, occurring only in Late Pleistocene deposits, while *S. elliptica* has only rarely been recorded as a fossil. Both species tend to occur in deeper water towards the southern end of their geographical ranges, a phenomenon seen particularly in the heavy-shelled, long-lived bivalve *Arctica islandica* (Fig. 155). This species has a boreal, amphiatlantic distribution, and in the east Atlantic is most common, and occurs at shallowest depths, around Iceland and the Faroe Islands, off Norway and around Scotland. It is increasingly rare, and found only in deep offshore waters, in the North Sea and off southern Britain. *Arctica islandica* seems to have been common in the Mediterranean through much of the Pleistocene, and its fossils are found today around most of the Mediterranean basin. *Buccinum undatum*, the Common Whelk (Fig. 149), is today abundant in shallow shelf habitats from northern Norway to the Bay of

FIG 157. Shells of the small scallop *Pseudamussium peslutrae*. Scale bar: 5 cm.

Biscay, and also had a wider geographical range through the Mediterranean of the Late Pleistocene.

The historical biogeography of northwest European shelf benthos thus shows distribution patterns of individual species that change through geological time, and appear to relate to long-term climatic cycles. However, such patterns need to be carefully examined: it is often difficult to confirm the equivalence of living and fossil specimens, the latter having lost all traces of non-calcified structural components, and possibly suffered abrasion and other post-mortem effects. Also, while the present-day geographical distributions of some species may embrace isolated, relictual populations in the Mediterranean, evidence for their existence must be based on records of living specimens. Modern geographical ranges founded on records of dead shells may prove to be partly based on Pleistocene fossils, either eroded from their stratigraphical context, or simply dislodged by dredges. The evolutionary history of a species may also cloud its biogeographical history. For example, the Great Scallop, *Pecten maximus* (Fig. 128), seems to have originated in the Late Pliocene, is found in Pleistocene deposits of northern Europe, and has been reported from a few fossil horizons of the same age in the Mediterranean. Its present-day distribution is centred on the northwest European shelf, with a northern limit at north Norway, and a southern limit off Portugal. Outlying populations have been recorded further south and from the western Mediterranean, but the latter, at least, might be considered doubtful, as the common Mediterranean scallop is *Pecten jacobeus*, which is probably genetically close to *P. maximus*.

Marine benthic species with extensive latitudinal ranges are distributed along latitudinal environmental gradients, the most important being the temperature gradient. Biological and ecological characteristics of such species tend to change continuously in response to the environmental gradient, and this gradient of continuous variation, with no obvious discontinuity between its two extremes, is termed a cline. At any point along the cline, populations of a species will be adapted to local environmental conditions, and differences in life-history characteristics between populations of a species may represent clinal variation along a latitudinal gradient. Contraction or expansion of a species' geographical range through geological time might result in dislocation of a cline, isolation of populations and possibly further local adaptation, perhaps leading to speciation, and to some interesting present-day distribution patterns. The Striped Venus, *Chamelea striatula* (Fig. 158), is a small infaunal bivalve, inhabiting clean, sandy habitats around the whole of the British Isles except for the extreme southeast, and distributed from Shetland to Brittany, and southwards to Portugal, southern Spain and the Atlantic coast of Morocco. Another species, the Mediterranean Striped Venus, *C. gallina*, is common in similar habitats throughout the

FIG 158. Shells of the Striped Venus, *Chamelea* (formerly *Venus*) *striatula*. Scale bar: 5 cm.

Mediterranean and the Black Sea, and so resembles *C. striatula* morphologically that, 200 years after Linnaeus first distinguished between them, it was concluded that they must represent just a single, variable species, and should be referred to *C. gallina*. Differences in life-history characteristics between populations were perhaps attributable to latitudinal environmental gradients, and represented a cline. In warm Mediterranean habitats *C. gallina* is a fast-growing species with a comparatively short life span of around three years; longevity increases and growth rate decreases in successive populations northwards, and around Shetland individuals may live for more than ten years. However, these two species of *Chamelea* may still be distinguished by the finest morphological details (Fig. 159), and firmly differentiated by genetic structure (Backeljau *et al.*, 1994). *Chamelea striatula* ranges no further into the Mediterranean than the Costa del Sol; *C. gallina* is the Mediterranean Striped Venus, and its range overlaps that of *C. striatula* only as far west as southern Portugal. Mixed populations do not interbreed, confirming their separate genetic identities, and vindicating Linnaeus' judgement.

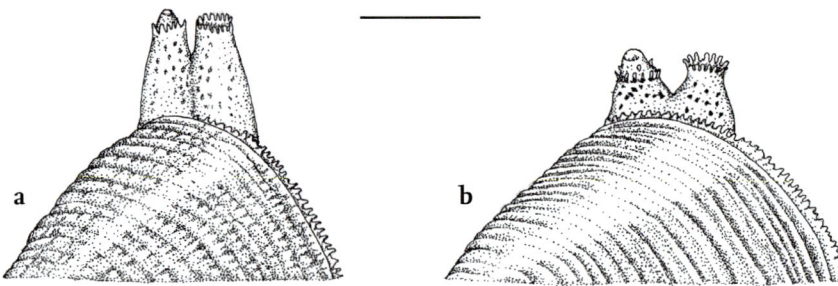

FIG 159. Posterior shell margin, mantle edge and pallial siphons of (a) *Chamelea striatula* and (b) *Chamelea gallina*. Scale bar: 5 mm. (After Backeljau *et al.*, 1994)

THE SHORTER TERM

The present warm interglacial period began about 18,000 years ago as the ice sheets began to retreat and the sea to rise again; the last glaciers in Scotland are thought to have melted by 10,000 years before the present, and the past 8,000 years have been marked by relative climatic stability. These great climatic cycles occur at scales measured in millennia, while small-scale environmental fluctuations may have periodicities of decades to centuries and may be masked by seasonal or interannual variation. In the eastern temperate North Atlantic, climatic fluctuations, evidenced by changes in sea surface temperature and correlated with changes in plankton communities, have been documented through much of the past hundred years and related to hydrographic influences, which seemed to explain alternating cycles in zooplankton and pelagic fish populations in the western English Channel, monitored from the 1930s onward (Southward *et al.*, 2005). Herring fisheries off the south Devon and southeast Cornish coasts began to decline around 1900, and by 1930 Herring stocks had failed, but stocks of Pilchard had grown sufficiently to support a new fishery that grew to a peak in the 1960s. Pilchard then decreased, and eventually disappeared by 1985, by which time Herring stocks had grown again. The two species have overlapping geographical ranges: the Herring is a cold-water fish that reaches its southern distributional limit off southwestern Britain, while Pilchard range from Morocco to a northern limit in the same region of the British Isles. Cold-water species of zooplankton also declined with the Herring, and catches of bottom-dwelling, cold-water fish species dropped, and the advent of Pilchard was marked by a change in zooplankton composition to a warm-water community. These cyclical ecological changes became known as Russell Cycles, and are now seen to have a periodicity that seems to shadow that of the Atlantic Multidecadal Oscillation (page 20), with cold-water communities correlating with low values of the AMO index, and warm waters corresponding to high positive values. The North Atlantic Oscillation (page 19) may be correlated with sea temperature fluctuations over a much shorter timescale than the AMO, and effects on marine ecosystems are probably less easily identified; sea surface temperatures also reflect sunspot activity, which has 11-year and $c.$100-year periodicities, and the shorter period may be reflected in population cycles in some littoral invertebrates. In particular, among barnacle communities of southwest England, the proportions of the warm-water *Chthamalus* species and the cold-water *Semibalanus balanoides* fluctuate with an 11-year periodicity (Southward, 1991).

The ecological effects of both short- and longer-term environmental cycles on marine ecosystems are still being explored, but it is apparent that a new phase of climatic change began in the last decades of the twentieth century, with further potential effects that may be difficult initially to distinguish from those arising from natural, cyclical change. Global temperatures may now be increasing at an accelerating rate, although the forces driving the change, its extent, and its potential consequences for the marine environment are still argued over. The possibility that such consequences might have economic and political implications perhaps explains the renewed interest in marine environmental monitoring, which was largely abandoned by 1988 as being of minor scientific merit (Southward et al., 2005). Monitoring and quantifying change in the marine environment may become important in documenting the rate of climate change and its effects on marine ecosystems, although significant caveats must be acknowledged when analysing and interpreting the data collected. Time series data often fail to reveal anthropogenic change in the benthic environment. Sampling protocols may be sufficiently rigorous with regard to sample size and replication, frequency and controls, and the monitoring programme may be sufficiently long to cover the average generation time for the most important framework species, but it is always too short. In order to distinguish natural environmental oscillation from human-induced change, benthic monitoring programmes need to be sufficiently protracted to embrace the natural cycles observed and recorded for Herring and Pilchard populations in the western English Channel. With the possible exception of a few remote oceanic islands that have never been settled and rarely visited, no part of the world's shore and shelf habitats has been unaffected by human activity, and the shallow seas of Europe have been continuously exploited for thousands of years. Time series data can only reveal change relative to environmental conditions at the commencement of a monitoring programme. Historical data are occasionally useful – alternation between Herring and Pilchard fisheries off southwest England has been recorded since the sixteenth century – but for the most part it is impossible to know how European coastal habitats have changed since humans first began commercial utilisation of marine resources. That change might have been profound.

Western North Atlantic and Caribbean coastal marine resources only came under pressure with the arrival and settlement of Europeans at the beginning of the sixteenth century, and there are good contemporary accounts of the profusion of marine life encountered by the first colonists, and subsequent records of its catastrophic decline (Jackson, 2001). Large animals were the first to be severely affected. In the Caribbean and on the Gulf coast of Florida Green Turtle, *Chelonia mydas*, populations that might have numbered tens of millions were practically exterminated. The consequences of this depredation have only recently been appreciated, and indeed the most severe effects might only have become manifest in comparatively recent times. Taxonomically diverse ecosystems are said to possess a degree of ecological redundancy, meaning that if one ecologically important species is removed others will come to occupy the vacant niche, and it is only when successor species succumb to further environmental stress that the impact of the initial disturbance is seen, and the ensuing environmental change may be complex and unexpected. Green Turtles feed on seagrasses, cropping the leaves to a height of around 4 cm, their waste products are dispersed in the water column and carried away from the seagrass bed, and little dead leaf litter accumulates among the living shoots. Unchecked by turtle grazing, seagrasses grow to heights of 30 cm, the leaves are shed annually and accumulate within the bed, boosting nitrogen and sulphide levels, creating periodic hypoxia, and a changed microbial environment. The slime moulds that gave rise to the wasting disease that eventually destroyed huge tracts of seagrass bed in the western North Atlantic and Caribbean may have been stimulated by the changed detrital communities within the soil of the beds.

The seagrass beds that floored much of Chesapeake Bay, to the north, decreased drastically in extent through the middle decades of the last century, destroyed by the same wasting disease suffered by the subtropical beds of Florida and the Caribbean, and it may be that this resulted from a profound change in microbial and plankton communities caused by the virtual extinction of native oysters. Chesapeake Bay is enormous; its main extent, from Norfolk, Virginia, to Baltimore, Maryland, covers about 2.5 degrees of latitude. In pre-Columbian times the oyster population of the bay totalled billions of individuals, with the capacity to filter the entire water body contained within the bay in one week; commercial exploitation and overfishing led to the almost complete destruction of the oyster beds by the 1870s, and filtration turnover time is now 48 weeks.

Eelgrass habitats are especially vulnerable to anthropogenic degradation, through physical disturbance, overfishing and eutrophication. Coastal engineering and development are significant sources of physical disturbance, leading to increased turbidity and changing hydrodynamic environments. Overfishing causes physical damage to *Zostera* beds but, more importantly, affects community structure and interactions by removal of large organisms. Eutrophication is often associated with environmental contamination, and while consequent nutrient enhancement promotes epiphyte and phytoplankton growth, increasing both shading and turbidity, pollutants may kill mesograzers and predators, causing further habitat impoverishment. Wholesale loss of eelgrass meadows may have striking ecological consequences, which have been well documented for some Danish habitats. The epidemic wasting disease did not result in the predicted collapse of coastal food webs, but led to some dramatic changes in inshore benthic habitats. Subtidal *Zostera* meadows slow and dissipate tidal currents and wave energy, accumulate fine sediments and create a buffer between the often sharply fluctuating physical environment of the littoral zone and the more stable sublittoral. Loss of eelgrass results in an immediate loss of sediment stability, and the development of a mixed substratum of coarse muddy sands and gravels, with patches of fucoid seaweeds. Habitat complexity is reduced, biodiversity, production and biomass all slump, and the character of the shallow benthic environment is profoundly altered.

The *Zostera* beds of Chesapeake Bay suffered their steepest decline in the twentieth century, approximately 50 years after the initial stress arising from the destruction of the oyster beds. European seagrass populations underwent an equally sharp decline, also as a consequence of slime mould attack, during the same period, perhaps also reflecting a lagged response to ecological stress. Turtles, of course, were never part of European coastal ecosystems, but Brent Geese and Wigeon, a small whistling duck, both feed on *Zostera*, and might have had significant roles in the ecology of seagrass beds before their populations declined, or moved, in response to human hunting and disturbance. It is not known whether northwest European eelgrass beds had associated communities of mussels, oysters and large clams, but it must be considered a possibility that there was once an indigenous eelgrass bivalve fauna, fatally accessible and fished to extinction centuries ago. Oysters, and possibly other large long-lived bivalves, were certainly much more abundant in coastal habitats of northwest Europe in the past, and must have already been severely depleted when commercial exploitation began, and coastal pollution grew, in the nineteenth century; and any significance they might have had for seagrass ecology would have been obscured. It would be a fascinating natural experiment to stock a *Z. marina* bed with mature clams and oysters, close it to exploitation and observe the consequences. It is just possible that a unique recovery might result. However, one insidious anthropogenic insult must first be controlled, and that is hydrocarbon contamination. Eelgrass communities in the proximity of oil facilities are always drastically depleted of crustacean herbivores, and suffer from persistent, strong trophic cascades.

Environmental change may have direct, and immediate, effects on marine ecosystems, or indirect effects arising from interactions within systems. A direct effect may be a strong year class, which might lead to expansion of a population, or a weak year class, poor recruitment and local extinction. The clearest examples of the direct effects of natural fluctuation in temperature are seen in the population dynamics of most shallow-water species of starfish and sea urchin. Indirect effects of temperature fluctuation include what are described as ecological mismatches, when, for example, populations of grazing zooplankton peak earlier or later than the phytoplankton blooms on which they depend. Progressive change in population and community structures in response to progressive environmental change is described as a 'linear response', but environmental change may be marked by an increasing frequency of severe events, or non-linearity. The frequency of severe events – prolonged storms, excessive temperature fluctuation – may have greater ecological significance than progressive change in mean values for environmental parameters, and biological responses may be abrupt, with irreversible regime shifts (page 121).

All stages in the life cycles of benthic organisms are mediated by temperature, and the distribution and density of each species are determined by the temperature tolerances of each population. At suboptimal temperatures gamete maturation may not be initiated, or completed; threshold temperatures for spawning may occur too early or too late in the reproductive cycle; fertilisation, larval development and growth rates, metamorphosis, and ultimately settlement and recruitment, are all dependent on temperature. At the extremities of its latitudinal range, populations of a species are likely to be small, limited by suboptimal conditions for growth and reproduction, and probably representing only a proportion of its genetic variability. Recruitment to these populations may depend upon dispersal of larvae or juveniles from within the breeding range of the species, or on sporadic immigration of adults during favourable environmental periods. The Common Octopus, *Octopus vulgaris*, for example, reaches a northerly distributional limit in the east Atlantic on the Brittany coast. Individuals have been recorded off the extreme southwestern coasts of Britain, but populations there are inconstant, originating from sporadic larval drift northwards from Brittany. During those decades in the first half of the twentieth century when warm-water assemblages, including Pilchard, became established in the western English Channel, *O. vulgaris* populations reached unusually high numbers in some years. It is probable that the species did not breed north of Brittany, but rather the benign environmental conditions enabled a high proportion of drifted larvae to settle and metamorphose successfully. Similarly, the small hermit crab *Clibanarius erythropus* is widely distributed in the Mediterranean, and on east Atlantic shores ranges from Morocco to northwest Brittany. In 1955 it was found to have rounded the corner, eastwards into the Gulf of St Malo, and from 1959 small populations were discovered at a number of localities from Wembury, south Devon, to Lizard Point, and at Trevone, north Cornwall (Southward & Southward, 1977, 1988). All English populations had disappeared by 1987, possibly partly as a consequence of the *Torrey Canyon* oil spill on the north Cornish coast, but probably mostly through reproductive failure of the north Brittany populations. Mediterranean populations are known to require a minimum seawater temperature of 18 °C to reproduce successfully, and it is unlikely that those at the northern edges of the species' range ever bred. However, at suboptimal temperatures *C. erythropus* grows slowly and has a long life span, and these factors, together with sporadic recruitment of drifted larvae, would account for the persistence of the northerly outliers.

Contraction or expansion in the latitudinal range of a species may occur over as brief a timespan as a single annual cycle, and the new distribution may persist as long as environmental conditions are tolerable. Long-term climatic cycles may have determined the present-day distributions of many benthic invertebrates and demersal fish through continuous range shifts over long periods of time. In the western English Channel short-term Russell Cycles were shown to drive range expansion and contraction in pelagic assemblages, and probably also among the benthos, while single, extreme environmental events may also have effects on the relative abundance and geographical distribution of some benthic species. A well-documented example of an extreme climatic event is the winter of 1962/63 (Crisp, 1964), when air temperatures dropped as much as 6 °C below the long-term average for a period of two months. Sea temperatures on the Essex coast dropped from 0 °C on 1 January 1963 to −1.5 °C on 12 January, and to −1.8 °C on 9 February, but rose from 0 to 5 °C through the first ten days of March. Intertidal habitats were the most severely affected, with up to 100% mortality recorded in some barnacle populations, and for top shells, sea anemones, bivalves and small crustaceans. Mortalities among benthic communities could not be quantified, but on North Sea, Channel and western coasts of Britain Conger Eels, wrasses and sole were killed in large numbers, and numerous benthic crustaceans were also washed ashore dead or moribund, including *Nephrops norvegicus*, *Cancer pagurus*, *Maja brachydactyla* and species of *Liocarcinus*. Disproportionate mortality was recorded among southern, Lusitanian, species, but the effect was less severe than might have been expected, possibly because northern range limits for these species are limited by summer temperatures during the breeding season rather than by the winter minimum. However, the relative abundance of these species would have been greatly reduced, and for some their latitudinal range would have been restricted, either through local extinction of peripheral populations or through subsequent reproductive failure.

The littoral top shell *Phorcus lineatus* (Fig. 160) is an especially good example of a Lusitanian species strongly affected by a single, severe climatic event. Prior to the winter of 1962/63, *P. lineatus* reached a northern latitudinal limit in the vicinity of Malin Head, Ireland, and Anglesey, north Wales, and ranged

FIG 160. The warm-water top shell *Phorcus lineatus* is recognised by the distinct 'tooth' on the inner border (the columella) of the shell aperture. Scale bar: 2 cm.

into the English Channel as far east as Lyme Regis (Mieszkowska *et al.*, 2007). Populations on the Welsh coast were wiped out by the severe winter cold, and those on the English Channel coast were greatly reduced. The consequence was that the geographical range of *P. lineatus* contracted sharply southwards and westwards, establishing new limits that persisted until the 1980s, when mean annual seawater temperatures began to rise again. The range of the species then began to expand once more to the north and east; by 2002 *P. lineatus* had re-occupied the whole of its previous range, and even extended its distribution in the English Channel as far east as the Isle of Wight. Populations surveyed at sites encompassing the newly recolonised area showed a predominance of year classes 0 to 4, and with individuals of every annual cohort up to 10, indicating successful recruitment in every year. *Phorcus lineatus* is extremely sedentary and can only increase its distribution by larval dispersal during a brief pelagic phase, and the demographic structure of the newly established populations was evidence of successful reproduction and recruitment.

THE PRESENT

The nature of the marine biota of northwest Europe has changed continuously in response to fluctuating environmental characteristics, cycling with periodicities of decades to centuries. Change is likely to accelerate through the next 50 years as global temperatures rise by a predicted 2–3.5 °C, a striking contrast with the 1.8 °C increase in mean annual SST recorded for the western English Channel through the past century (Genner *et al.*, 2004), possibly presaging a cycle of unpredictable periodicity, and seeming permanent and irreversible. That change began from 1980 onwards, marked by records of unusual species of fish off southwest England (Stebbing *et al.*, 2002). Fish are conspicuous and easily caught, and well-organised recording schemes ensure that individuals of unfamiliar species are submitted to specialists to confirm identifications, and all field data are accurately logged. Records for the north Cornwall sea area show that from 1960 to 1980 no species of warm-water fish were newly reported, but from 1980 to 2000 one or two new records were noted annually (Fig. 161). The total number of warm-water species new to the region reached 18 by 2000, and the cumulative curve correlated with that for the annual SST anomaly for the period, and represented significant northward range shifts for a group of Lusitanian species. From 1988 a number of tropical fish species showed similar northward range extensions, to the coast of Portugal; between 1972 and 1992 the north-flowing European shelf current showed a 2 °C mean temperature increase, and five species of deep-water fish, occurring at depths of 200–600 m, spread northwards as a consequence. One of these, the Sailfin Dory, *Zenopsis conchifer*, was estimated to have extended its range northwards at a rate of 60 km per decade from 1960 to 1995 (Swaby & Potts, 1999).

Increasing annual mean SSTs are driving rapid change in the marine assemblages of the northwest European shelf, and the responses of the biota are likely to be differential. Lusitanian species will show northerly range shifts;

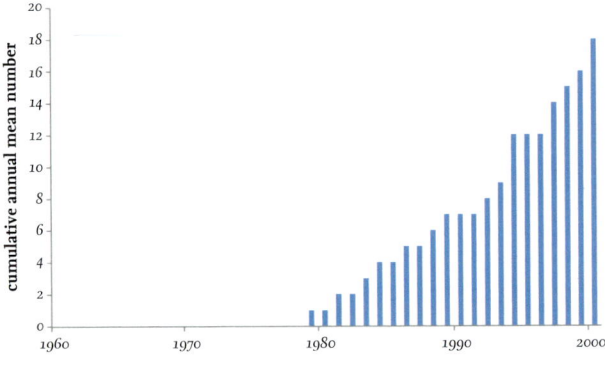

FIG 161. Numbers of southern, warm-water fish species newly recorded in Cornish waters, 1960 to 2000. (After Stebbing *et al.*, 2002)

for motile species such shifts may initially represent an extension of seasonal migration, until self-sustaining resident populations become established within the new range limits. For example, the trigger fish *Balistes carolinensis* was an uncommon vagrant off southwest Britain prior to 1980, but certainly by 2002 resident breeding populations were present on the coasts of southwest Wales, and the Sponge Crab, *Dromia personata* (Fig. 162), is now found commonly in crab pots off the Gower coast (John E. Lancaster, personal communication, 2013). Latitudinal ranges of boreal species might contract northwards, or populations might retreat offshore to cooler waters, during a modified migratory period. More sedentary species might be expected to show a slower rate of range shift,

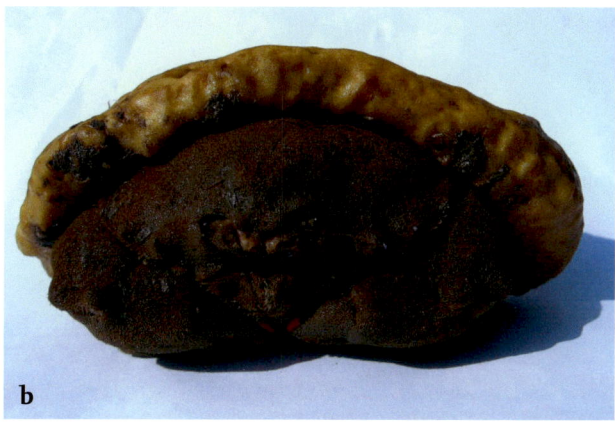

FIG 162. The Sponge Crab, *Dromia personata*, from a Gower crab pot. (a) On the move; (b) within its sponge domicile. (J. E. Lancaster)

with southernmost populations simply displaying reduced reproductive success, declining densities and ageing populations.

The large acorn barnacle *Perforatus perforatus* (Fig. 163) is a perfect exemplar of a Lusitanian species. It encrusts hard substrata in the lower intertidal and shallow subtidal, with a geographical range from tropical West Africa to the Mediterranean, and northwards along the European Atlantic coast to a limit off southwest Britain. It is found commonly off southwest Wales, but does not reach Ireland, and on the English Channel coast formerly occurred as far east as Chichester, where it was rare. The bitter winter of 1962/63 decimated these southern English populations, the geographical range of the species contracted sharply, and it became effectively extinct east of Swanage. Welsh populations were also severely reduced, although range contraction was not recorded. The geographical distribution of the species in the English Channel was subsequently surveyed between 1993 and 2001, and compared with distributions mapped in earlier surveys, in the 1940s and 1960s (Herbert *et al.*, 2003). It was apparent that *P. perforatus* had recovered most of the territory it had occupied prior to 1962/63, and had extended eastwards by a further 120 km on the English coast and 190 km on the French coast. Rising mean annual sea

FIG 163. An intertidal barnacle assemblage. The large volcano-shaped tests are those of the Lusitanian species *Perforatus perforatus*.

temperature seems to have been the principal factor responsible for the spread of this barnacle after 1980. Populations close to the former northern limit of the species, on the Welsh coast, did not breed; individual animals tended to be small and none achieved sexual maturity. With a mean summer seawater temperature of 15 °C, 5% of the population were found to be brooding embryos, but the proportion rose to 90% at 19 °C. Mature barnacles translocated eastwards, from Lyme Bay to Bembridge, on the east coast of the Isle of Wight, showed lower reproductive output than the source population, producing fewer embryos per milligram of body weight. *Perforatus perforatus* has a discontinuous distribution through its newly established range on the English coast of the Channel, partly because of the type of substratum available for settlement, chalk not being an optimum habitat, but possibly also because of seawater temperature. It tends to be absent, or only sparsely present, on exposed headlands, occurring instead in sheltered embayments where the water column is stratified through most of the summer, and the warmer upper levels provide appropriate temperatures for embryo development.

Perforatus perforatus is illustrative of fluctuating geographical distribution patterns that may prove to be characteristic of many Lusitanian species, but another warm-water barnacle species, *Solidobalanus fallax* (Fig. 164), has shown an extraordinary range extension over a very short period of time. The known range of *S. fallax* encompassed the coast of tropical and subtropical West Africa, from Angola to Morocco, a distribution termed Senegalian; it had been recorded from Algeria but not elsewhere in the Mediterranean, and probably reached its extreme northerly limit off southwest Portugal, where it was rare. In 1994 it was discovered at Plymouth, encrusting live individuals of the Queen Scallop, and the following year was collected at numerous stations in the Plymouth region, as well from Lundy Island in the Bristol Channel and on the Gower coast of south Wales (Southward *et al.*, 2004). All of these localities had been surveyed and sampled by several generations of marine biologists, and the almost simultaneous occurrence at each of an exotic species of barnacle suggests either an extremely rapid spread, and perhaps several successive introductions, or that it had previously been mistakenly recorded as the native species *Balanus crenatus*. Re-examination of dredge samples collected off west Cornwall in 1988 revealed specimens of *S. fallax* attached to hydroid clumps, proving that it had been present off southwest England at least six years earlier than the Plymouth records, but it seems improbable that it would have been consistently misidentified, especially by barnacle specialists, as individuals often have distinctively pink-patterned shell plates, and yellow and brown bands on the tissue flaps lining the operculum. *Solidobalanus fallax* occurs naturally on

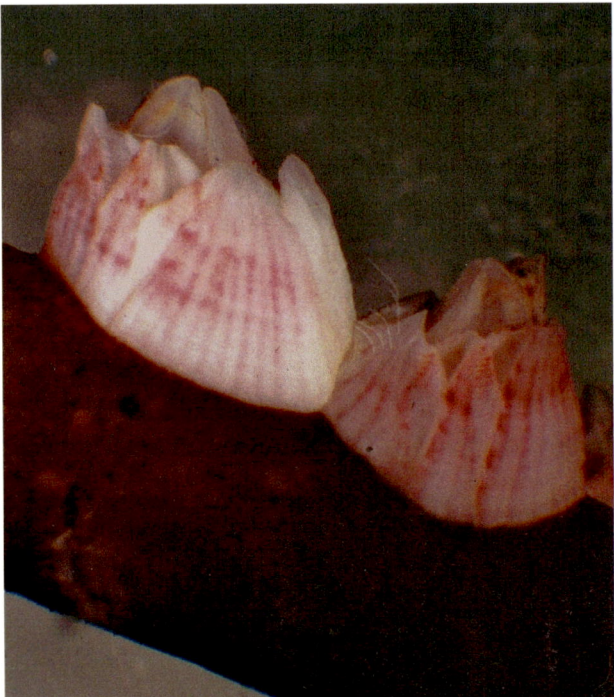

FIG 164. The warm-water barnacle *Solidobalanus fallax*, encrusting a kelp hapteron, from the Gower coast.

biological substrata, such as Queen Scallops, sea fans (*Eunicella verrucosa*), spider crabs (*Maja brachydactyla*), shells occupied by hermit crabs (*Pagurus bernhardus*), and a number of other large sedentary bivalves. It does not occur on rock or on the Great Scallop; while it will settle on inert surfaces, it does not seem adapted to high-energy habitats, and the mobility of the scallop may result in hydrodynamics at the shell surface/water interface similar to those associated with rock. The present range of *S. fallax* now extends to the southern North Sea and to Anglesey, and it is abundant along Atlantic coasts from Portugal to Brittany. A warming environment may have been an important factor, enabling new colonists to establish breeding populations, but its rapid spread might also be partly attributable to its tendency to settle on artificial substrata. Fishing nets, ropes, nylon lines, buoys and every kind of stationary fishing gear are readily colonised by barnacle larvae, and most small fishing harbours on the Atlantic coasts of Portugal, Spain and France are population centres of *S. fallax*. Discarded fishing gear, as well as every kind of plastic flotsam, would also have provided further means of dispersal; the barnacle has been reported as beached on Dutch shores, and it is probable that self-sustaining populations will eventually be found around much of the southern North Sea coast.

Biological responses to climate change, in the marine environment, will mostly be more subtle than simple range shifts by individual species. Those species with broad latitudinal ranges are likely to show correspondingly broad ecological tolerances, with perhaps little mixing between populations. An increase in SST across a species' range may thus affect populations differentially, depending on the degree of local adaptation shown by each, with different consequences for fecundity, recruitment and adult mortality. Responses to change are also likely to be mediated by interactions within benthic assemblages. The species composition of a community may change following an influx of species new to the region, and the loss of others; its structure may also change as formerly abundant species decline in density, to be supplanted by a new suite of numerically dominant species. Such impacts were well shown in a 25-year sampling programme in Bridgwater Bay, Somerset, through a period of climatic change characterised by a large swing in the NAO index, increasing mean SST and fluctuating salinity (Henderson, 2007). Samples of fish and decapods were collected monthly, from 1980, from the filter screens of the Hinckley Point power station cooling-water inlets. The number of fish species recorded each year showed a gradual increase, from 33 in 1983 to a maximum 46 in 1998, and a total of 81 species was logged through the period of the study. Maximum summer SST varied annually, showing no trend, but minimum winter SST increased steadily from below 5 °C in the early 1980s (1.6 °C in 1987) to annual values generally above 5 °C after 1987. Salinity also varied annually, in relation to rainfall and river outflow, between 17.5 and 33 psu, but mean values declined each year from 1991 to 2000, and increased thereafter. The structure of the fish community in Bridgwater Bay changed abruptly in 1985, when the relative frequencies of its permanent components decreased sharply. Numbers of Sea Snail, *Liparis liparis*, Poor Cod and Dab recovered from the filter screens dropped, while Bass, *Dicentrarchus labrax*, Pout, Whiting and Sole, all of which had been present formerly in low numbers, became the numerically dominant species. This sharp switch in community structure could be related to a swing in the NAO winter index in the mid-1980s, which, together with rising mean annual SST, stimulated increased phytoplankton productivity, higher densities of zooplankton grazers and enhanced survival of fish larvae. A second sharp change in species composition was recorded after 1993; it was not related to NAO-enhanced primary productivity, but rather to a change in seasonal visitors: cold-water fish species were replaced by a suite of southern, warm-water migrants as a consequence of rising annual mean SSTs.

The changing environment also had an impact on the population of the Brown Shrimp, *Crangon crangon*, a key species in the trophic web of Bridgwater Bay, but these effects were less immediately obvious (Henderson et al., 2006). Spring abundance varied little between years, and the adult population seemed stable, but recruitment increased annually, and tended to show greater variability from year to year. Brooding female shrimps move offshore in July, newly metamorphosed juveniles return in late summer, and the population peaks in September and October as recruitment is completed. Abundance of recruits was positively correlated with mean monthly SST for the period January to August preceding recruitment; average relative recruitment (i.e. number of individuals captured over a six-hour ebb) for the years 1994 to 2004 was more than double that for the previous decade (Fig. 165a), with anomalously high autumn abundance in some years – recruitment in 2002 was 25 times that of 1985 – but spring adult densities did not vary significantly, suggesting a strong density-dependent effect on mortality. The most important source of mortality was predation. Brown Shrimp move across mudflats, feeding with the inflowing tide, and retreat with the ebb; they require muddy substrata in which to burrow, and at low tide the shrimps are vulnerable to predators. Predatory fish limit the adult shrimp population, as increasing shrimp recruitment leads to an increasing abundance and diversity of predators (Fig. 165b), and anomalous peaks in shrimp recruitment are matched by equally strong peaks in the abundance of predators. If mean annual SST for the Bristol Channel region continues to rise it might be hypothesised that as it approaches the upper thermal limit of *Crangon crangon*, the structure of the benthic community must change. That threshold is not known with precision, although recruitment was very low in two years when summer SST exceeded 22 °C. Decreasing shrimp recruitment, declining shrimp populations and increasing competition for food from warm-water immigrant species might result in the development of a high-diversity ecosystem. Alternatively, at a critical threshold temperature, Brown Shrimp recruitment might fail, adult populations might collapse, and the ecosystem of the region might undergo an abrupt regime shift. Higher seawater temperatures and NAO-induced increases in primary productivity will together lead to profound changes in the ecology of large estuarine systems such as Bridgwater Bay, and the wider Bristol Channel region, as fish recruitment and growth also increase in consequence, but the stability of the system decreases. High SST, low winter NAO indices and high salinity might result in a sharp drop in primary productivity, and the collapse of consumer populations.

Comparing the data for fish species abundance in Bridgwater Bay with records for the English Channel revealed that while fish communities in both

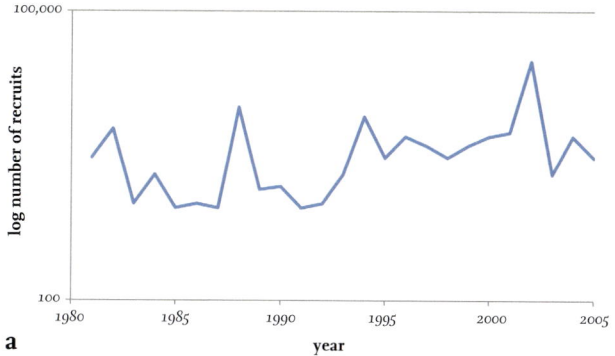
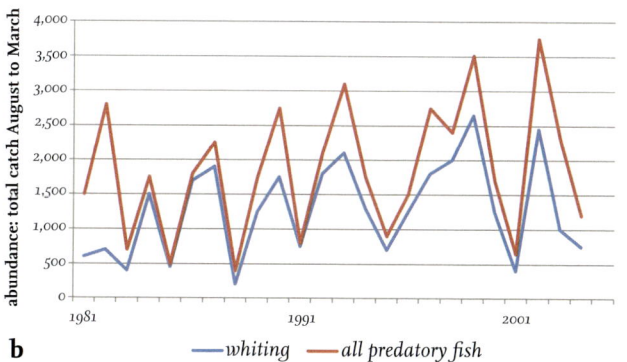
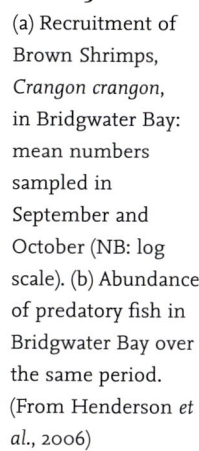

FIG 165. (a) Recruitment of Brown Shrimps, *Crangon crangon*, in Bridgwater Bay: mean numbers sampled in September and October (NB: log scale). (b) Abundance of predatory fish in Bridgwater Bay over the same period. (From Henderson et al., 2006)

regions were strongly influenced by annual mean sea surface temperatures, responses at both community and species level were not identical (Genner et al., 2004). In each dataset 33 demersal fish species were identified as 'core species', according to their abundance and persistence in the samples from year to year. For Bridgwater Bay these 33 taxa, of a total 81, comprised 99.42% of all individuals caught, and for the English Channel samples, 99.43% of a total 72. Nine of the English Channel core species increased in abundance during warm years, showing a significant positive response to increasing SST; these represented an average of 57.43% of all individuals caught in each sampling year. The Bridgwater Bay fish community also showed a significant positive response to increasing SST by nine species, but another, the Sea Snail, declined in abundance in warm years. Community responses to warming were thus similar in the two regions, but individual species responded differently. There was no species in common to the English Channel and Bridgwater Bay that showed a positive response to increased SST, and there was no relationship between changing abundance and latitudinal range: widely distributed and narrowly distributed species were equally likely to change in abundance in relation to rising sea surface temperatures. Changes in the composition of fish communities, and in the abundance of individual species, as a consequence of a warming environment, cannot be predicted on the bases of the known geographical range and physiological characteristics of each species. Life-history characteristics and migratory behaviour of each population are likely to be attuned to local environmental variables, and exchange between populations might be limited. However, perceived differences in community- and species-level responses to warming seas at different locations might also relate to the demographic structures of each population. Sheltered coastal habitats in, or near to, large estuarine systems frequently serve as nursery areas during the first few years of an individual's life, while adult populations may be distributed in deeper, offshore habitats.

Ecological interactions may be significant factors affecting population and community responses to environmental change on a local scale, as particular species, or age classes, may be dependent upon food resources that may also be affected by changing environmental variables. The Dover Sole, *Solea solea*, was one of the nine core species in the Bridgwater Bay assemblage that showed a positive response to increasing mean annual SST. Abundance of Sole sampled at the Hinckley Point water intakes rose each year after 1985, increasing by almost an order of magnitude by 2003 (Henderson & Seaby, 2005). Bridgwater Bay is a nursery area; young Sole move inshore through July, at an age of 2–3 months, migrating offshore with the onset of winter and returning the following spring. Beyond three years of age seasonal inshore migration ceases and the fish remain offshore close to the spawning grounds. The annually increasing numbers of Sole in Bridgwater Bay thus reflected increasing recruitment and higher juvenile survival. A high mean annual SST probably leads to accelerated larval development and a shorter planktonic phase, and thus a reduced risk of mortality through predation. Juvenile survival, however, is more closely related to the winter NAO index, which has a positive correlation with higher primary production, more food and rapid growth. The mean length of juvenile Sole was shown to have a significant positive correlation with abundance, suggesting that rapid growth to a size refuge from predation results in increased survival. A positive correlation between growth and the winter NAO index has also been demonstrated for populations of Sole in the Thames estuary, but this congruence does not permit extrapolation to all populations, for all environmental variables.

Southern North Sea Sole populations have been reported to show a negative correlation between recruitment and winter SST, and the causal basis of this is not fully understood.

Long-term records for North Sea fisheries for Sole and Plaice reveal that both species have shown significant shifts in distribution over a 30-year period, since 1980. They are usually part of a mixed-species assemblage targeted by beam trawlers, but are also an important bycatch of otter trawlers fishing for demersal roundfish, and landings by British otter trawlers have been recorded, as catch per unit effort (CPUE), for all ICES statistical rectangles (see page 8) between 51° and 62° N since 1920 (Engelhard et al., 2011), with a maximum 77 years of records for most rectangles. Plaice have a more northerly distribution than Sole; they are most abundant in the southern and central sectors of the North Sea, but occur in most of the northern ICES boxes, within the 200 m isobaths. In the 1920s, 1930s and 1940s the greatest catches of Plaice were recorded in the southern sector, along the southeast coast of England, and in the area bounded by the Netherlands and Germany, referred to as the German Bight, while landings decreased sharply to the north and northwest. From 1960 through the 1980s the CPUE of Plaice declined for the southeastern ICES boxes, but remained high for the central and eastern boxes. After 1990 catches recorded for the eastern region also declined, but records for the central North Sea, between 54° and 56° N, were unchanged, while an abrupt increase in CPUE was recorded for northeast Scotland, Orkney and Shetland (Fig. 166). Sole are absent from the northern North Sea, and most abundant in the south and southeast. In the 1920s and 1930s highest catches were recorded for the most southerly ICES boxes, and for those in the German Bight, and in the 1950s and 1960s boxes in the latter area yielded the greatest CPUE. This pattern was reversed after 1970 as catches from coastal southeast England increased sharply, while those from the German Bight declined equally sharply (Fig. 167).

These extensive datasets encompass the alternating periods of warming and cooling documented in the northeast Atlantic through the twentieth century, and the more sustained period of warming of the latter decades, and demonstrate changes in the spatial distribution of the two species that correlate significantly with fluctuations in SST for the period. The distribution of Plaice shifted northwards in response to rising temperatures, by as much as 2 degrees of latitude from 1940 to 2000, equivalent to a range extension of 142 km, and offshore into deeper water, at a rate of +3.96 m per decade. Its longitudinal range drifted eastwards from 1950 into the 1980s, but after 1988 showed an abrupt switch westwards of 3 degrees, reflecting the collapse of Plaice catches in the German Bight, and a sudden appearance of large

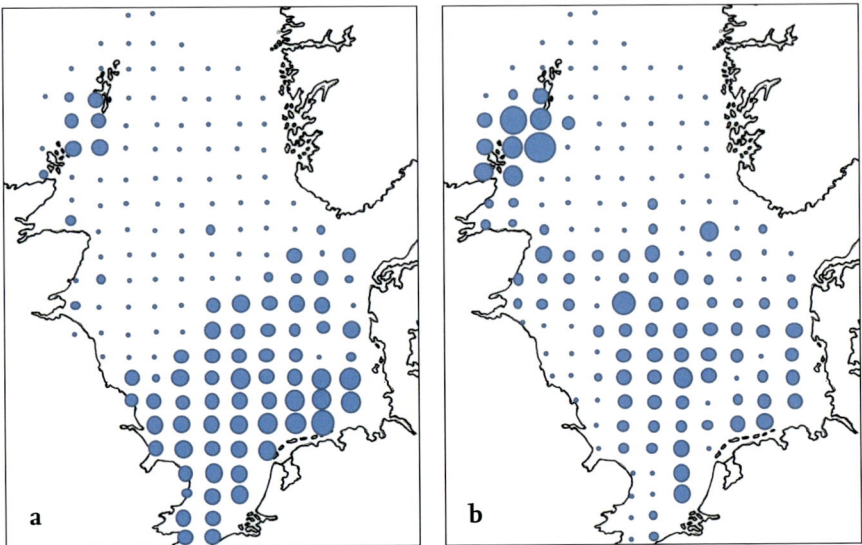

FIG 166. Temporal change in distribution and density of North Sea Plaice, shown by relative catch per unit effort (CPUE) records for (a) late 1940s and (b) 2000s. (After Engelhard et al., 2011)

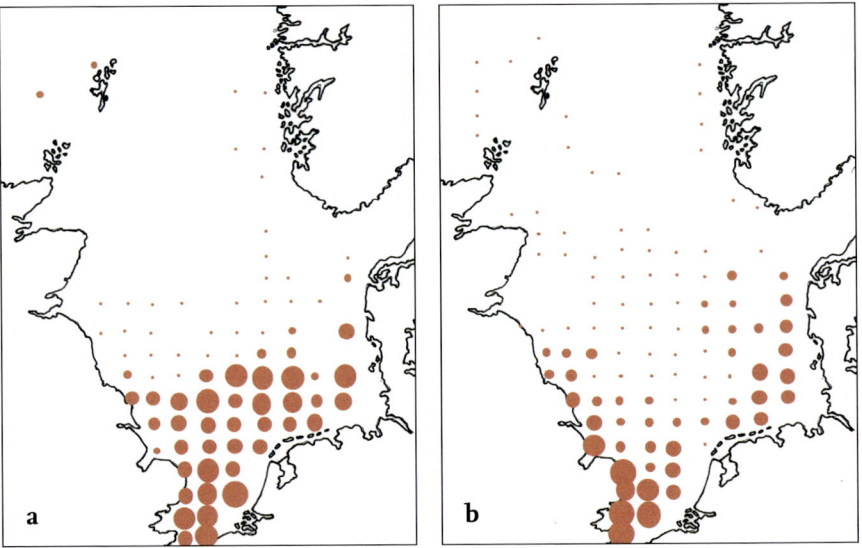

FIG 167. Temporal change in distribution and density of North Sea Sole, shown by relative catch per unit effort (CPUE) records for (a) late 1940s and (b) 2000s. (After Engelhard et al., 2011)

stocks off northeast Scotland and the Northern Isles. Sole distribution shifted southwards and westwards, a range extension of 63 km, into shallower, warmer water, and its mean depth distribution decreased by 10 m; earlier spawning and longer residence in shallow coastal waters were stimulated by rising winter temperatures.

Long-term changes in the spatial distribution of Plaice and Sole in the North Sea were considered to have been independent of fishing mortality (as measured by commercial otter trawl landings, i.e. CPUE) for Plaice (Engelhard et al., 2011), while for Sole there was a significant positive correlation. However, in comparing fish communities through time it is often difficult to distinguish temperature-related change from that induced by fishing pressure. Four benthic surveys conducted in a limited area of the southern North Sea, from 51.5° to 54.5° N and between 2° and 5° E, over seven-year periods through two episodes of cooling and two of warming, yielded abundance and body length data for 48 demersal fish species (ter Hofstede & Rijnsdorp, 2011). The two earlier periods, one cool and one warm, were characterised by low fishing pressure and those later by high fishing pressure. Each species was categorised as either Lusitanian or Boreal, and both species richness, which would have been positively correlated with trawl duration, and the mean ratio of Lusitanian/Boreal species, which would have been unbiased, were analysed in relation to mean winter SST (Table 9). Species richness differed significantly between cool and warm periods, at both low and high fishing intensity. Mean winter SST showed no significant difference between the two cool periods, but fish community structure differed significantly

TABLE 9. Comparison of demersal fish assemblages in the southern North Sea (51.5–54.5° N, 2–5° E), in alternating cold and warm periods with different fishing intensities: mean species richness, ratio of Lusitanian to Boreal species, maximum length (Lmax) for each of the two categories. Mean values ± 1 standard error. (Data from ter Hofstede & Rijnsdorp, 2011)

Period	1902–1908	1950–1956	1978–1984	2002–2008
Mean winter SST (°C)	6.02 ± 0.04	6.32 ± 0.1	6.25 ± 0.07	7.66 ± 0.09
Fishing intensity	low	low	high	high
Species richness	19.2 ± 2.2	24.7 ± 1.0	19.6 ± 1.2	32.4 ± 0.8
Ratio L/B	1.91 ± 0.23	2.61 ± 0.24	1.20 ± 0.13	1.57 ± 0.09
L_{max} Boreal	78.1 ± 3.9	60.7 ± 2.2	67.7 ± 3.0	53.7 ± 1.7
L_{max} Lusitanian	65.0 ± 1.3	58.6 ± 1.9	66.0 ± 0.4	46.1 ± 2.4

in the second period, 1978–1984, which was characterised by high fishing pressure. The annual L/B ratio increased significantly from the cool period, 1978–1984, to the latest warm period, 2002–2008, and the mean L/B ratios for the two periods, both with high fishing pressure, also differed significantly, as higher temperatures stimulated an increased proportion of warm-water Lusitanian species. Mean body size was significantly smaller, for both Boreal and Lusitanian species, in warm than in cool periods, and body size of Boreal species was largest at times of low fishing intensity; during 2002–2008, a warm period with high fishing pressure, decreased body size was probably as much attributable to an increased proportion of small Lusitanian species as to fishing pressure.

ALIEN INVASIONS

Natural range extension by exotic, warm-water species is likely to change the composition of both benthic and pelagic communities of northwest Europe, and further change may result from an increase in the number of species introduced accidentally through human agency. With the possible exception of Antarctica, shelf environments worldwide now support a complement of alien species, originating from the fouling assemblages on ships' hulls, from ballast water, and from communities associated with translocated oysters. A warming environment is likely to enable easier establishment of breeding populations, and consequent spread, of newly introduced species. Coastal pleasure craft provide for further dispersal, while marinas and other coastal engineering installations provide convenient way stations. An interesting possibility is that a number of species introduced into the northwest European coastal environment during the latter half of the last century, such as the Hard Clam, *Mercenaria mercenaria*, and the American Oyster Drill, *Urosalpinx cinerea* (Fig. 168), might begin to spread at a far greater rate than previously recorded. Several of these, such as the American Slipper Limpet (Fig. 70), are already regarded as 'invasive' species, with broad ecological tolerances, and capable of displacing native species across a wide habitat range. Present ranges of such species may be temperature-limited, either directly through suboptimal reproductive conditions, or indirectly, with ambient temperatures favouring reproductive success and competitive ability of native species. An even more intriguing possibility is that certain species, non-native but not presently recognised as alien, and introduced so long ago, and so well-established that their origin, and time and method of introduction, cannot now be known, may gain ecological advantage from a changing environment, and begin to increase in both distribution and abundance.

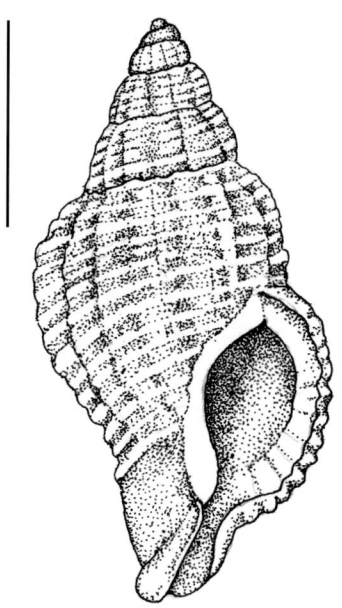

FIG 168. The predatory American Oyster Drill, *Urosalpinx cinerea*, employs an acid-secreting gland on the sole of its foot, and a toothed radula, to bore into the shell of its prey. Scale bar: 1 cm.

Ships have been significant agents for the dispersal of marine organisms, probably since the times of the earliest seafarers. Slow-moving coastal craft, following coastlines, mooring for protracted periods in estuaries and sheltered bays, would have accumulated communities of sessile or sedentary plants and animals, and redistributed them, and their propagules, along trade routes. Mariners would have been aware of the biological burdens their ships carried, and careening was a frequent necessity to remove potentially damaging colonists such as the shipworm *Teredo navalis*. The geographical origins and patterns of dispersal of many ship-fouling organisms are unknown, but early anthropogenic dispersal has been suggested as the explanation for some curious modern-day distributions. For example, the large Sand Gaper clam, *Mya arenaria*, has a circumboreal distribution, but in the northeast Atlantic became extinct at the beginning of the Pleistocene. It has since recolonised the region, and is found as far south as the Bay of Biscay, but the oldest shells recovered from Holocene sediments have been aged at around 700 years, suggesting a very recent reintroduction, perhaps through human agency (Lasota *et al.*, 2004). The larvae of *M. arenaria* have a pelagic phase of just 2–3 weeks, insufficient for long-range natural dispersal, but it is possible that adult animals were transported aboard ship, perhaps as part of a food store, or amongst sand and gravel ballast. Comparison of genetic structure in populations of the clam from Poland, Sweden, the Netherlands and Biscay showed little variation between them, suggesting a recent introduction to European shelf habitats and a rapid dispersal.

Another puzzle concerns a rather dull bryozoan, *Victorella pavida*, so named for its discovery, in 1870, in the Victoria Docks, London. It is now no longer found at its original, or type, locality, and in Britain exists only as a single small population in a lagoon at Swanpool, Cornwall, a scale of rarity that earns it conservation status. Yet *V. pavida* is found in estuarine habitats practically worldwide, a distribution that can only have been established by ship-borne dispersal over many decades. Its original distribution is unknown, but it is common in Italian estuaries and lagoons, and in similar habitats around the Black Sea, and it is probable that the species began its dispersal attached to the hulls of trading vessels in Classical times. Species such as *V. pavida*, which do not appear to be native to a particular region, or fauna, and yet cannot be shown unequivocally to have been introduced by human agency, are termed cryptogenic, and in any coastal marine fauna, in regions long populated by humanity, the proportion of cryptogenic species may prove to be substantial.

The small amphipod *Corophium sextonae* is a more recent enigma. It was discovered at Plymouth, and described as a new species in 1937. It had not been collected in the Plymouth area previously, and was assumed to be a newly introduced alien species. Subsequently, it was established that material had been collected in New Zealand, and then from Portugal, prior to its discovery at Plymouth. However, it is now known to occur also in Tasmania, so while it is not part of the indigenous British marine fauna, its origin cannot be known with certainty, and it should be regarded as a cryptogenic species.

Most of the known alien species present amongst the northwest European marine biota have a documented history, comprising a dated first record, and a sequence of later records charting their colonisation and spread. Alien status is confirmed by the lack of historical evidence for its presence prior to the first record, which does not, of course, correspond to first introduction, although that may be known in some cases; it may take several generations before an introduced species becomes apparent amongst its new host community.

By the end of the twentieth century around 50 species of marine invertebrates and plants had been recognised as non-indigenous to northwest European shelf habitats, and thus exotic or alien, introduced to the region by human agency. The American Slipper Limpet, *Crepidula fornicata* (Fig. 70), was among the earliest invaders to be documented, recorded in Liverpool Bay in 1872, but it did not establish permanent populations until after 1877, on the Essex coast (Eno *et al.*, 1997). Most of the marine species now naturalised in northwest shelf habitats seem to have been first noted in the middle years of the twentieth century, perhaps as a result of a renewed interest in marine ecology, but the number of newly discovered aliens began to rise each year after 1950, and seemed to plateau from 1980. A review of non-indigenous marine species recorded from the British sea area listed five diatoms, 15 marine algae, one angiosperm and 30 macroinvertebrates reported up to that date (Eno *et al.*, 1997). The predominant invertebrate invaders were polychaetes, with eight species, crustaceans, with seven, and nine species of mollusc. The roll of non-indigenous marine species, for all European seas, had reached 851 by 2003, and while this included planktonic as well as benthic species, and while the Mediterranean was the recipient of the majority of introductions, a striking total of 114 species of macrozoobenthos had been recognised amongst the faunas of the European Atlantic coasts and the northwest shelf (Streftaris *et al.*, 2005). Eleven of these species were found to occur along Atlantic coasts, through the southern North Sea, and into the Baltic, and this group included species that had been early introductions, such as *C. fornicata*, and were characterised by broad ecological tolerances, or rapid habituation to new habitats. Thirty-two species were noted as occurring only on Atlantic coastlines. These included a number of relatively recent additions to the European marine fauna, such as the barnacle *Solidobalanus fallax* (Fig. 164), the bryozoan *Watersipora aterrima*, and the tunicate *Perophora japonica*, as well as several species of oyster, and their associates, introduced for aquaculture. The 21 species present on Atlantic coasts and in the North Sea perhaps have rather limited distributions within the area, such as *Mercenaria mercenaria*, or have been introduced to a few sites as contaminants of oyster stock, or perhaps have simply dispersed slowly from a single point of introduction. Twenty-six non-native marine invertebrates are presently known only from the North Sea, and 14 only from the Baltic, while a further ten are established in both regions.

The potential for new translocations has grown enormously over the past century, as total world shipping has boomed (Mineur *et al.*, 2012). Encrusting, fouling, communities of plants and animals have been carried from port to port and around the world for centuries, but faster ships, shorter residence time in port, and effective antifouling treatments all reduce the probability of successful transportation, and translocation, of fouling faunas and floras. However, larger ships, and larger ballast tanks, provide an alternative mode of transport, and millions of marine organisms are now carried around the world annually in ballast water. Entire pelagic communities are easily shipped between oceans, thriving in the ballast water of giant tankers: of the 851 non-indigenous species recorded in European seas by 2003, around 100 were species of phytoplankton and zooplankton (Streftaris *et al.*, 2005). The number of newly introduced species reported annually had seemed to level off around 1980, but is certain to have begun to increase again thereafter as a consequence of rising sea surface temperatures from 1985 on.

Aquaculture has been another cause of alien introductions, through the importation of edible bivalves, especially exotic species of oyster, and their undetected associates, such as *Urosalpinx cinerea* (Fig. 168) and *C. fornicata*.

The latter is already an environmentally damaging pest, and the former has the potential also to become a nuisance, but the small, sacciform phaeophyte *Colpomenia peregrina* is an interesting and inoffensive addition to the native macroalgal flora. It is probable that translocation of oyster stock is now conducted with more care in order to limit the spread of further pest species, but it is probably practically and economically impossible to prevent the dispersal of viruses, bacteria and endoparasites associated with oysters.

Improved monitoring of marine habitats for the presence and spread of exotic species has probably been a factor in the increase in numbers of newly introduced species reported from 1960 onwards, and it will also contribute to the increasing number of species reported in the years to come. A further factor is a significant new vector in the translocation of alien species, namely pleasure craft (Mineur et al., 2012). Coastal pleasure craft are ideal agents for the dispersal of sessile and sedentary marine organisms: they spend a large proportion of each year moored in marinas, accumulating thriving fouling assemblages; they tend to travel more slowly, and over shorter distances, than commercial vessels, and their biological burdens are less likely to be damaged as a result. Leisure craft usually spend more time at each destination, allowing encrusting plants and animals to shed spores, eggs and larvae into the new surroundings, or for portions of the assemblage, and its non-sessile component, to detach and occupy new habitat space. Coastal yacht havens tend to be sheltered, warm and clean, less polluted than commercial docks, but with sufficient suspended material to maintain food resources. Jetties and pontoons may support permanent communities of sessile and sedentary organisms supplying a source of spores and larvae that are likely to colonise any clear space on adjacent hulls. The increasing development of marinas, and coastal engineering projects such as sea defences, provide new islands of habitat allowing stepping-stone dispersal of sessile marine species (Mineur et al., 2012) and may be of especial ecological importance on sedimentary coasts. The gregarious tubeworm *Ficopotamus enigmaticus* (Fig. 100) is the perfect example of a long-range migrant subsequently dispersing gradually through its newly extended range. It probably originated in the tropical eastern Atlantic, is tolerant of low and fluctuating salinity, and is capable of non-sexual reproduction, through fission, quickly establishing new self-sustaining colonies in each new habitat it colonises. *Ficopotamus enigmaticus* has been transported around the world by commercial shipping, and has expanded along each newly settled coastline on the hulls of coastal craft. Its original point of introduction to northwest European coasts is not recorded, but it can probably now be found in almost every port, harbour and marina on the coasts of England and Wales.

Successful colonisation by a non-indigenous species initially depends upon it encountering suitable ecological conditions, in which it is able to resist resource competition and predation by native species, and to establish a self-sustaining population. Temperature is especially important: all stages in the reproductive cycle, in all species of marine invertebrate and fish, are regulated by temperature. Species with narrow ecological tolerances, and in particular, narrow temperature tolerances, are generally able to colonise only a narrow range of new habitats, and usually display a slow rate of dispersal through their newly extended geographical range. For example, Hard Clams were deliberately introduced into Southampton Water in 1925 and dispersed only slowly, scarcely ranging beyond the eastern entrance to the Solent, despite possessing a pelagic larval stage. It requires a minimum water temperature of 18–19 °C for spawning, lower than experienced by native populations in the western Atlantic, but still higher than mean summer temperatures on the south coast of England; it probably only reproduces successfully in the vicinity of warm-water outflows from power stations.

Currents are also critical to the establishment and further dispersal of newly introduced non-indigenous species (Dunstan & Bax, 2007). A prolonged pelagic larval stage will allow a rapid dispersal, and an increasing range, provided that the larvae are carried by coastal currents over areas of suitable habitat. However, for a population to become self- sustaining, hydrodynamic regimes must also result in a degree of larval retention, as over-dispersal of significant proportions of the annual reproductive output is likely to lead to local extinction in years of reproductive failure. *Urolsalpinx cinerea* was introduced with stock of the oyster *Crassostrea virginica*, on which it feeds, and has dispersed very slowly since it became established in Essex coast oyster beds. It lacks a pelagic larval stage, its egg capsules discharging crawling juveniles, and it only achieves any significant range extension when attached to translocated oysters. *Crepidula fornicata*, a suspension feeder and also a co-immigrant of *C. virginica*, has a pelagic larva and broad ecological tolerances, and is not dependent on an oyster host. It has achieved a spectacular dispersal, and is now abundant around the southern coasts of Britain, from Yorkshire to Pembrokeshire, and on much of the mainland European coast. Human agency has been significant in aiding this dispersal. In the decades following its first introduction *C. fornicata* would have been steadily redistributed among the oyster beds of western Europe with every translocation of oyster stock. Once established in shallow bays and estuaries the limpet populations would have been further dispersed by dredges, and as part of the bycatch dumped overboard along the route of each oyster smack. In the Bay of Brest, Brittany, the biomass of *C. fornicata* exceeded 12,000 tonnes by 1995 (page 130), but in the Bay of Arcachon, where it was first recorded in 1969, a more modest biomass of 150 tonnes was estimated by 2000. There, it had colonised only about 5% of the subtidal habitat of the bay and its distribution was partly determined by hydrodynamic and sediment characteristics. It occurred principally on muddy substrata, but was scarce where these were occupied by *Zostera* meadows. Discrete patches of *C. fornicata* had densities of up to 2,000 individuals per square metre, with biomass of up to 250 g dry weight per square metre (dwt/m^2), but the mean biomass for the bay as a whole was only 1.6–22 g dwt/m^2, rather low compared to values for other oyster culture sites on the French Atlantic coast, where mean slipper limpet biomass ranged from 60 to 300 g/m^2 (de Montaudouin *et al.*, 2001). Significantly, dredging is not permitted in Arcachon Bay, so the aggregations of *C. fornicata* are not disturbed, scattered and continuously dispersed, and the benthic environment is not homogenised.

The best-documented, and most impressive, dispersal by a newly introduced non-indigenous species is that of the American Razor Clam, *Ensis directus* (Fig. 169). A population of this plump razor fish was first found in the Elbe estuary, Germany, in 1978 (Armonies, 2001). It consisted of one-year-old animals, which were assumed to represent the recruits of the first strong year class produced by a recently established, and successfully breeding, alien species.

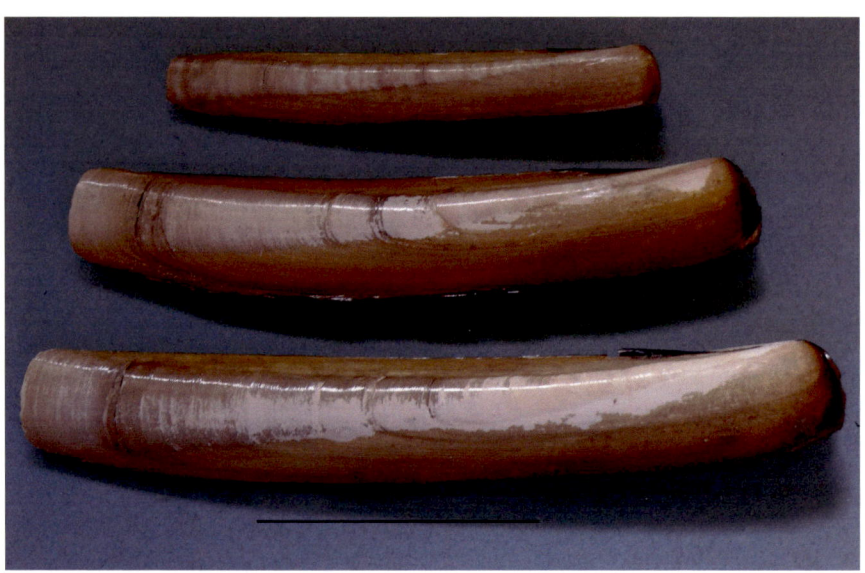

FIG 169. Shell of the invasive American Razor Clam, *Ensis directus*. Scale bar: 5 cm.

Ensis directus has a native range along the eastern coasts of North America, from Labrador to Florida, and pelagic larvae are thought to have been released in ballast-water discharges by transatlantic shipping, in the southern North Sea region, prior to 1978. The species spread rapidly around the southern borders of the North Sea. It reached the Danish coast by 1982, and Belgium by 1984, and the French coast of the English Channel by 1986. It was first found on the coast of Norfolk in 1989, is now abundant in the Wash, and occurs as far north, at least, as the Humber estuary. Unlike the three native species of *Ensis*, each of which is limited to sand of a particular grade, *E. directus* is fairly eurytopic, and will colonise a wide range of sediments, from very fine, sandy mud to coarse, well-sorted sands with little organic content. Also unlike European species, it will live in both estuarine and fully marine conditions. In the Dutch Wadden Sea it colonised lower littoral and shallow sublittoral sand flats from 1981, burrowing into coarse, current-swept sands with low organic content, and with a low biomass and diversity of macroinfauna. It seemed to have occupied a largely vacant niche, and became the dominant infaunal species (Beukema & Dekker, 1995). However, densities seemed rather low: although recruits might number tens to hundreds per square metre, mortality was high, and average spring juvenile densities were around six per square metre. Through a 12-year study of the Wadden Sea populations up to 1994, good recruitment occurred in only one year. In the Wash, densities of adult *E. directus* have frequently exceeded 2,000 per square metre, but while in the Wadden Sea it had a maximum life span of seven years, in the Wash the average age was only three years, suggesting a density-dependent effect (Palmer, 2004). The American Razor Clam releases a planktonic larva that has a pelagic phase of 10–29 days, and its rapid spread around the southern shores of the North Sea is attributable to larval dispersal. Coastal currents of the North Sea have an average velocity of 0.02–0.1 m/s (= 1.7–8.6 km per day) in calm weather, but may exceed 0.3 m/s (25 km per day) in stormy weather (Armonies, 2001), and with a 20-day pelagic phase this could equate to a dispersal of 34–170, or up to 500 km. In addition, post-recruits may disperse further through byssal drifting in benthic boundary currents. The spread of migrant *E. directus* from the Elbe estuary to the Skagerrak, Denmark, in four years would demand a dispersal rate of 125 km per year, a distance that could be travelled within a single generation. With such a potential for gene flow between the foci of *E. directus* colonisation in the southern North Sea, all populations in the region should be viewed as a single, continuous population (Armonies, 2001), with a predictable low level of genetic variation. Also, with such widespread larval dispersal, poor recruitment in one area probably represents poor reproductive output elsewhere, within a 125 km radius. Thus, the Wadden Sea subpopulation is perhaps not self-sustaining, but instead surviving at the edge of the species' ecological tolerances, and with its continued existence dependent upon recruitment of immigrant larvae.

Environmental contamination, overexploitation of marine biotic resources, habitat destruction and the spread of invasive species are the four apocalyptic threats to the marine environment. All have anthropogenic origins, but while the effects of the first three can be mitigated to some extent, and perhaps in some cases reversed, the consequences following the introduction of fast-spreading invasive species may be irreversible. There is little possibility that any invasive species can be eliminated, or its further dispersal limited by anything other than natural ecological processes. The naturalisation of invasive species is an example of biological disturbance caused by human agency. The anthropogenic dispersal of fouling communities is another, more recent, type of biological disturbance, and while translocation of biota in ballast water is possibly the major source of such disturbance, it has the best potential for control, by treating water prior to discharge, while limiting the spread of sessile fouling communities is much more difficult. All temperate and tropical shelf environments, worldwide, now have a complement of non-indigenous species, some with the potential to develop invasive characteristics. The spread of non-indigenous species has been documented for many coastal regions, but in only few cases have the ecological consequences of marine invasions been demonstrated. The introduction and establishment of non-indigenous species may have profound effects on coastal ecological processes; native species may be outcompeted and ultimately displaced, and biodiversity and habitat heterogeneity may decrease.

Some alien species may appear to achieve a degree of population stability without undue ecological dislocation. For example, the Manila Clam, *Ruditapes philippinarum*, is native to the northwest Pacific, but has long been favoured for aquaculture, and transplanted into lagoonal habitats along the Pacific coast of North America, Atlantic European coasts, and in the eastern Mediterranean, and everywhere it has acclimatised and naturalised. Its most northerly population in western Europe is in Poole Harbour, Dorset, where it was introduced in 1988 to investigate the possibility of its cultivation there (Jensen *et al.*, 2004; Humphreys *et al.*, 2007). Introductions elsewhere on the south coast of England failed, but in Poole Harbour the species naturalised, establishing a self-sustaining wild population with densities above 100 individuals per square metre and an annual production of 17.04 g AFDW/m^2, sufficient to support a fishery that by 2006 was yielding 250–500 tonnes per year. *Ruditapes philippinarum* is adapted to eutrophic lagoonal habitats, occupying muddy substrata rich in benthic diatoms, and its further dispersal to similar habitats in northwest Europe will probably depend upon transport for aquaculture developments, and it perhaps has only limited invasive potential. However, elsewhere across its geographical range wild populations of the clam have been recorded at densities above 5,000/m², and in Italy, where it extended its range by 30 km annually from its initial site of introduction in the Venice Lagoon, densities of 2,000–2,500/m² were shown to have deleterious effects on the abundance of zooplankton and macroalgae. Cold winter weather seems to be one factor presently limiting the populations in Poole Harbour, but that constraint might disappear as climatic warming continues, and the clam populations might show sharp increases in density as a consequence. Two common native infaunal bivalves, *Scrobicularia plana* and *Macoma balthica*, seemed unaffected by the naturalisation of this species in Poole Harbour, initial declines in the density of both being related to tributyl tin contamination, but should populations of the Manila Clam reach similar densities to those described for Italian lagoons, both the infaunal and epifaunal ecology of the area could be seriously affected.

Non-specialist predators are especially damaging as invasive species, and may disturb entire ecological networks. A particularly notorious invasive predator is the ctenophore, or comb jelly, *Mnemiopsis leidyi*, a native of the temperate northwest Atlantic, which was recorded in the Black Sea for the first time in 1982, probably transported there in the ballast water of a large tanker. It is eurytopic, tolerating a salinity range of 2–38 psu, and temperatures of 2–32 °C, and is a euryphagous, pelagic carnivore, consuming a wide range of zooplankton, including the eggs and larval stages of pelagic fish. At its peak abundance, in 1989, *M. leidyi* reached densities of 400 per cubic metre of seawater; its effects were catastrophic, resulting in the collapse of commercially important fisheries, and a hypoxic benthic environment. In 2006 *M. leidyi* was found in the North Sea, and in the western Baltic (Kube *et al.*, 2007) where densities of 100 per cubic metre were reported. In the Black Sea population, densities of *M. leidyi* dropped sharply, perhaps partly through the depletion of zooplankton resources, but certainly in part through the appearance in 1999 of a Mediterranean ctenophore, *Beroe ovata*, a specialist predator of other comb jellies. A new equilibrium seems to have been established in the Black Sea marine ecosystem, but probably with a quite different ecological structure than existed prior to 1982. In the North Sea and the Baltic *M. leidyi* does not seem so far to have been the cause of significant ecological dislocation, although its effects have not been investigated closely, but there are several native northeast Atlantic species of *Beroe*, any of which might be important in the ecological interactions of the invading species.

The European Green Crab, or Shore Crab, *Carcinus maenas*, is another generalist predator, and it is now established on both the Atlantic and Pacific coasts of the United States and Canada, in South Africa, and along the southern coasts of Australia. It feeds on many different species of mollusc, and on northwest European coasts it is an important factor regulating populations of infaunal bivalves and epilithic gastropods, and thus structuring both soft sediment and rocky shore communities. It was carried to the west Atlantic in the early nineteenth century, and is now distributed from Virginia to Nova Scotia, and was present in Australia by early in the twentieth century, but was not recorded in the eastern Pacific until 1989, when it was discovered in San Francisco Bay. In 1993 it was reported 100 km north, in Bodega Bay, and continued to expand northwards at a rate of around 55 km per year until 1997, when it was found on the Oregon coast, 300 km further north, probably transported by ship (Jamieson et al., 1998). In the Pacific northwest *C. maenas* is truly invasive: its mean carapace width is greater than that of its European ancestors; it will occupy a range of habitats, from muddy grounds to gravel and rock, and aggressively outcompetes native decapods, while consuming the widest range of molluscan prey. Oregon, Washington and British Columbia support commercially important shellfisheries, based on both wild and cultured species, including Pacific Oysters, *Crassostrea gigas*, and Manila Clams, and the naturalisation of an invasive crab predator is a serious ecological and management problem.

It is a matter of increasing concern that aggressively invasive species are now shown to have the potential for ecological change that approaches the catastrophic, as continuing research on the American Slipper Limpet, *Crepidula fornicata*, reveals. It has long been regarded as a nuisance. Its dense populations filter huge volumes of water, and the accumulation of faeces and pseudofaeces (particles filtered and rejected, bound with mucus) among the clumps of slipper limpets, creates a muddy environment that may benefit small detritivorous crustaceans and polychaetes, but smothers and supresses populations of other suspension feeders. Slipper limpet beds also have wider effects on the marine environment. In the Bay of Brest populations of *C. fornicata* were shown to be so dense that they were effective in depressing chorophyll *a* concentrations during the early stages of the spring phytoplankton bloom, reflecting a reduction in densities of phytoplankton cells in the water column. They also promoted rapid cycling of nutrients and silica, and thus prolonged diatom blooms into early summer. Subsequently, the potential environmental effects of such a huge biomass of slipper limpets were shown to be even more far reaching (Martin *et al.*, 2007). Measuring the flux of dissolved oxygen and dissolved inorganic carbon at the sediment/water interface, and deriving estimates of benthic community metabolism, emphasised the relation between slipper limpet biomass and carbon production. At a low density, with an average of 62 ± 77 *Crepidula* per square metre, carbon production for the entire benthic community was 180 g per square metre per year, while at a higher average density of $1,224 \pm 239/m^2$ this figure rose to 440 g. Slipper limpet density averaged over the whole 180 km² of the Bay of Brest was $260/m^2$, which was estimated to produce an annual yield of $220 \text{ g C}/m^2$. This figure proved to be higher than that for the combined production of the phytoplankton and benthic microflora communities, and was equivalent to an annual release of 10,000 tonnes of carbon. Carbon cycling by the *C. fornicata* population significantly increased the partial pressure of dissolved carbon dioxide, to the extent that the bay contributed to a positive flux of carbon dioxide to the atmosphere. A further consequence of *C. fornicata* metabolism is that the accumulation of biodeposits – faeces and pseudofaeces – stimulates bacterial metabolism, leading to mineralisation of organic residues and sediment hypoxia.

CHANGING HABITATS

The introduction of invasive marine species and their subsequent spread, especially when native species are displaced and communities dislocated, may be regarded as an environmental stress, or pressure, generated by human activity. Other such pressures can be recognised, and have been categorised (Table 10).

Toxic contaminants may have effects throughout the entire marine environment, resulting in toxic pollution that may be relatively short-term and acute, in the case of large-scale spillages of hydrocarbons, or long-term and chronic, as in the case of tributyl tin leached from antifouling paints. These types of environmental pressure, together with organic and non-organic nutrient enrichment, affect both biological and ecological processes, but ecosystems usually repair in time, even from such catastrophic incidents as the *Torrey Canyon* and *Braer* oil spills. Marine pollution is a subject of considerable environmental and commercial importance, with an extensive literature, and several scientific journals and a number of authoritative textbooks (e.g. Clark, 2001) devoted to it, and it has not been addressed at length in this book.

Similarly, fisheries science is a subject with very great commercial and environmental significance, and attended by difficult political questions. Fishing constitutes a profound biological disturbance to shallow shelf habitats, and the overexploitation of fish populations – overfishing – is one problem for marine biologists that almost everybody will be aware of. That fish populations are severely depleted everywhere is evident from the prices fresh fish commands, and the fact that fish species that were once steamed for the cat are now presented in elaborate dishes by culinary celebrities. However,

TABLE 10. Categories of marine environmental stress resulting from human activities, with some effects and sources. Stress resulting in direct physical effects marked in bold. (After Eastwood *et al.*, 2007)

Category	Effect	Source
Physical loss of habitat	**Smothering**	Artificial obstructions, spoil and waste disposal
	Obstruction	Permanent constructions
Physical damage to habitat	**Siltation**	Industrial runoff, outfalls, dredging
	Abrasion	Boats and anchors
	Extraction	Aggregate dredging, benthic fishing gear
Non-physical disturbance	Noise	Shipping and leisure craft, sonar
	Visual	Shipping and leisure craft
Toxic contamination	Release of synthetic compounds	Pesticides, antifoulants
	Release of non-synthetic compounds	Heavy metals, hydrocarbons
	Release of radionuclides	Nuclear industries
Non-toxic contamination	Nutrient enrichment	Agricultural runoff, outfalls
	Organic enrichment	Aquaculture, outfalls
	Temperature changes	Power stations, outfalls
	Turbidity changes	Dredging, runoff
	Salinity changes	Water extraction, outfalls
Biological disturbance	Release of microbial pathogens	Aquaculture
	Introduction of alien species	Ballast water discharges, ship-borne fouling, aquaculture
	Removal of selected species	Fishing

it is difficult to demonstrate effects of fishing on fish communities through experimental techniques because, in the northeast Atlantic region, fishing has been conducted across shallow shelf environments for at least a century and there are no undisturbed areas that might serve as controls. Evidence from elsewhere suggests that the most profound effects occur when fishing is first begun, and that further effects may be undetectable when fishing intensity is low to moderate. Fishing has many consequences for marine ecosystem function: removal of target species affects the composition of fish assemblages, and selective fishing for the larger size classes changes the population structure of each target species. These might include large predatory species, smaller planktivores, and various omnivores and scavengers, and changing the relative abundance of species and size classes in a mixed-species assemblage might result in changing interactions within the fish community, or between fish communities and benthic invertebrate assemblages. Fisheries biology was established and developed as an applied science, focused on the reproductive cycles, growth, population structure and dynamics of commercially important species, usually as single, defined population, or 'stocks'. Into the twenty-first century the most important task of fisheries science is to assess stock levels of target species, with the objective of designing management strategies that result in the maximum yield of fish protein to humankind while maintaining the resource, and only comparatively recently have fish population studies been integrated into ecosystem-based marine ecology (Jennings & Kaiser, 1998). The effects of fishing on fish populations will not be considered further here, but the physical damage caused by fishing, together with that resulting from aggregate dredging, offshore engineering and coastal development, are particularly significant pressures affecting benthic habitats, and are reviewed in the final sections of this book.

Extraction of sand and gravel –'marine aggregates'– for the use of the construction industry has especially damaging impacts on the benthic environment. Dredging by stationary vessels, or anchor dredging, creates a cratered benthic topography of large pits, often more than 50 m diameter and many metres deep, while trailer dredging, by slow-moving vessels, leaves crisscrossed tracks several metres wide and perhaps 0.5 m deep (Newell *et al.*, 1998). Licences to extract aggregates usually have long leases, and the volumes of material removed from a dredge site over successive years must have significant cumulative effects: from a site off the Essex coast dredged over a period of 25 years, more than 100,000 tonnes of sand and gravel were removed annually, with a record 872,000 tonnes in a peak year (Boyd *et al.*, 2005). Suction pumps employed in aggregate extraction create plumes of suspended fine material, causing a further physical effect as the silt is dispersed, settling on and blanketing adjacent communities. The extent of dispersal and siltation will depend upon current regimes. In low-energy environments re-suspended silt is likely to settle close to the disturbed site, although it was estimated, in

2004, that while 134.5 km² of the coastal waters of England and Wales were subject to aggregate extraction, 2,994.8 km² were affected by plume dispersal of silt (Eastwood et al., 2007); however, this figure probably includes silt plumes generated by bottom-fishing gear. Commercial demands for marine aggregates require a sand/gravel ratio of from 50:50 to 65:35, but dredged sediments most often comprise 15–55% gravel (Newell et al., 1998). Ratios are adjusted on board the dredger by screening, a process by means of which material too fine or too coarse for commercial purposes is discarded; as much as 60% of the material rejected may be in the particle size range 0.25–1.00 mm – fine, medium and coarse sand according to the Wentworth scale (page 72). Material returned to the sea creates further new topographic features, and long-term dredging results in a permanent change in the sediment profile over the dredged area, partly as a consequence of the proportional increase in the volume of fine to coarse sand, and the winnowing out of silt, but possibly also as a result of the exposure of deeply buried material, finer or coarser in grade than the overlying sands and gravels. The scars left by marine dredgers may be visible for years following removal of aggregates, although persistence will depend upon such factors as water depth, bottom current velocity and sediment mobility. High-velocity currents may lead to pits infilling within a year, as the edges slump and the central depression acts as a sediment trap. On sea floors subject to the slowest current flow, and on tidal flats, dredge scars have been noted to remain visible for up to ten years. However, in shallow, inshore environments sediment overturn caused by seasonal storms may obliterate dredge tracks and pits in one winter, and on the coarse gravelly banks of the central Bristol Channel it may take just three tidal cycles for all traces of a dredging operation to disappear (Newell et al., 1998).

Removing a substantial proportion of a soft-sediment habitat has severe consequences for its epifauna and infauna, and post-dredging surveys show an immediate reduction in the number of species, and in the density of individual species, and sharp drops in the total number of individuals and total biomass. Large-bodied species, such as epifaunal molluscs, decapods and echinoderms, are likely to be damaged, dislodged and either swept away or made vulnerable to predators and scavengers. Similarly, the large-bodied infauna, mostly *K*-selected, tube-building and burrow-dwelling species, will be destroyed, and small free-living infaunal species washed away; community structure is dislocated, and complexity and stability are lost. Even the most profoundly disturbed, defaunated sedimentary habitat will eventually be recolonised, and new benthic communities will develop, but the process may be very slow, and the composition of the fauna may not be the same as that which existed prior to the disturbance. In time the new community will develop increasing diversity and complexity, and thus approach a stable equilibrium state equivalent to that of the original, but this may not represent a 'recovered' but rather a 'replacement' community. The rate of development of the new assemblage will be determined by current regime and sediment grade, and is likely to be swifter in current-swept sands than in low-energy muddy habitats.

Recolonisation of perturbed sediments follows a familiar pattern (Fig. 170): initial colonists will be fast-growing, r-selected species, usually small polychaetes, with high reproductive output and widely dispersing larvae. Total number of individuals may reach a peak early in the succession, but the number of species grows more slowly, reaching a peak after the early successional species have begun to decline and to be gradually displaced by long-lived species towards the K-selected end of the r–K spectrum. The total number of individuals then declines, but biomass rises and all three parameters approach equilibrium. There have been numerous studies monitoring the recolonisation of sediments disturbed by dredging, and a consensus suggested that early-colonising polychaetes were likely to appear 5–10 months after the cessation of dredging activity, and that benthic assemblages, and total biomass would be restored within 2–4 years. However, most of these investigations have focused upon sedimentary habitats that had been subjected to relatively short-term perturbation, usually over a period of just one year, and sand and gravel extraction from the same area over many years might lead to a quite different recovery response once dredging activity ceased.

A site off the coast of southeast England, 900 m by 300 m in area, at 27–35 m depth, licensed for aggregate extraction over a period of 25 years, was monitored at four, five and six years following the end of dredging activity, and showed a pattern of recovery that was in contrast to all previous surveys (Boyd et al., 2005). Grab samples from two areas within the extraction site, one of which

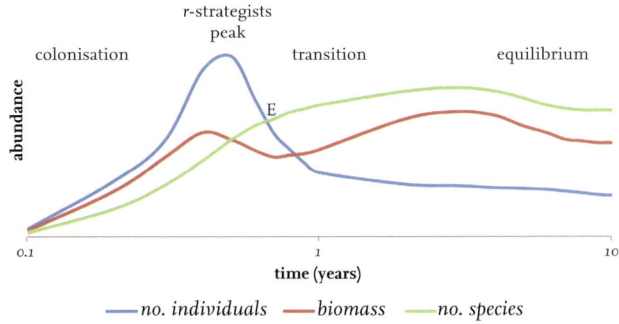

FIG 170. Ecological succession in the recolonisation of marine sediments following disturbance. E = ecotone point, at which r-selected taxa begin to be displaced by K-strategists. (After Newell et al., 1998)

had been subject to high-intensity dredging and the other dredged at much lower intensity, were compared with samples from two reference sites that had not suffered dredging activity. Replicate samples from all four sites showed considerable variability both in sediment particle size frequency, and in the composition of the macrofauna. There was especially large variation in the proportions of sand and gravel among the samples from the high-intensity dredge site. Particle size frequency distributions (page 73) of samples from the low-intensity site were more similar to those from the non-dredged reference sites, but the latter had a higher proportion of silt and clay particles. Thus, dredging activity perhaps increases variability of sediment characteristics through screening, and enhances substratum patchiness. Numbers of macrofaunal invertebrate species, and population densities, were lowest in the samples from the most heavily dredged area, as might be expected, but of greater interest was the composition of the faunal assemblages at each sample area. The sediment at the most intensively dredged area was inhabited by small infaunal species which even six years beyond the end of dredging activity included a high proportion of juveniles, suggesting that continued habitat instability precluded each species from growing to maturity and establishing self-sustaining populations. The undredged reference sites, and the site that had been subjected to only low-intensity dredging, supported communities of individually larger species, both infaunal and epifaunal, and with a broader phyletic range.

Dredging intensity – the duration of each dredging episode in relation to the area dredged – thus seems to be a significant factor determining the rate of recolonisation of defaunated sedimentary habitats, and the composition of the developing fauna. Intensive dredging across a limited area, for a protracted period of time, results in a complete change in the nature of the sedimentary habitat, from coarse sandy gravels to gravelly sands, or from coarse sands to fine sands, which may not achieve physical stability for many years. The benthic communities that ultimately develop in these new habitats will reflect their new sedimentary characteristics, rather than those that pertained prior to dredging. The recovery of benthic communities in intensively dredged areas will only result in the re-establishment of the original communities if the topography and sediment profiles of such areas are returned to their original state. Low-intensity dredging has the least damaging and shortest-term effects: the fauna might be damaged, dispersed or removed, but recolonisation will follow the predicted sequence from r-selected pioneers to K-selected dominants, and the equilibrium community might resemble that which existed prior to dredging. Recolonisation following intensive dredging is likely to result in a change from gravel-associated to fine-sand communities, and perhaps an attenuated succession, and a decrease in diversity.

The physical disturbance caused by benthic fishing gear is much less severe than that arising from aggregate dredging, and while the top layers of sediment may be profoundly disturbed, none of it is removed and its particle size profile is not changed so abruptly. However, demersal fishing, employing trawls and dredges, has been conducted over much of the south and central North Sea, and in the coastal waters of the British Isles and adjacent mainland Europe, for more than a century, with continually increasing intensity. This long-term, low-level pressure, a chronic physical disturbance, has led to fundamental changes in the composition and structure of benthic communities. Sediment structures are destroyed by overturn, together with the burrows and tubes of the infauna. Much of the larger epifauna may be removed, together with the target species, and much more may be damaged or quite destroyed. Deep-burrowing infauna may be relatively unaffected, but species living in the upper layers of the sediment are exposed to predators and scavengers. Despite these dramatic consequences, it has been claimed that soft-sediment communities recover from the effects of demersal fishing quite swiftly, and that there are no obvious long-term effects. A contrary opinion avers that contemporary soft-sediment communities are the result of ecological structuring by a century of dredging and trawling, and that natural, unaffected communities cannot be known.

Each of the three principal types of towed demersal fishing gear has a particular impact on the sea floor, with effects that vary according to the type of substratum. Otter trawls have weighted ground ropes that churn up the sediment as the net is towed, and the two trawl doors, or otter boards, that act to spread the wings of the net widely apart leave deep furrows, around 2 m in width. Beam trawls are equipped with a pair of steel runners, and heavy 'tickler' chains in front of the trawl mouth specifically to drive bottom-dwelling fish and crustaceans out of the sediment and into the path of the net, which may be underlain by a chain mat. Beam trawls are more damaging than otter trawls, but the many types of dredge used to fish for scallops and clams are usually equipped with toothed lower edges designed to rake the sediment to depths of 10 cm or more, causing the greatest damage to the sea floor.

Ecologists began to investigate the effects of trawling on benthic habitats in the North Sea, in particular, in the last few decades of the twentieth century. Apart from the need to monitor population densities and demographic structure of commercial target species, conservation and management policies demand data on the entire benthic community, to assess to what extent species richness, individual species densities, and total individual densities, as well as individual species biomass and total biomass, are affected by towed benthic fishing gear. It is important to discover how benthic communities recover following trawling and dredging, and at what rate, and in assessing and comparing data from similar areas of sea floor it is essential to know the intensity of trawling – frequency, trawl duration and the speed of tow – to which each area has been subjected. While many studies into the effects of fishing disturbance on benthic assemblages have yielded valuable insights, it is often difficult to make valid comparisons between them because the methods, as well as the objectives, are often not consistent. Analysis and interpretation are further hampered by the practical impossibility of obtaining data from pristine habitats that would provide controls for natural, spatial and temporal, variation in community structures. One approach is to compare benthic assemblages from an experimentally trawled area with those from a similar but undisturbed site; another is to compare samples from similar habitats known to have been fished at differing intensities, while others have begun to stress the importance of historical data for understanding the degree and rate of change in shallow-shelf benthic communities as a consequence of benthic fishing disturbance.

Effects of trawling disturbance on the benthos have been investigated by an experimental technique referred to as 'before – after – control – impact', or BACI (Schratzberger et al., 2002). Areas of sea floor with comparable depth and sediment characteristics, and known to have experienced only minor disturbance, are sampled, and the structure of their benthic communities recorded. One area is then subjected to a measured degree of trawling while the other is left as the undisturbed control. Both areas are then sampled at successive intervals, with the objective of comparing seasonal and temporal change in benthic communities, in an area disturbed by a known intensity of trawling and another experiencing only natural perturbation. For example, commercial fishing in Loch Gareloch, southwest Scotland, was restricted from 1967 and banned entirely in 1989. Its enclosed northern end was undisturbed for at least 27 years, and provided an opportunity to examine the effects of otter trawling on a relatively non-impacted muddy benthic sea floor (Tuck et al., 1998). An experimental track was created by trawling for one day, at monthly intervals, for a period of 16 months. The infauna of the trawl track was sampled 5, 10 and 16 months after trawling began, and 6, 12 and 18 months after it ended, and compared with samples from an undisturbed control site sampled at the same frequency. Polychaetes comprised 50% of the 147 infaunal invertebrate species recorded for the two sites, and molluscs 27%. On the trawled track 84% of all individuals were polychaetes and 14% molluscs, while at the control site the figures were 77% and 22% respectively. The relative proportions of polychaetes and molluscs remained consistent, at each site, for the duration of the experiment, but the structure of the infaunal community at the disturbed site had changed significantly after the first five months of disturbance, and the change was maintained through the 18-month post-trawling recovery period. The number of species at the disturbed site was greater than at the control site after the first 16 months, and remained so through the recovery period. Numbers of individuals were higher at the fished site than at the control site at the beginning of the experiment; there was no significant difference between the two sites after five months, but significantly greater numbers at the disturbed site at 10 months, and no significant difference, again, at the end of the 18-month recovery period. However, the three most informative diversity indices (page 62) all showed significant differences between the two sites after just five months, and these differences were sustained until 12 months into the recovery period. Trawling thus reduced infaunal diversity: numbers of species and numbers of individuals may not have differed significantly at the end of the recovery period, but the most common species became more abundant, the least abundant species declined even more, and rare species were especially sensitive.

The impacts of towed fishing gear on infaunal and epifaunal communities vary in relation to the physical characteristics of the trawled ground, and to the extent and frequency of previous disturbance. For, example, a 4 km long experimental track established in Liverpool Bay, trending southeast, at 26 m depth, to northwest, at 34 m, encompassed a transition from rippled, coarse sand to a low-energy, fine-sand habitat (Kaiser & Spencer, 1996). The coarse mobile sediment to the southeast supported a poor fauna characterised by motile, free-living polychaetes, such as species of *Glycera* and *Nephtys*, while the finer, stable sediment to the northwest was occupied by a richer, more abundant fauna, particularly of sedentary tube- and burrow-dwelling species, including maldanid polychaetes and large amphipods such as species of *Ampelisca* and *Urothoe*. Beam trawling had the least effect on the communities of the coarse, mobile sediments, but on fine-sand habitats it led to a loss of rare species and a general decline in diversity (Fig. 171).

Trawling for demersal fish has had especially damaging effects on populations of the Stone Coral, *Lophelia pertusa*, which is found along the entire West Norwegian Shelf edge, at depths of 200–400 m, and is especially abundant between 62° 30' and 65° 30' N (Fosså et al., 2002). Here, rich fish resources had long been exploited using passive techniques, specifically long lines and static gill nets. The development of heavy 'rock-hopper' trawls, able to bounce over hard grounds, and the deployment of more powerful ships enabled bottom

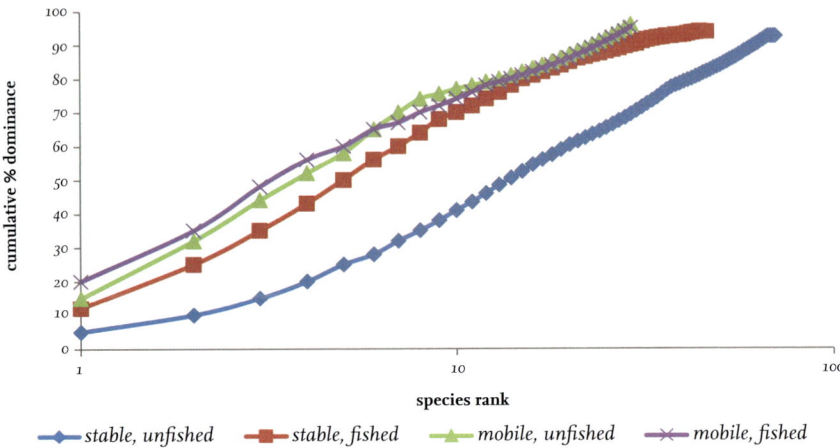

FIG 171. *K*-dominance curves for faunal assemblages in mobile and stable sediments before and after trawling. (After Kaiser & Spencer, 1996)

trawling of *Lophelia* banks for the first time, with catastrophic consequences. By the early 1990s fishermen were reporting steep declines in landings from the area, prompting a survey which showed that as much as 50% of the coral communities had been severely damaged. Long lines and static nets might damage colonies, but the fragments were not dispersed; they provided fresh surfaces for colonisation, and were able to regenerate new growth of coral. Conversely, bottom trawling smashes coral banks and distributes the fragments of coral across the whole track of the trawl, reducing the density of coral colonies, and effectively diluting its high-diversity focus. Frequent trawling presumably leads to continual re-suspension of fine materials, with a resultant smothering effect that suppresses both coral regeneration and recolonisation by associated species. Individual *L. pertusa* corallites grow slowly, with an annual elongation of 5–10 mm. A colony 1.5–2.0 m in diameter may represent several centuries of undisturbed growth, while the huge banks, up to 30 m thick, that were formerly distributed along the West Norwegian Shelf edge may have taken several thousand years to develop. It is likely that the damage to this important foundation species created by benthic fishing gear is effectively permanent, in relation to human timespans.

Bottom fishing for crustaceans and molluscs is as wasteful as trawling for demersal fish, and this is strikingly demonstrated by data for the Clyde Sea fishery for the Norway Lobster, *Nephrops norvegicus* (Bergmann *et al.*, 2001).

This is not a traditional fishery; it developed from the late 1960s to supply an increasing taste for 'Dublin Bay prawns', but by the turn of the century it was the most commercially important shellfish industry in the United Kingdom. Each trawl hauled inevitably contains a bycatch of many other invertebrate species, especially epifaunal echinoderms and decapods, which is discarded overboard. In the Clyde Sea as much as 50–90% of the volume of each haul may be discarded, and 90% of the total individuals discarded are likely to be epifaunal invertebrates. For each kilogram of *N. norvegicus* landed, 9 kg of other benthos is rejected, and it was estimated that the Clyde Sea *Nephrops* fishery was generating 25,000 tonnes of discards annually (Bergmann et al., 2002). Most discarded animals are either damaged or moribund, or dead, by the time they are dumped overboard. The degree of damage to, and the initial mortality rate of, discards are related to such factors as towing speed, water depth and substratum; coarse grounds and vigorous movement of the fishing gear increase the likelihood of the catch being crushed, and of individual animals losing appendages. The duration of the tow and the total volume of material in the net are also significant factors: long hauls and large catches also result in the more delicate organisms being crushed. Further physical damage occurs on deck as the cod end of the net is emptied, and a proportion of the target species, *N. norvegicus*, is likely to be too damaged to be marketable, adding to the discard bulk. Dehydration, hypoxia and temperature have additional inimical effects during the processing of the catch, and even relatively undamaged animals may be dead or dying before they are returned to the sea. Perhaps as much as 25% of this rejected benthos is eaten at the surface or within the water column by seabirds and pelagic fish, while the rest sinks to the bottom, scattered along the course of the moving trawler, across habitats that are possibly quite different from those from which it was taken.

The vulnerability of bycatch species to trawl damage differs according to body size, morphology and functional mode: large individuals are more frequently damaged than small ones; hard-shelled molluscs suffer less than slender, leggy decapods (Fig. 172). Epifaunal echinoderms are especially vulnerable: brittlestars lose arms readily. More than 70% of the smooth-armed *Ophiura ophiura* may suffer severe damage as part of the bycatch, and mortality of damaged individuals returned to the sea is usually total (Bergmann et al., 2001). The stiff-bodied starfish *Astropecten irregularis* is almost as vulnerable, with as much as 50% of a trawled sample showing damage, but the lax-bodied *Asterias rubens* is more resilient, with only 30% of a beam trawl bycatch damaged. Epifaunal decapods, especially swimming crabs (*Liocarcinus* species) and the long-clawed squat lobster *Munida rugosa*, are particularly liable to damage, with around 60% of individuals

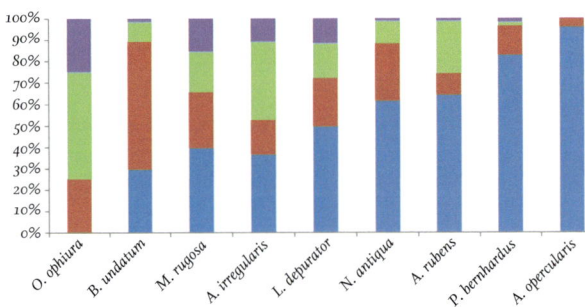

FIG 172. Degree and frequency of damage sustained by invertebrate bycatch of the Clyde *Nephrops* fishery. (Data from Bergmann *et al.*, 2001)

losing one or more limbs. Hard-shelled molluscs, as might be expected, are less susceptible: Queen Scallops, *Aequipecten opercularis*, seem to be largely unaffected by encounters with towed gear, and while high proportions of the whelks *Neptunea antiqua* (50%) and *Buccinum undatum* (64%) may experience damage, this mostly consists of chipped shell aperture lips, which the animals are capable of repairing. A preponderance of echinoderm and decapod discards injured during their transit from benthos to ship's deck and back will eventually die, through increased risk of predation and disease, inability to forage and feed efficiently, and perhaps as a result of the energetic cost of repairing or replacing damaged body parts. Yet it has been noted that the most commonly occurring bycatch species in the Clyde Sea *Nephrops* fishery seem still to be abundant despite the high intensity of trawling in the region (Bergmann *et al.*, 2001), which might suggest that some species benefit from fishing disturbance, as predators and competitors are depleted, while the discards provide additional sources of food for the community. However, regularly fished areas appeared to support a greatly impoverished fauna, and the bycatch from such areas rarely included large epibenthic species, such as the starfish *Luidia ciliaris* and *Marthasterias glacialis*, sea pens and large sea anemones.

Species richness is soon reduced by high-intensity trawling, but the common species will persist so long as the loss of biomass suffered as a consequence of fishing is outweighed by the energetic gain derived from discarded bycatch. Field experiments in the Clyde Sea seem to support this hypothesis (Bergmann *et al.*, 2002), showing that the principal scavenging species were also the largest proportion of the bycatch. Bait was prepared from the most abundant discard species, each in the same proportion as it occurred in the bycatch (Table 11). Thus, the largest proportion of the bait consisted of detached cephalothoraces – 'heads' – of *N. norvegicus*, while undersized *Nephrops*, *Ophiura ophiura*, *Crangon*

TABLE 11. Proportional composition of discard species, characteristic of north Clyde Sea *Nephrops* fishery, in bait used for time-lapse recording of scavenger activity. (From Bergmann et al., 2002.)

Species	Number
Nephrops 'heads'	40
Nephrops, whole animals	10
Munida rugosa	2
Liocarcinus depurator	9
Pagurus bernhardus, without shell	1
Crangon allmani	9
Pandalus sp.	1
Asterias rubens	7
Astropecten irregularis	1
Ophiura ophiura	26
Sepietta oweniana	1
Merlangius merlangus	3
Eutrigla gurnardus	1
Pleuronectes platessa	4
Limanda limanda	1
Hippoglossoides platessoides	1
Glyptocephalus cynoglossus	1
Total	108

allmani, *Liocarcinus depurator* (Fig. 173), *Asterias rubens*, *Munida rugosa*, six small fish species and five other invertebrates comprised successively smaller proportions. The bait was laid by divers and its fate recorded by time-lapse photography. All was consumed, within 28 hours in one experiment and 48 hours in a second, by a succession of scavengers, the most predominant of which were *A. rubens* and the Green Crab, *Carcinus maenas*. The shrimp *Crangon allmani* (Fig. 51) and brachyuran crabs were the first scavengers to arrive at the bait, and the density of decapods peaked six hours after its deployment, but the slow-moving Edible Crab, *Cancer pagurus*, was among the last to arrive.

In the northern part of the Clyde Sea the most abundant discard species were the swimming crab *L. depurator*, the squat lobster *M. rugosa*, the starfish *A. rubens* and the brittlestar *O. ophiura*, and these were presented as individual

FIG 173. The swimming crab *Liocarcinus depurator*. (J. S. Porter)

baits in standard crab pots – creels – to test the preferences of each scavenger species. Each creel incorporated an inner funnel trap that retained the smaller scavengers. The most consistent scavenger was *A. rubens*, occurring in 27–44% of 60 creels deployed for 48 hours, at bimonthly intervals, through a complete year (Fig. 174a). The hermit crab *Pagurus bernhardus* was the next most abundant scavenger species (Fig. 174b), found in 10–27% of the creels, while the three crabs *L. depurator*, *C. maenas* and *Necora puber* together occurred in 16–37% of the traps, and the whelks N. *antiqua* and B. *undatum* in 5–12%. Two amphipods, *Scopelocheirus hopei* and *Orchomene nana*, were abundant in the funnel traps, and through the year comprised 15–43% of the total number of individuals trapped. Crustacean

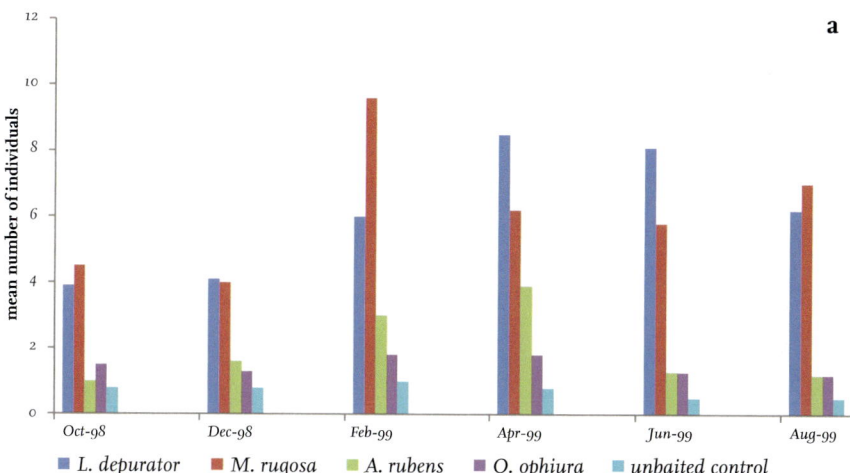

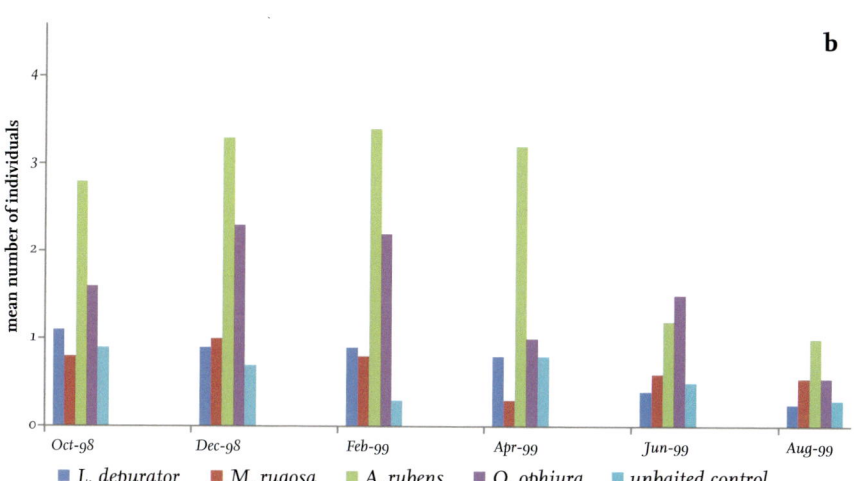

FIG 174. Mean numbers of (a) the starfish *Asterias rubens* and (b) the hermit crab *Pagurus bernhardus* recovered from *Nephrops* creels baited with four different invertebrate discard species, and from unbaited control creels. (Data from Bergmann et al., 2002)

bait was favoured by all scavenger species, with *O. ophiura* a close second, with the exception of *P. bernhardus*, which showed an overwhelming preference for *A. rubens*. In the deeper southern region of the Clyde Sea, undersized *N. norvegicus* and *Nephrops* 'heads' are a large part of the discard, and the experiment was repeated utilising these two bait categories, together with *L. depurator* and *A. rubens*. A single 48-hour deployment of baited creels trapped far fewer scavengers, representing only four species, but *N. norvegicus* was the most abundant of these, favouring *L. depurator* only a little more than undersized conspecifics. *Nephrops norvegicus* populations in the southern Clyde Sea, although dense, consist of small, slow-growing animals, and may be limited by food availability, and fishery discards may thus provide an important source of additional energy.

The question of whether bottom fishing with trawls and dredges results in increased densities of scavenging species has also been addressed experimentally. Discarded fish and invertebrates are likely to be widely strewn from moving trawlers, and further dispersed by currents and waves. They may not necessarily land in habitats similar to those from which they were removed, and are unlikely to support local concentrations – communities – of scavenging species. Yet the passage of towed benthic gear disturbs sediments and leaves a trail of damaged and exposed epifaunal and infaunal organisms, and this non-catch mortality provides a potentially substantial source of food. Scavenging is a facultative trophic mode: apart from various small crustaceans, especially amphipods, that feed entirely on carrion, the principal benthic scavengers tend to be eurytopic species, of both fish and invertebrate, with broad dietary niches, and a tendency to track odour trails towards food sources. *Asterias rubens*, *B. undatum*, *P. bernhardus* and some bottom-feeding fish, including gurnards (*Aspitrigla cuculus* and *Eutrigla gurnardus*), Whiting and Dab, have all been observed moving onto newly disturbed grounds, and feeding on disabled epifauna and infauna, and this response to benthic fishing has been investigated by field experiments off the coast of Anglesey (Ramsay *et al.*, 1996). A 4 m commercial beam trawl was used to create a disturbed track, 30–40 m wide, on an infrequently fished coarse-sand bottom. Numbers of the two hermit crab species, *P. bernhardus* and *P. prideaux*, sampled with a 2.8 m beam trawl, were recorded prior to deploying the commercial trawl, and then daily for the four following days. The small beam trawl was also used to sample two unfished control sites at the same frequency. The response by *P. bernhardus* to trawl disturbance was immediate: prior to trawling, it was present on the experimental track at densities of 4.5 individuals per 1,000 square metres, rising to 12.8 one day after trawling; on day 2 it was recorded at 81.8 individuals per 1,000 square metres, and at 60.5 on day 3. By day 4 densities had fallen again and were not significantly different from the pre-trawling densities, or from those recorded at the control sites, which were 9.7 individuals per 1,000 square metres at the beginning of the experiment, and from 5.0 to 7.6 on subsequent sampling occasions. Following the initial trawling, the size frequency distribution of *P. bernhardus* shifted to the right, on each subsequent sampling day, as larger animals moved onto the experimental track, reflecting perhaps a size-selective competition for the carrion, or simply faster movement by the larger crabs. Gut contents of these animals included higher proportions of crustacean and polychaete material than was found in hermit crabs from the control sites, as they took advantage of newly exposed tube and burrow dwellers, such as the polychaete *Lagis koreni*. *Pagurus prideaux* showed a less positive response to trawl disturbance than *P. bernhardus*. It appeared not to migrate, population densities showed no significant effect of trawling, and it did not change its diet, which consistently included a higher proportion of molluscs and crustaceans, and far fewer polychaetes, than that of *P. bernhardus*. It perhaps occupies a narrower niche than *P. bernhardus*, and is thus less likely to benefit from additional food sources provided by trawl disturbance, or is perhaps excluded by competitive pressure from its more eurytopic congener. Field experiments may fail to reveal any effect of beam trawling on grounds that have been intensively fished over long periods of time, because during this time benthic communities will have been extensively restructured by fishing pressure (Kaiser & Spencer, 1996). The populations of each scavenger species may be severely reduced by high fishing intensity, or there may be such a constant supply of carrion created that scavengers do not need to migrate towards each new source.

Scallop dredges are the most damaging of the principal types of towed benthic fishing gear. The Newhaven dredge, employed by UK scallop dredgers, was especially designed for bottom fishing on rough grounds. It has a span of 77 cm, with the steel edge bearing nine spring-loaded teeth, each 10 cm long and 0.8 cm wide, and spaced 8 cm apart, and is used to dredge for the Great Scallop. A modified version, with slightly shorter, more closely spaced teeth, is designed to catch the smaller Queen Scallop. Both species live on coarse, mixed grounds, bedded into the sediment; the toothed dredge rakes the substratum, dislodging both the quarry species and assorted large objects, including pebbles and cobbles. Large epifaunal animals are likewise forked above the substratum and engulfed by the net. There is considerable potential for substantial damage as target and bycatch species are tumbled together in the dredge bag with rocks and shells. Brittle-bodied animals are at risk of crushing in full dredge bags, or may be smashed through vigorous movement in partly filled bags, and in either case further damage is likely as the dredge contents are discharged onto the trawler's deck.

FIG 175. The Curled Octopus, *Eledone cirrhosa*. (J. S. Porter)

A survey of scallop-fishing grounds around the Isle of Man employing both Great Scallop and Queen Scallop dredges caught a total 87 species of macroinvertebrate and fish (Veale *et al.*, 2001). The number of species recorded in each haul, and the relative abundance of each, varied from site to site around the island, in relation to the type of substratum and the history – intensity and duration – of scallop dredging on each fishing ground, but in all areas the same eight species were predominant among the bycatch, with both types of gear. These were the two whelks *B. undatum* and *N. antiqua*, the Curled Octopus, *Eledone cirrhosa* (Fig. 175), the Edible Crab, and the echinoderms *Asterias rubens*, *Echinus esculentus* and *Spatangus purpureus* (Fig. 176). The two scallop species

FIG 176. The Purple Heart Urchin, *Spatangus purpureus*, a shallow-burrowing species vulnerable to trawl damage. (J. E. Lancaster)

frequently occur in the same habitat, and the Queen was a significant part of the bycatch in Great Scallop dredges. *Spatangus purpureus* was the species most susceptible to damage, with both types of dredge, while the molluscs were the most robust. Diver observations of tow tracks, 20 minutes after the passage of the dredge, revealed the expected trail of crushed and disabled animals, and quantifying this showed that the degree of damage to the larger epifaunal organisms was as great among those left on the bottom as among those swept into the dredge bag. In the case of the crabs, C. *pagurus* and *Liocarcinus* species, the latter was the safer option as many of those individuals escaping the mouth of the dredge were crushed by the steel mesh underlying the bag (Veale et al., 2001). Most depressing was the revelation that despite the huge scale of the damage caused by scallop dredges, in this study the capture efficiency of the target species, the Great Scallop, was just 19%, about midway through the range, 6–41%, reported for other studies.

Scallop fishing in Manx waters began in 1937, with a fleet of just nine small boats, no longer than 10.5 m, each deploying up to three hand-hauled dredges, 1.1–1.8 m wide, and the quarry was the Great Scallop (Bradshaw et al., 2002). After 1950 boats increased in size and number, powered winches were introduced to haul larger dredges, and in 1972 the Newhaven dredge was first used, by when the Queen Scallop had become the second target species. By 1983 the Manx fleet of scallop dredgers topped 70 vessels, up to 25 m long, and was augmented by trawlers from mainland Britain and Ireland. Such a huge increase in fishing effort around the Isle of Man must have led to significant changes in the physical and biological characteristics of the scallop grounds. Sediment profile may change substantially when fishing intensity is high, fine material is winnowed away and large pieces of substratum, especially shell, are crushed and reduced to small fragments. Long-term effects of continuous scallop dredging are very difficult to predict, as usual, because of a lack of undisturbed grounds to serve as experimental controls. However, some insight may be gained from the analysis of historical data, such as the faunal surveys conducted from the Port Erin marine station prior to 1952 (Bradshaw et al., 2002). Comparing species lists from localities off the south end of the Isle of Man, with differing sediment characteristics, and subsequently dredged at differing intensities for up to 60 years, new quantitative samples revealed considerable changes in the benthic communities at all sites. Many taxa had decreased in abundance, or disappeared, including the suspension-feeding brittlestars *Ophiothrix fragilis* and *Ophiopholis aculeata*, erect, sedentary species, such as the hydroids *Lafoea dumosa* and *Sertularella* species, and tufted bryozoans in the genera *Scrupocellaria* and *Crisia*, as well as many encrusting modular taxa. Continual overturning of the substrata, and enhanced turbidity, would have led to the decline of sessile taxa, and their associated fauna, with a consequent decrease in habitat heterogeneity and species richness. Motile, scavenging species had increased in abundance, *Asterias rubens*, *Buccinum undatum*, *Neptunea antiqua* and *Pagurus bernhardus* in particular, together with shrimps, squat lobsters and spider crabs, and the brittlestars *Ophiocomina nigra*, *Ophiura albida* and *Amphiura filiformis*.

In the Manx study, life-history traits were good predictors of probable responses to disturbance: fast-moving epifaunal species and deep burrowers are more likely to survive dredge impacts than sessile, sedentary or shallow-burrowing species. Suspension feeders are most vulnerable, while scavengers benefit from disturbance, and muddy-sand communities suffer less damage than those inhabiting coarse sand and gravel. However, results contrasting with those of the Isle of Man scallop grounds survey were obtained in a study comparing numbers of species and individuals, and total biomass, of bivalves, brittlestars and echinoids from five deep-water sites in the east and central English Channel, and five shallow-water sites in the Gulf of St Malo, sampled in 1998, with similar data for the same precise localities sampled more than 40 years earlier (Kaiser & Spence, 2002). At all ten sites the sediment was coarse, consisting largely of gravel and pebbles, and in the 1950s the benthic communities were characterised by species with individually large body size that might have been predicted to be especially vulnerable to high-intensity bottom fishing. Many more species and individuals were recorded in the later survey, as a result of a far greater sampling effort, and numbers of species and individuals, and total biomass, varied significantly between the ten sites. There was also significant variation through time, as benthic assemblages changed through the 40-year interval between the two surveys, yet the same, individually large species which had characterised each site in the 1950s were still abundant in 1998. The brittlestar *Ophiothrix fragilis* had declined in abundance, though perhaps for causes other than fishing intensity, but the large bivalves *Glycymeris glycymeris*, *Tapes rhomboides*,

Timoclea ovata and *Venus verrucosa* were still dominant, and the small-bodied taxa expected to have increased as a consequence of chronic fishing disturbance since the 1950s were not a significant proportion of the benthic communities. Depth-related environmental gradients may have provided a partial explanation for the persistence of large, vulnerable species, but differences in the type of fishing gear employed, and fishing intensity, are probably also significant, suggesting that even in the otherwise heavily exploited benthic realm of the eastern English Channel it is possible to identify areas that have been relatively lightly impacted by human activity.

The effects of scallop dredges on the structure and ecology of maerl beds have been examined in detail in the Clyde Sea region (Hall-Spencer and Moore, 2000b, 2000c). Collections of maerl made by naturalists off the Clyde Sea island of Great Cumbrae in the late nineteenth century, and preserved in museums, demonstrate how profoundly the expansion of scallop dredging has affected the habitat (Hall-Spencer and Moore, 2000c). In the long term, the effects of regular dredging are seen in a decreased proportion of living maerl in bottom samples and a sharp reduction in the average size of individual thalli. A sample collected in 1885 contained more than 100 thalli up to 58 mm in length, while three samples collected in 1891 consisted of more than 100 thalli up to 38 mm long, 65 up to 50 mm and 18 up to 43 mm, and all had been alive when collected. A core sample obtained during a survey in the period 1995–1997 yielded just 16 live thalli, all less than 20 mm in length (Kamenos *et al.*, 2003). During the latter survey maerl thalli collected from previously undredged areas had a median length of 12.8 mm and a median diameter of 1.7 mm; thalli from dredged beds were similar in diameter but had a median length of just 6 mm. Frequent dredging creates a habitat similar in structure to that of coarse gravel, with low heterogeneity and consequently low species diversity, of which live corallines are just a minor component.

While some of the maerl beds have been dredged regularly and frequently for more than 40 years, there are less accessible areas of relatively undamaged maerl habitat that appear to have suffered little anthropogenic impact, and experimental dredging of an unfished site demonstrated profound short- and long-term impacts (Hall-Spencer and Moore, 2000b, 2000c). The immediate effects of a Newhaven scallop dredge are catastrophic: a modern boat might deploy up to 16 such dredges, and so equipped would rake 6.6 km^2 of sea floor for every 100 hours of fishing. The maerl bed is ploughed to a depth of 10 cm, the live upper layers are overturned and buried, the interlinked coralline thalli are dislocated and the stable lattice of the bed is destroyed, tubes and burrows of the associated community are smashed and the heterogeneity of the habitat is drastically reduced. In the wake of the dredges, sand and silt disinterred from beneath the maerl bed spreads and blankets the community. It has been estimated that up to 70% of the live maerl overturned by the dredge remains buried, and dies. The clear water above undisturbed maerl was found to carry no more than 0.5 g of suspended silt for every square metre of maerl. Dredging liberated 340 g of settled sediment for every square metre of the dredge track, and the water column carried 25 g of suspended sediment per square metre of sea bed at 2 m distant on each side of the track, and 5 g per square metre at 10 m from the track. Grab samples from undredged maerl yielded an average 25% living maerl for every 0.1 m^2 sample, while for dredged and damaged sites the proportion was less than 2%. Dredge tracks in previously unfished maerl beds were still visible four years after the dredging, and recovery of the associated fauna took from six to 30 months, depending upon the physical characteristics, especially depth and current regime, of the experimental site. Populations of scavengers, notably whelks, decapods and flatfish, increased sharply, and very soon, after dredging; mobile epifauna destroyed or displaced by the dredge returned or was renewed fairly quickly, and deeply burrowed mud shrimps, *Upogebia deltaura*, repaired their burrow openings within days. However, the species diversity and the individual species abundance of the maerl fauna were both much lower at sites long exposed to dredging than at unaffected sites; in particular, scallops were younger and smaller, and less frequent, and the nest-building file shell *Limaria hians* was especially badly affected.

    The most important of the commercially exploited bivalve species in northwest European seas, traditionally, have been oysters, scallops, mussels and cockles. Continually increasing demand for esculent molluscs could not be met simply by increasing the harvest of wild populations. The market for oysters and mussels is now largely supplied from cultured stocks, and it is probable that a broader variety of species will be cultivated in the future, including native species, especially scallops, introduced species such as the Manila Clam, *Ruditapes philippinarum*, and even successful invaders such as the American Razor Clam, *Ensis directus*. Management of wild stocks is essential: most Britons' experience of cockles is still of dishes of vinegary morsels consumed at seaside shellfish stalls, but the huge demand from mainland Europe, where they feature in main course cookery, has led to expanding fisheries in tidal waters of the northern nations, producing thousands of tonnes of cockles annually. On the Atlantic and Mediterranean coasts of Europe a wider selection of species has been collected by traditional, artisanal methods, among others venus shells such as *Chamelea gallina*, the wedge shells *Donax trunculus* and *D. vittatus*, trough shells such as *Spisula solida*, and several species of *Ensis*, razor shells. In the last few decades of the twentieth century many such species became the focus of commercial fisheries, regulated by local, national and EU regulations that impose minimum size at landing for each species, catch quotas, close seasons and closed areas, and dictate the numbers of boats licensed to fish. Some of these fisheries have grown so rapidly that wild stocks have been quite depleted; proximate factors are not always clear, but, as well as overexploitation, probably involve habitat degradation, and declining viability of stressed populations. However, the demand for edible species of clam continues to rise, and in comparatively recent times fisheries have developed in northwest Europe for previously unexploited species, to supply burgeoning export markets. Ireland, for example, set a world record for landings of wild-caught *Ensis* species for a brief period, 1998–2000 (Hauton et al., 2007), although razor shells had never figured conspicuously in the traditional Irish diet. Most of these small clam species are infaunal in habit, and some live deeply burrowed within the sediment. A variety of small toothed dredges has been developed for commercial clam fishing, operating on the same principle as the Newhaven scallop dredge, but fishing efficiency has been greatly increased by the introduction of hydraulic dredges. These are rigid steel-framed boxes towed on steel runners, with a chain-mesh collecting bag at the rear end, and a wide blade, in proportions resembling a sharp spade, which overturns the sediment as the dredge is towed. Ahead of the blade the sediment is liquefied and suspended by pressurised jets of seawater, driven by a pump aboard the fishing boat. Small hydraulic dredges, with a blade less than 0.5 m wide, are used in coastal waters, and huge designs, weighing a tonne or more, are deployed in deeper shelf environments. Dredged material is usually mechanically sieved; clams above the legal minimum size are retained on the sieve, and those passing through it are returned to the sea. Cockle beds are harvested by hand raking, at low water, and at high water by suction dredging, a technique similar to that employed in the removal of aggregates – the sediment and its infauna are simply vacuumed up, and the marketable clams removed on board.

Small mechanical clam dredges and hydraulic dredges have the same effects on benthic habitats as the large Newhaven dredges, but at different scales, and varying in degree according to the type of gear and the substratum it is deployed upon. Structure and composition of the sediment are altered, a proportion of the epifauna and infauna is killed, or is damaged to an extent that it dies soon after, and there is an inevitable bycatch, which is likely to suffer differential mortality according to body size, morphology and taxonomic category. However, it is difficult to distinguish the physical and biological effects of clam dredging from those caused by natural environmental disturbance, which may be reflected in frequent short-term fluctuations in the abundance and composition of benthic assemblages. It is often suggested that benthic communities in shallow inshore sedimentary habitats must be adapted to withstand frequent disturbance from winter storms, that clam dredging does no more damage to coastal benthic habitats than that caused by winter storms, and that benthic communities must thus recover just as swiftly as they do from natural perturbation. Contrary opinion holds that frequent hydraulic dredging in shallow waters has damaging consequences for benthic communities, especially in low-energy environments, and suction dredging for cockles is banned on many coastlines, and limited by licence on others. Cockle populations filter fine materials from the water column that accumulate in the beds, as faeces and pseudofaeces, and contribute to those finest fractions of the sediment that are winnowed out by the dredging procedure, and in regularly fished beds there is a significant decrease in the proportion of the smallest particles, and in mean particle size of the sediment. There is no indication that recruitment is adversely affected by suction dredging, as recruits seem to be attracted by qualities of the sediment, while high adult densities may depress numbers of recruits, but damage to juveniles below the legal minimum marketable length, 14 mm, is a significant source of mortality. The associated fauna is also affected, in particular the thin-shelled bivalves *Macoma balthica* and *Tellina tenuis*, which have no commercial value but are important food resources for wading birds and sea duck. The catch quota for commercial cockle fisheries is usually set at around 30% of the biomass of adults (i.e. individuals > 14 mm length), which must be determined each year, and which will vary in relation to sediment type and hydrodynamic conditions.

There is little experimental evidence for the consequences of clam dredging, for the usual reasons that all shallow coastal habitats have been subject to anthropogenic disturbance over long periods of time, and undisturbed natural habitats are largely unknown. Yet, it is predictable that the severity of dredge damage, and the time required for damaged habitats to recover, will reflect habitat heterogeneity and complexity: benthic communities in well-sorted sands seem likely to recover quickly, while seagrasses, maerl beds and other structurally complex habitats suffer long-term damage that is only slowly repaired. A maerl bed in the Clyde Sea, already degraded by beam trawling, was sampled with a hydraulic dredge, with a blade 0.39 m wide that ploughed to a depth of 0.34 m, following pressurised water jets with a combined flow of 320 cubic metres per minute (Hauton *et al.*, 2003). It was towed at a depth of 10 m for 8 minutes, creating a dredge track 127 m long and 1 m wide, and gathering 366 kg of material, 65% of which consisted of largely dead maerl thalli. The sample included 60 species of large, macrofaunal invertebrates, and 94% of the biomass consisted of large infaunal bivalves. Most abundant were *Lutraria angustior, Ensis arcuatus, Clausinella fasciata, Tapes rhomboides* and, outnumbering all of these, *Dosinia exoleta* (Fig. 132), which comprised 84% of the bivalve biomass and 79% of the total biomass of the sample. Sediment composition of the dredge track was changed: plumes of suspended fine sediment were dispersed up to 20 m from the track, and the coarsest fraction – pebbles, maerl thalli and shell – were gathered by the dredge and effectively redistributed as they were thrown back. Dead thalli stained with a fluorescent dye and used to create an artificial maerl bed along the experimental track were collected by the dredge at a rate of 5.2 kg/m^2, or 183,000 individual thalli. Most strikingly, only a minor proportion of the stained maerl was recovered, the bulk being ploughed up and scattered, demonstrating that the most damaging effect of hydraulic dredging on maerl beds is the disruption of its three-dimensional structure, and the breaking, dispersal and burial of the coralline thalli. There is less evidence of damaging consequences of hydraulic dredging on sand, and it has been demonstrated that negative effects on benthic communities of sand habitats could be reduced by modifying the type of gear employed. In particular, large mesh sizes are more effective in reducing the volume of bycatch and the proportion of undersized target species collected (Fig. 177), regardless of the spacing between the dredge teeth (Gaspar *et al.*, 1999), while rigid collecting boxes allow a greater proportion to escape and reduce the risk of crushing when the catch is hauled (Gaspar *et al.*, 2003).

However, confidence that shallow-water dredging for clams has little or no adverse effect on sand communities may be misplaced. Artisanal harvesting of small bivalve species on the southern coast of Portugal developed from 1969 into a licensed fishery that deployed mechanical dredges from small inshore vessels, exploiting populations of *Spisula solida, Donax trunculus, Chamelea striatula, Pharus legumen* and *Ensis siliqua* (Chícharo *et al.*, 2002). Three decades later, ecological consequences of the fishery could not be distinguished from natural fluctuations in benthic community structure, which would also have varied in relation to depth and sediment grade. But comparison of infaunal communities in dredged areas with those in areas that had been closed to clam dredging four years earlier showed that macrofaunal abundance and biomass were significantly lower in the dredged area, in which populations of large K-selected species, the commercially important clams, had been depleted, and replaced by communities dominated by small, r-selected species, principally small polychaetes.

Even more striking, and focusing for a while longer on southern Europe, is the history of the Adriatic fishery for the Mediterranean Striped Venus, *Chamelea*

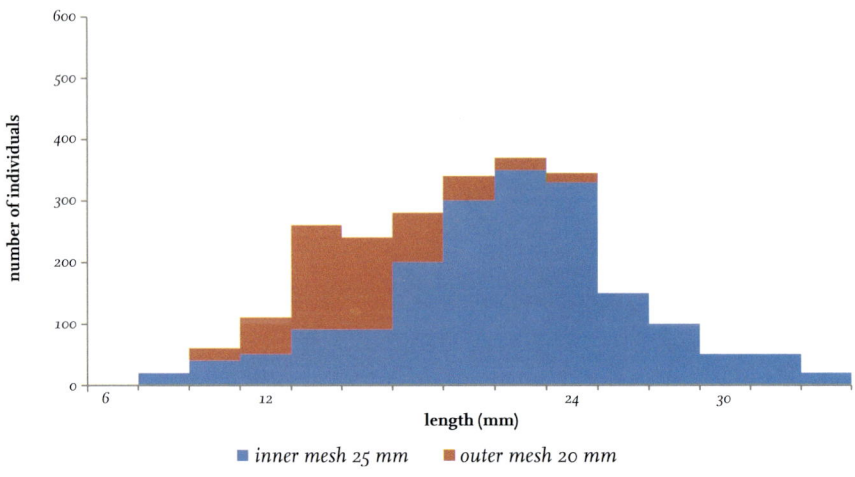

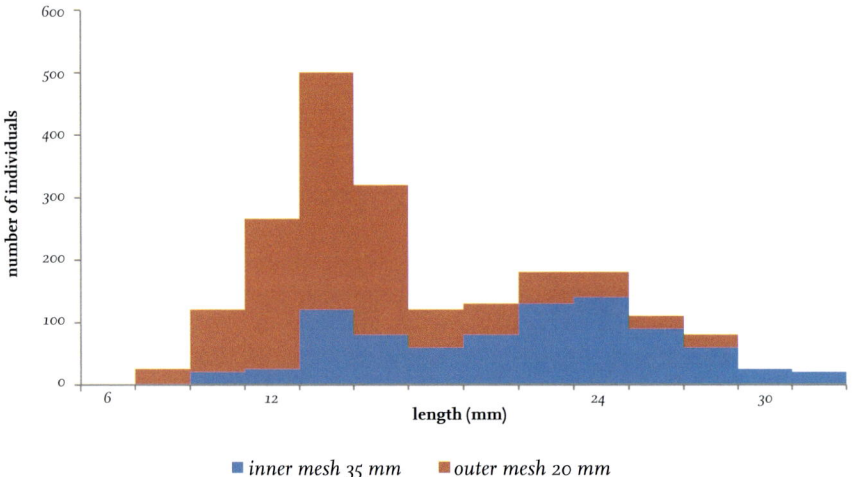

ABOVE AND OPPOSITE: **FIG 177**. Size frequency distributions of *Chamelea gallina* sampled with three experimental mesh sizes : 25, 35 and 50 mm. The catch retained by an outer net with 20 mm mesh represents the potential bycatch of unmarketable clams below the 25 mm minimum legal size. Fewer of these small clams are retained by the coarsest commercial nets. (After Gaspar *et al.*, 1999)

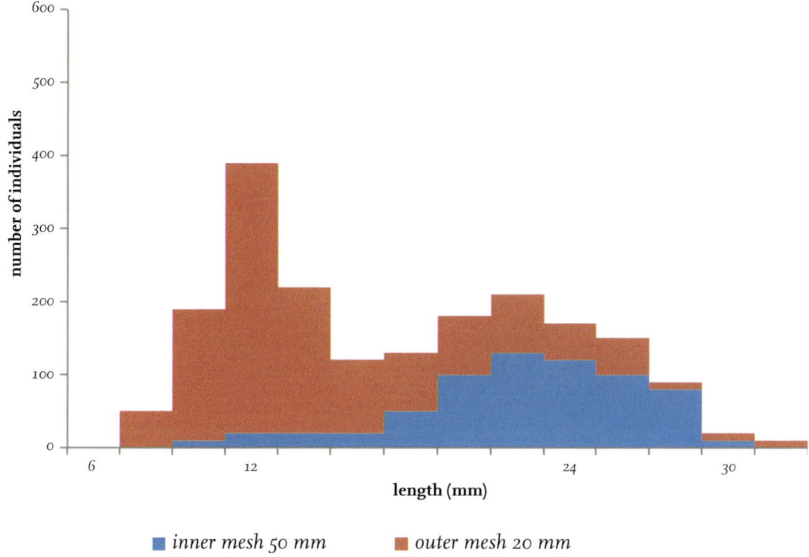

gallina. This fishery began in the early 1970s, using hydraulic dredging methods, and by 1974 was landing 80,000 tonnes of clams each year (Romanelli *et al.*, 2009). Landings increased to a record 108,000 tonnes in 1984, after which the fishery began a sustained decline, decreasing by about 3,500 tonnes each successive year, to a low of 12,000 tonnes in 2002, levelling off at 21,000 tonnes in 2005. Such a steady decrease, continuous through two decades, remains difficult to explain. In the northwest Adriatic *Chamelea gallina* inhabits fine, well-sorted sands, with low organic content, within a depth range of 3–12 m. In contrast to many north Mediterranean coastal regions, this is not an oligotrophic environment; the outflow of the Po river system provides abundant organic nutrients, boosting primary productivity, which in turn supports huge populations of suspension-feeding bivalves, among which *C. gallina* is predominant. Densities of marketable clams, longer than 25 mm, exceed 70 individuals per square metre, and the fishery was considered commercially viable, with catches of 12.5 kg per 1,000 square metres of sea floor dredged. The decline in catch was accompanied by a decrease in the mean length of clams, within each marketable size class, and a drop in mean individual body weight, and the mean biomass of exploited populations fell from 20 g/m² to 5–10 g/m² in just ten years.

Overexploitation may have been partly responsible for the initial decline in
C. gallina populations, but measures to regulate and limit the fishery failed to
halt it. Hydraulic dredging for C. gallina had been subject to regulation since its
inception: minimum size at landing was set at 25 mm, so that marketable clams
would be at least two years old, and would have reproduced at least once (in the
northern Adriatic the species lives for little more than three years). Catch quotas
were imposed from 1979, and the number of licensed vessels was reduced from
1993. Environmental factors were certainly partly responsible for decreasing
exploitable populations of C. gallina: mass mortalities in some summers were
caused by prolonged benthic hypoxia, and others might have been related to
fluctuations in the Po river outflow, resulting in lower nutrient levels and a drop
in primary production. Changes in the composition of the benthic communities
might also have been a factor: predator numbers perhaps increased as frequent
dredging exposed the clams and increased their vulnerability, in particular to
naticid gastropods – necklace shells – that specialise on bivalve prey. However,
a significant pressure on C. gallina populations may have been the physical
damage and physiological stress inflicted on the non-marketable component by
hydraulic dredging (Marin et al., 2003). This was tested through comparison of
the physiological condition of two samples of C. gallina collected by hydraulic
dredging, one subjected to commercial fishing techniques, employing high-
pressure water jets followed by mechanical sieving, the other using much
lower water pressure and manual sorting. The two samples were maintained in
laboratory aquaria, and filtration and respiration rates were measured for each
individual. These data allow calculation of the scope for growth, the difference
between the energy derived from filtered food and the energy expended in
respiration: a high value indicates a healthy surplus available for somatic
growth. Concentrations of free phagocytic cells, haemocytes, in the blood,
haemolymph, of the clams were also measured as indicators of the health of
their immune systems.

Filtration rate was greater in summer samples than in those collected in
winter, for both groups of clams, but in both seasons it was significantly lower
in clams collected with high-power hydraulics and mechanically sieved, and
which also had significantly higher respiration rates and thus low scope for
growth values. Lower concentrations of haemocytes in the haemolymph of
the latter group was considered a further indication of physiological stress.
Undersized individuals of C. gallina may have suffered some physical damage,
more frequent among those collected by high-power dredging and mechanical
sieving (28.87%) than those less roughly handled (14.8%), and such damage may
impair burrowing ability and increase their vulnerability to predation, but these

effects are likely to be compounded by severe physiological stress. This leads to reduced survival time and increased mortality among clams exposed to air, especially marked at high summer temperatures. In some areas of the Adriatic clam beds are dredged almost daily through the open season, and the stress inflicted on undersized animals must have cumulative effects on growth and reproduction, and ultimately on the viability of the whole population. The most damaging attribute of the hydraulic dredge, perhaps not fully acknowledged, is its efficiency. At optimum towing speeds hydraulic dredging of Adriatic *C. gallina* was shown to be 100% efficient, removing all clams above the legal minimum length in the area dredged; this probably represents the greater proportion of the adult breeding stock, and continued dredging at such efficiency would seem to threaten apocalyptic consequences for the clam population. It must also affect population structure through physical damage and physiological stress, and increased mortality, inflicted on undersized individuals, and the loss of an unknowable proportion of new recruits. Close seasons might reduce the mortality rates of recruits and juveniles, but reproductive cycles of bivalves are cued by temperature, and the timing of spawning, settlement and recruitment is likely to vary between years, and may extend beyond the end of any close season. Bivalve species typically show great variation in recruitment across their latitudinal ranges, with some species often failing to recruit for several years in succession, and so year class strengths may also vary markedly.

Razor clams are now fished widely on northwest European coasts, but while landings, and exports, continue to rise, there is very little information on the sustainability of these fisheries. The elongate, thin shells of all *Ensis* species are brittle and readily damaged, and it is likely that a proportion of each catch is unmarketable. In the Clyde Sea two species, *Ensis siliqua* (Fig. 178a) and *E. arcuatus* (Fig. 178b), are the focus of expanding commercial interest, although the use of hydraulic and suction dredges is still limited in Scottish waters. Experimental sampling of a mixed population of the two species, living at 4–6 m depth in coarse sand off Hunterston, and of a population of *E. siliqua* in fine sands, 5–6 m deep, in Irvine Bay emphasised the efficiency of hydraulic dredging (Hauton *et al.*, 2007). Within an effective fishing track 0.45 m wide a single haul could collect 90% of the *Ensis* biomass, representing almost 100% of the population. The dredge collected individuals as small as 5 cm long, half the EU legal minimum landing size. The densities of the two species were low: the biomass of *E. siliqua* at Hunterston was 139.5 g AFDW per 10 m^2 with an annual production of 18.08 g AFDW per 10 m^2; *E. arcuatus* biomass was 192.11 g AFDW per 10 m^2 and had an even smaller annual production of just 1.82 g AFDW per 10 m^2. The production/biomass (P/B) ratio was thus very small, for both species, at 0.11 for *E. siliqua*

FIG 178. The razor clams (a) *Ensis siliqua* and (b) *Ensis arcuatus*. Scale bar: 5 cm.

and 0.01 for *E. arcuatus*. It must be stressed that these are *K*-selected species, long-lived and probably very slow-growing; beyond two years they cannot be accurately aged in the hand, and larger size classes comprise individuals of different ages. In the Irish sea *E. siliqua* requires three years of slow growth to reach the minimum landing size, and off the southwest Scottish coast it is likely to take even longer, and it is not certain that Clyde Sea populations reach sexual maturity at that length.

At present, it seems that northwest European clam fisheries, for at least some species, are simply depleting the standing stock of each population, rather than harvesting sustainably a surplus annual production, and presumably once a population, or species, is reduced to a level at which it is no longer commercially viable to fish, then the fishery moves on to the next population, or species. *Dosinia exoleta* comprised 84% of the bivalve biomass in a degraded maerl bed (Hauton *et al.*, 2003), with a density and mean individual size that made it of potential commercial interest. Apart from the physical impacts of intensive clam dredging, the removal of the larger part of the population of so many species of large suspension-feeding bivalves must lead eventually to profound changes in the structure and function of benthic infaunal assemblages, with further indirect effects on epifaunal and demersal communities. One small bivalve tastes very much like another, and mostly of very little, and it must be wondered, when oysters, mussels, and probably scallops, can be so easily cultivated, why they need be fished at all.

CHANGING REGIMES

Towed benthic fishing gear is creating catastrophic damage to shelf environments throughout most of the world's seas. Initial impacts result in increased mortality, and eventual elimination of large, long-lived epifaunal and infaunal organisms, with further severe effects dependent upon the nature of the benthic communities impacted, the type of gear employed, and fishing frequency. Sediment is flattened and smoothed by the passage of otter trawls over sandy grounds, or churned up to a depth of several centimetres by beam trawls, and overturned to a greater depth by toothed dredges. The structure of the sediment is thereby broken down, biological and physical characteristics are altered and the chemical environment is changed, the degree of change depending upon sediment grade and hydrodynamic regime. The overturn of fine muddy sands is likely to carry organic material to the surface, and to re-oxygenate deeper levels, perhaps with positive effects on production and carbon cycling by some components of the benthic community. In certain cases chronic fishing disturbance has been shown to change particle size distributions in sandy environments, and selective removal of elements of the fauna, naturally, changes the composition of the fauna. Mean size, and thus age, of target species decreases, relative abundances change, and defaunated patches of habitat may be colonised by a suite of species different from the original occupants. Changes in faunal composition, including both demographic and taxonomic characteristics, must be reflected in changing interactions, especially with regard to trophic relationships, within the benthos, and between benthic and demersal assemblages. For example, on frequently fished grounds, populations of large bivalves and shallow-burrowing heart urchins are replaced by assemblages of smaller bivalves and small polychaetes (Jennings *et al.*, 2001a); these are the favoured prey of flatfish species such as Sole and Plaice, which accordingly increase in abundance. As benthic assemblages change in response to fishing disturbance, so fish assemblages may also change, and all benthic and demersal assemblages are affected by changes in predation and competition pressures. These consequences of benthic fishing employing towed gear pose some very interesting questions. In particular, does it result in changing rates of production, and do any such changes result in new stable communities?

Steam-powered trawlers began to replace sailed vessels in the last decades of the nineteenth century, and by the middle of the twentieth century these had been supplanted by the larger, diesel-engined boats that powered the expansion of the European fishing fleets through the next 50 years. Total annual fish landings provide a useful proxy for demonstrating the increasing fishing effort of the seven European nations exploiting North Sea demersal fish stocks (Callaway *et al.*, 2007). In 1906 some 250,000 tonnes of Cod, Haddock and Whiting – 'roundfish' – were landed, and 50,000 tonnes of flatfish: Plaice, Sole, Turbot, *Scophthalmus maximus*, and Brill, *Scophthalmus rhombus*. These together with the three roundfish species account for 80–84% of the total European catch of demersal fish. Landings of both categories rose steadily through the century (Fig. 179), with roundfish peaking at around one million tonnes by 1970, and then declining to below the 1906 level by 2000. Flatfish catches peaked by 1990 and dropped through the 1990s, but by 2000 had begun to rise again.

The development and expansion of benthic fishing techniques in the North Sea would have initiated changes in the composition, structure and function of benthic communities from their inception. All anthropogenic pressure on the benthic environment may be expected to increase through the present century, and it is essential that the health and status of benthic habitats are continually evaluated in order to maintain sustainable management policies and appropriate conservation measures. However, a century of fishing disturbance, and a lack of pristine habitat with which to make comparison, precludes a simple assessment of the degree of damage inflicted by current benthic fishing practices, although, as the Manx scallop ground survey showed (page 334), historical data may be helpful. The North Sea has been the most intensively fished area of northwest European waters, and the test ground for developing benthic fishing methods.

It has also been the subject of many faunal surveys through much of the last century, and the results of these surveys, as specimens or data records, are to be found in national scientific institutions. Material collected between 1902 and

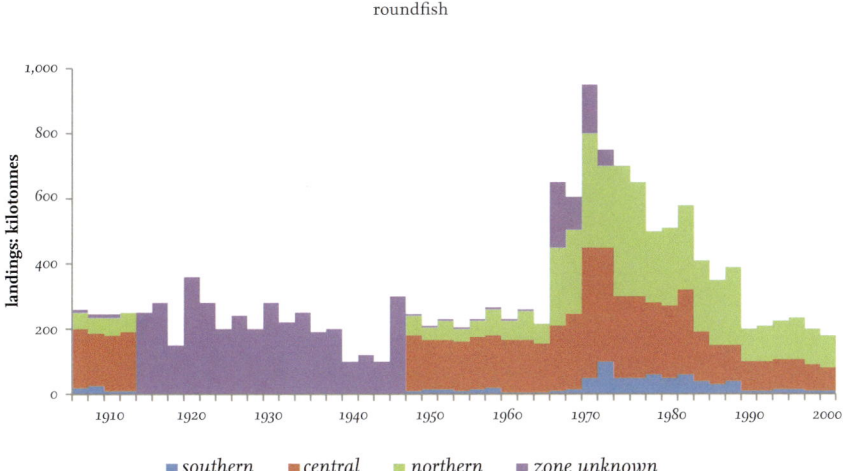

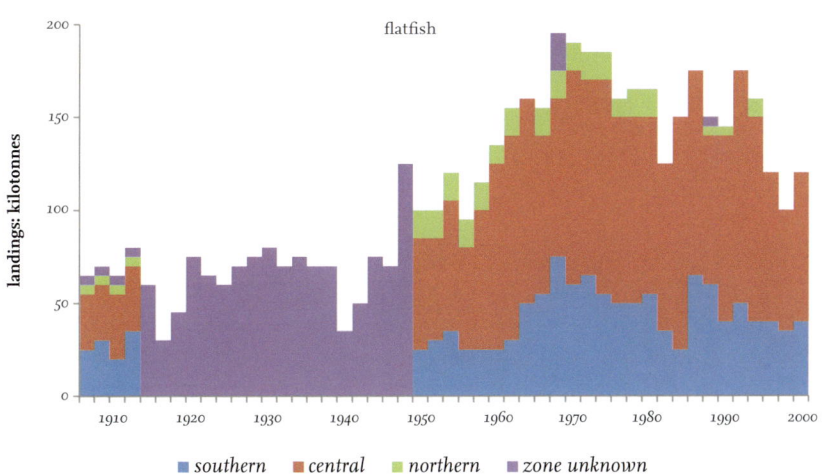

FIG 179. Fish landings from the southern (< 53.5° N), central (< 57.5° N) and northern (< 62° N) regions of the North Sea, 1906–2000; regions were not recorded during the war and inter-war years. (After Callaway et al., 2007)

1912, in the course of epibenthic surveys by marine biological stations at Kiel and on Heligoland, is stored in Kiel Zoological Museum, and comparison with the results of a benthic survey conducted in 1986 revealed significant changes in the composition of epibenthic communities, and the frequency of occurrence of particular species, between the two periods (Rumohr & Kujawski, 2000). Some species had decreased in abundance, others had increased, while some species that occurred commonly in the earlier surveys were not recorded at all in 1986. The brittlestar *Amphiura filiformis* was present at almost all of 56 stations sampled in 1902–1912, but at only 5% of 40 stations sampled in 1986; the two starfish *Leptasterias muelleri* and *Henricia sanguinolenta* occurred at 40% and 20%, respectively, of stations sampled in the first surveys, but were absent from all the 1986 samples. Of three whelk species, *Neptunea antiqua* showed no change in occurrence, *Buccinum undatum* was much more frequent in 1986 than previously, and *Colus gracilis*, which had not been recorded in the first surveys, was found at almost 80% of the 1986 stations. The most striking changes were found among the bivalve data (Fig. 180). While three species collected in 1986 were previously unrecorded, 11 species present in the earlier surveys, mostly at fewer than 40% of the stations, were not found again in 1986, and most of the rest showed sharply decreased occurrence, some, such as *Arctica islandica*, occurring frequently only at the deeper stations. These differences in occurrence of a selection of epifaunal animals over an interval of 74–84 years cannot be unequivocally attributed to

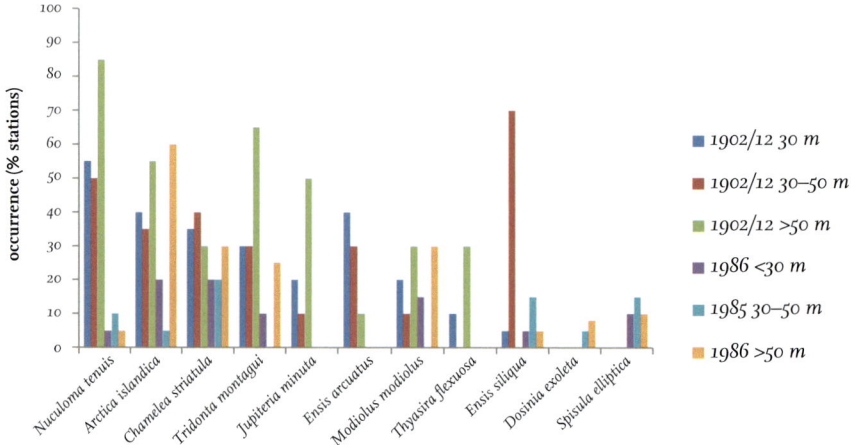

FIG 180. Changes in the recorded occurrence of selected bivalve species at different depth zones in the North Sea. (Examples from Rumohr & Kujawski, 2000)

benthic fishing pressure, although it is noteworthy that bivalve communities showed the greatest change, perhaps being the least motile and the most vulnerable to physical damage.

The 1902–1912 specimen records have been compared subsequently with data from 1982 to 1985, and from 2000, when epibenthic species were recorded in the course of demersal fish surveys (Callaway et al., 2007) (page 00). Species records exist from all three time periods for 40 ICES rectangles: two lie in the southern sector of the North Sea, south of 53° 50' N, 24 in the central sector, south of 57° 50' N, and the rest in the northern sector, to 62° N. From these 40 ICES rectangles the early surveys recorded 80 epibenthic species, but sampling techniques were not standardised and there are no quantitative data; the two latter surveys recorded 66 (1982–1985) and 146 (2000) species from standardised beam trawl hauls, but it could not be known whether species present in these surveys but not in the first had actually been absent, or simply not recognised or not recorded. With these constraints, only those species present in all three surveys could be used in any analysis, and the only valid biodiversity measure was average taxonomic distinctness (D^*) (page 67). Distributions of species had changed with time. For 14 species the number of ICES rectangles in which they occurred in 1902–1912 had dropped by more than 50% in the later surveys; these included seven large echinoderm species and four large bivalve species, all considered especially vulnerable to trawling disturbance. Twelve species had increased their spatial distribution and by 2000 were present in as many as 2–4 times the number of rectangles in which they were found in the first surveys; perhaps significantly, these included a couple of 'mud urchins', the Sea Mouse, *Aphrodita aculeata*, and the pelican's foot gastropod, *Aporrhais* species, and three peripatetic decapod scavengers. Four species showed no change in spatial occurrence, notably the starfish *Asterias rubens* and *Luidia sarsi*, and for 11 others there was no clear trend, with spatial pattern fluctuating through time. Mean average D^* did not show any significant difference between the first survey period and the latter two, but it was significantly lower in 2000 than for the period 1982–1985. Benthic fishing activity in the 40 ICES rectangles examined has had an effect on benthic community structure, but much of the change had occurred prior to 1902, and spatial distribution changes had mostly occurred prior to 1980. Also, the significant difference in average D^* between 1982–1985 and 2000 might be attributable to factors other than fishing, such as climate change and increasing eutrophication.

Fish landings may be indicators of total annual fishing effort for the North Sea and other regions of the northwestern shelf, but they provide no information on the intensity of benthic fishing in relation to location and size of the main fishing grounds. Fishing intensity is patchily distributed: some grounds may be trawled just once or twice each year and others on an almost monthly basis. Intensity will also depend upon the size of the vessel, the type of gear employed, and, in the case of beam trawlers, number and size of trawls deployed. Aircraft have been used by the United Kingdom fishery protection agency to patrol the coastal waters of the British Isles, recording sightings of trawlers, and in the central and southern North Sea the four quarters of each ICES rectangle were each overflown about 100 times per year (Jennings et al., 2001a). The data collected may be expressed as sightings (of beam/otter trawlers) per unit effort (i.e. flight) – SPUE – and provide a measure of fishing intensity for each quarter rectangle.

The effects of trawling disturbance on epifaunal and infaunal communities, in relation to fishing intensity, was explored in a comparison of two fishing grounds in the central North Sea, between 54° and 55° N, and 0° to 3° E, referred to as the Hills and the Silver Pit (Jennings et al., 2001a). The Hills extends from approximately 100–200 km off the Yorkshire coast, with depths of 40–60 m and a substratum of fine sand; the Silver Pit, further to the east, has depths of 60–80 m, and fine muddy sands. The physical environments of the two grounds are thus similar, but the Silver Pit is subject to a far greater fishing intensity than the Hills. Sightings of trawlers were recorded between 1994 and 1998 for 103 km^2 (= 5 × 6 nautical miles, nm) blocks within each of the ICES quarter-rectangles encompassing the two grounds, and proportional fishing intensity, in terms of SPUE, calculated for 3.43 km^2 (1 nm^2) sample sites randomly positioned in the larger blocks. There was a 10-fold range in fishing intensity across the Hills grounds, and a 27-fold range at the Silver Pit, where the most intensively trawled site was trawled almost three times as often as the most frequently disturbed Hills site. At the Silver Pit sites the total biomass, of both epifauna and infauna, decreased significantly with increasing fishing intensity, while there were no significant differences in total biomass among the sample sites on the less disturbed Hills grounds (Fig. 181). The biomass of bivalves and heart urchins decreased significantly at higher SPUE measures among the Silver Pit sites, but polychaete biomass was unaffected; across the Hills sites polychaete biomass increased at higher levels of disturbance, a result possibly consistent with the 'intermediate disturbance' effect, but on both grounds bivalve spat decreased, as a proportion of total biomass, with increasing fishing disturbance. Intensive trawling pressure was thus shown to lead to a decline in total epifaunal and infaunal biomass and a change in the composition of the infauna, as large, long-lived organisms were replaced by small polychaetes. The total polychaete biomass did not change, but it comprised an increasing proportion of the total biomass per sample as bivalves and heart urchin biomass declined. In some other studies in the southern North sea, by contrast, similar changes in faunal composition in response to increasing trawling frequency were found to be accompanied by large increases in total polychaete biomass, although this may have been a consequence of factors other than trawling disturbance, such as rising nutrient levels and increasing primary production.

Fishing effort by beam trawlers is very patchily distributed, dependent initially on depth and substratum, which, obviously, also determine the distribution of the target species and their associated communities, and on the locations of obstacles, such as wrecks and rigs, and on the tracks of main shipping lanes. Broad-scale patterns of beam-trawling effort in the North Sea thus reflect the distribution of habitats and areas suitable for beam trawling (Callaway *et al.*, 2002), and densities of benthic megafaunal species, and mortality rates attributable to beam trawling, are likely to vary at scales far smaller than that of the standard 30 × 30 nm ICES rectangle. While they are fishing, locations of EU-registered trawlers are now continually recorded by automatic GPS technology with an accuracy of 180 m (Piet *et al.*, 2000), and the data allow fishing effort to be calculated at a resolution of 1 square nautical mile (nm). In the Dutch sector of the southern North Sea the abundance of 21 megafaunal species, both infauna and epifauna, showed less variation at the 1 nm^2 scale than at the scale of the ICES rectangles (Piet *et al.*, 2000). Direct mortality estimates were on average higher when calculated in relation to substratum and depth than according to the ICES rectangles, and in both cases differed according to the resolution at which fishing effort was recorded, reflecting the patchy distribution of both bottom benthic assemblages and fishing effort. For 20 species mortality estimates were lowest when calculated in relation to depth and sediment, and for fishing-effort data at the 1 nm^2 scale; for one species, *Spisula subtruncata*, mortality estimates were least when based on sediment/depth data. This region of the North Sea was the first to be subjected to offshore trawling, and it is probable that the invertebrate megafauna present comprises those species which by virtue of lifestyle and/or reproductive mode have proved resilient through 150 years of benthic fishing. An area closed to fishing for two decades, around a gas rig at 53° 35' and 4° 05' E, at 30–40 m depth off the Dutch coast, was shown to have a far richer benthic community than sample sites situated 1.5 nm along each compass point from the rig and subject to regular bottom fishing (Duineveld *et al.*, 2007). The infauna of the closed area included a great abundance of mud shrimps, *Upogebia deltaura* and *Callianassa subterranea*, as well as large bivalve species such as *Thracia convexa*, *Arctica islandica*, *Acanthocardia echinata* and *Dosinia lupinus*, all especially vulnerable to trawling disturbance, and smaller, thin-shelled species, including *Abra nitida* and *Cultellus pellucidus*.

354 · SHALLOW SEAS

The large epifaunal and infaunal animals especially vulnerable to benthic fishing gear are long-lived and slow-growing organisms, with the usual correlated *K*-selected attributes: low reproductive output, low recruitment and often limited capacity for dispersal. Populations have a long turnover time, and chronic fishing disturbance eventually eliminates such species from the

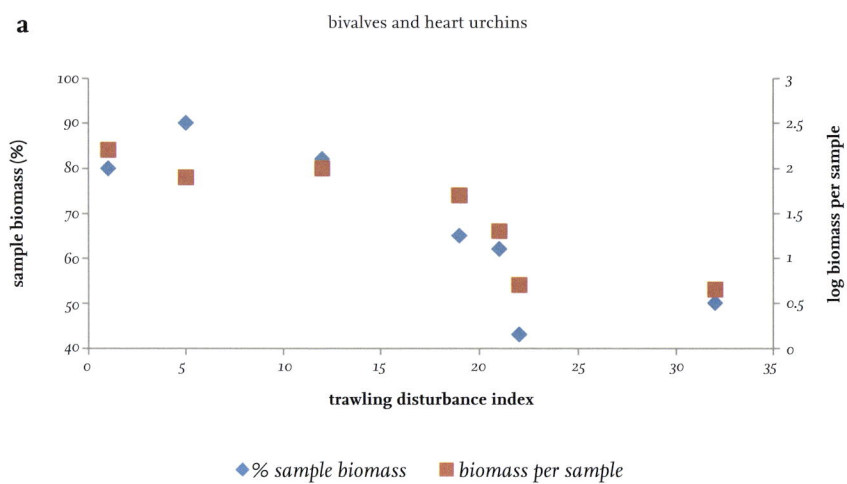

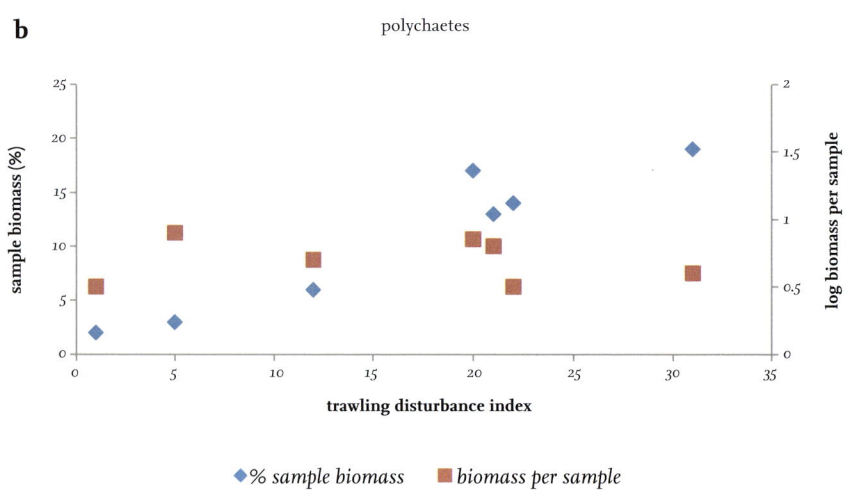

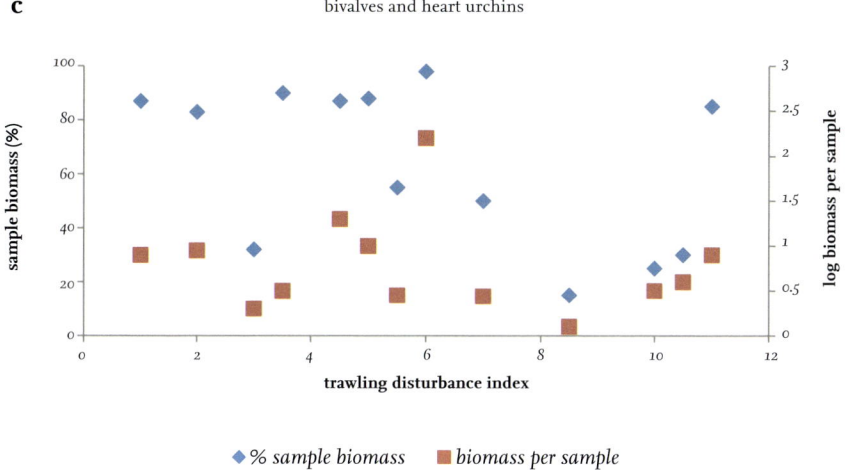

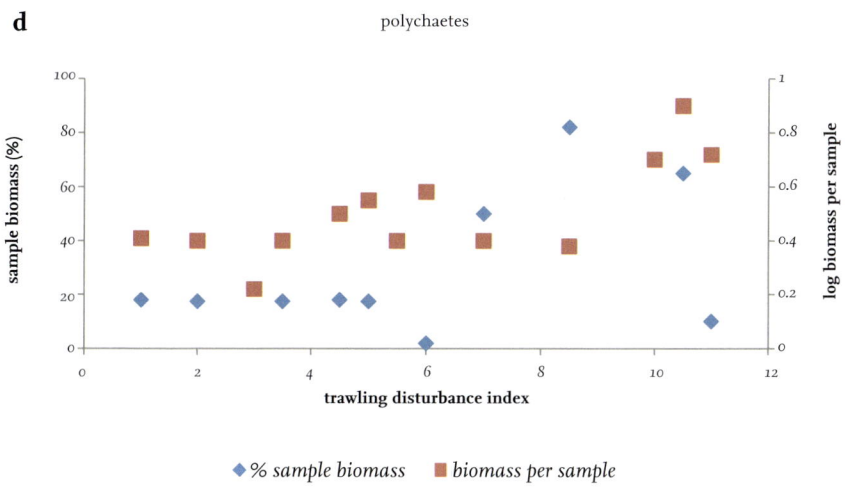

OPPOSITE AND ABOVE: **FIG 181.** Mean sample biomass and mean percent biomass of infaunal bivalves + heart urchins and polychaetes, in relation to trawling disturbance, in two areas of the North Sea: (a, b) the intensively fished Silver Pit, and (c, d) the little-impacted Hills grounds. (After Jennings et al., 2001a)

bottom community. Small-bodied animals are less vulnerable, and the degree of damage and overall mortality they are subject to may be sufficiently low that populations can withstand a greater frequency and severity of disturbance than larger species. Some small-bodied species may be relatively long-lived, with a slow generation time, but most are likely to display r-selected traits: fast growth, short life cycles, large reproductive output and significant dispersal capability. Once disturbed, a benthic habitat may be rapidly recolonised by r-selected species from adjacent communities, and frequent low-intensity disturbance may even favour such species, restricting the development of an equilibrium community, maintaining it in a constant early-successional state. The impact of benthic fishing gear has differential effects on the functional groups and habitat engineers present in each benthic community, and on the trophic structure of the community. Large-bodied bivalves and tube-building polychaetes are suspension feeders, the bulky mud shrimps, *Callianassa subterranea* and species of *Upogebia*, are variously suspension feeders and deposit feeders, while the larger, motile species of epifauna are generally predatory or scavenging in habit. Removal of these functional groups increases the quantity of suspended and sedimented food available to smaller animals, and reduces competitive pressures accordingly, and also results in a decrease in mortality from predation.

Along the sandy coast of the Algarve clam dredging, principally for *Spisula solida*, *Donax trunculus*, *Chamelea gallina* and *Ensis siliqua*, changes the trophic structure of macrofaunal benthic communities (Gaspar et al., 2009). At relatively undisturbed sites, unexploited for at least ten years, suspension-feeding animals are the dominant functional group, with deposit feeders also constituting a moderate proportion of the community (Fig. 182). At sites subject to moderate exploitation, where less than 12.5×10^6 m^2 of sea floor is dredged each year, suspension feeders decrease sharply and the number of carnivores increases, but deposit feeders are the dominant functional group. At heavily exploited sites, with more than 12.5×10^6 m^2 dredged annually, carnivory is dominant, perhaps reflecting immigration of scavengers into continually disturbed grounds, and deposit feeders, presumably being much of the prey, are significantly reduced. However, differences between sites will also reflect differences in sediment grade and mobility, as well as dredging frequency, and also sporadic recruitment events. For example, the suspension-feeding Lancelet, *Branchiostoma lanceolatum*, is adapted to mobile, medium-grade sands and was among the top most abundant species at some of the highly exploited sites.

As the proportions of each functional group comprising a benthic community change as a consequence of bottom-fishing disturbance, so biochemical

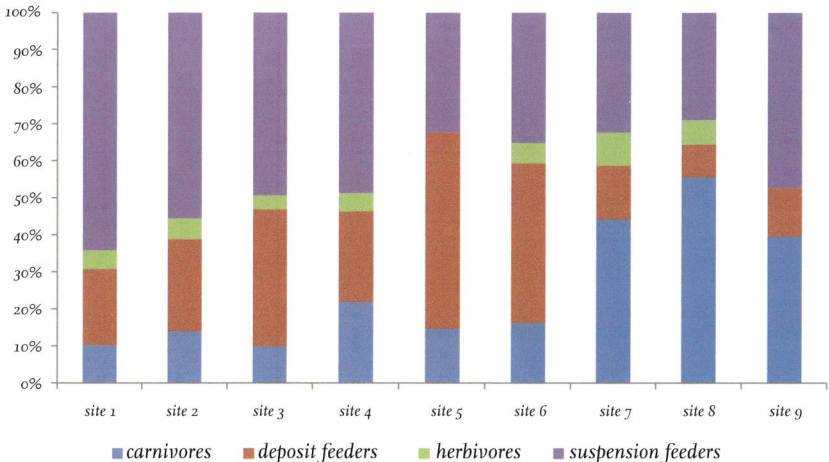

FIG 182. Trophic group analysis of soft-sediment macrobenthic communities at nine sites on the Algarve coast with contrasting degrees of disturbance by clam dredging. Sites 1–3, unexploited; sites 4–6, moderately exploited; sites 7–9, heavily exploited. (After Gaspar et al., 2009)

processes within the sediment will also change, in relation to the type of gear employed, sediment type, depth and fishing intensity (Allen & Clarke, 2007). Computer simulation based on more than 100 studies into the effects of trawling disturbance was employed to predict responses to trawling frequencies of twice and five times a year at two model North Sea sites, one at 85 m depth with a seasonally stratified water column, and a shallower site, 50 m, with a permanently mixed column. Beam trawling was shown to have the most destructive effect, and at the deep-water site twice-yearly trawling, for a five-year period, removed 88% of suspension-feeder and 67% of deposit-feeder biomass. After a following five-year period without trawling, the deposit-feeding community had recovered but the suspension-feeder biomass was still 40% lower than for untrawled control sites. At the unstratified shallow-water site suspension feeders were practically eliminated by five years of trawling, at twice-yearly frequency, and recovered to only 5% of their original biomass in the following five years, while the biomass of deposit feeders increased above the original level to become the dominant proportion of the total biomass. Removing suspension feeders reduced oxygen demand within the sediment, increasing the depth of the oxygenated layer. This stimulated greater nitrification and a sharp reduction in ammonium flux, which dropped by 20%; increased oxygenation led to a rise in the amount of phosphate adsorbed onto the sediment, and a decrease in the concentration of phosphate in interstitial water. The resulting fluctuations in nutrient levels, however, do not seem to have direct consequences for the production of pelagic communities, especially in shallow, unstratified waters where the removal of suspension feeders reduces grazing pressure.

A five-year moratorium on trawling was considered sufficient to allow complete recovery following disturbance by most gear types, on most sediment types, except in the case of high-frequency (five times a year) beam trawling on sand, resulting in 90% mortality of suspension feeders or deposit feeders, which would lead to a permanent change in ecosystem function. Production drops sharply following the initiation of benthic fishing: removal of large bivalves and heart urchins from some North Sea communities, for example, led to a six-fold decrease in total community production (Jennings et al., 2002). Biomass and production of small-bodied polychaetes may increase in areas trawled at frequencies of up to 2.3 times per year, but decrease at higher fishing frequency (Jennings et al., 2001b). However, any subsequent increase in the total biomass of the community is insufficient to balance the loss of biomass following from the extermination of the large-bodied, K-selected animals, although P/B ratios will increase. Chronic trawling disturbance of sedimentary habitats has little further effect on production rates of infaunal communities dominated by small-bodied polychaetes. For these species, reproductive output, recruitment and growth rates are sufficiently high, and mortality sufficiently low, that both biomass and production are largely unaffected by trawling impacts. Over much of the central and southern North Sea regular beam trawling is creating benthic communities quite different from those of pristine, unfished habitats, characterised by small-bodied epifauna and infauna, preyed upon by small demersal fish (Jennings et al., 2002). The diets of small Sole, 11–25 cm long, Dab, 11–30 cm, and Plaice, 11–35 cm, consist mostly of small polychaetes. Larger size classes may switch diet, Plaice, for example, favouring the siphons of bivalves, but the size structures of flatfish populations are also determined to a considerable extent by bottom trawling, and large individuals become increasingly rare on regularly fished grounds.

In recent decades there has been evidence of increasing growth rates among Plaice populations in the central and southern North Sea, and a coincident increase in primary production. Causal factors driving these phenomena are not entirely clear. Climatic amelioration may be one, while another may be the frequent re-suspension of nutrients by trawling, promoting more intense and prolonged phytoplankton blooms, and enhanced energy flow to the benthos. Primary production began to increase in 1985, coinciding with rising sea surface temperatures and positive North Atlantic Oscillation values, and correlated with increasing benthic biomass (Jennings *et al.*, 2001b). It has been suggested that regular sea floor disturbance could be regarded as 'marine farming', the sea floor being ploughed and aerated, and nutrients mixed and redistributed within the sediments, and in the water column, stimulating both primary and secondary production. In this one sense only, ploughing the sea floor may be compared with bioturbation, but in all other respects it is profoundly different. The large, bioturbating, habitat-structuring infauna, particularly mud shrimps, the echiuran *Maxmuelleria lankesteri* (page 89) and burrowing heart urchins such as *Brissopsis lyrifera* (page 86), rework huge volumes of sediment, but in addition to simply aerating their immediate environment they fulfil critically important roles in driving biogeochemical fluxes, including the cycling of carbon, nitrates, phosphates and silicate, through the degradation and remineralisation of organic detritus. The elimination of large suspension-feeding animals, and the larger deposit feeders, from benthic assemblages creates an excess of organic material – fresh and degraded phytoplankton, and detritus from every stage in the pelagic–benthic food web – that has to be incorporated into the benthic web by alternative pathways. Increasing numbers of small infauna will consume only a small proportion, and much may be consigned to a carbon sink in the sediments, and contribute to increasing eutrophication. However, 'benthic farming' may be apposite in one limited sense, in that overturn and aeration of the sediments may be a benefit to that component of the infauna that is least affected by trawling disturbance, namely the meiofauna.

The contribution of meiofaunal animals to benthic production and energy flow is strikingly disproportionate: as frequently noted, while the biomasss of soft-sediment meiofauna is typically a fraction of that of the macrofauna, its production may be an order of magnitude greater. The meiofauna constitutes a critical trophic link between bacterial, microfaunal and microfloral production, and the temporary meiofauna – post-larval stages of macrofaunal animals – and small-bodied macrofauna. Meiofaunal production thereby passes into the food webs of all benthic carnivores, a functional group encompassing most commercially important crustaceans and demersal fish. Meiofaunal animals are the least likely to suffer physical damage, or to be killed, by the passage of a trawl or dredge. They are carried into the water column among re-suspended sediment, and while a proportion may be dispersed by turbulent current flow, the majority will be returned to their habitat as the sediment resettles. Any loss to the population is probably swiftly restored. Individual reproductive output may be low – most meiofaunal animals produce very few eggs at each reproductive event – but reproduction tends to be continuous instead of seasonal, and development is direct: there is no pelagic larval phase and eggs hatch as advanced juveniles. Meiofaunal assemblages are remarkably resilient. Field experiments on intertidal mudflats on the Essex coast showed that meiofaunal communities buried by a layer of anoxic mud, thus simulating sediment overturn, recovered swiftly (Whomersley et al., 2009). The meiofauna consisted principally of nematodes, 46 species in total, and one month following experimental burial, species number, richness and diversity actually increased, another demonstration of the 'intermediate disturbance effect'. While macrofaunal abundance also increased within the same period, this was attributable to colonisation by two opportunistic species, and species number, richness and diversity all decreased. The structure of the nematode community changed slightly, reflecting the different capacity of each species for vertical migration through the sediment, and the dominance ranking of some species also changed, but no new colonising species appeared.

Meiofaunal animals have very short generation times, and disturbed communities are restored through rapid reproduction and growth by residual populations rather than by migration or recruitment from other sources. Experimental before – after – control – impact (BACI; page 324) trawling at two relatively undisturbed sites in the southern North Sea, a muddy sand habitat at 39 m depth, and a mud environment at 59 m, also demonstrated the resilience of the meiofauna (Schratzberger et al., 2002). Neither diversity nor biomass of meiofaunal nematode assemblages showed any effect of perturbation one day, one month, six months and one year following the experiment. Changes in the structure of meiofaunal communities attributable to trawling disturbance seemed to be minimal in comparison to natural, seasonal changes. However, trawling intensity may be an important factor determining the responses of meiofaunal populations to disturbance, and sampling from three locations on the Silver Pit grounds (page 352) suggested that trawling frequency might have a significant effect on the composition of nematode assemblages (Schratzberger & Jennings, 2002). Communities sampled in the same season, from areas subject to different degrees of trawling intensity showed significant differences in total abundance, number of species, diversity and richness, in both spring and autumn. Communities in areas trawled on average four times each year, at medium intensity, had a greater total abundance of nematodes than those in areas affected just once each year, low intensity, and those trawled with the highest average frequency, of six times per year. Number of species, diversity and richness were significantly lower in communities trawled with the highest intensity, and species composition showed greater variation in relation to trawling intensity than to any seasonal change.

CONCLUSION

The benthic ecosystems of the northwest European shelf seas are essentially dynamic. Benthic communities change in response to natural fluctuations in physical environmental factors, which may cycle with periods of decades, centuries or millennia. It is now apparent that a broader, and perhaps irreversible, climatic change is accelerating change in the marine benthic environment of northwest Europe. This is reflected in changes in the composition of benthic faunas, in the geographical distributions of species and communities, and in the establishment and spread of cryptogenic and alien invasive species. Human activity, especially bottom fishing with dredges and trawls, and aggregate extraction, has severely damaging effects on benthic habitats, some of which may not recover at a timescale appreciable by humankind. It is unlikely, for example, that some biogenic reefs, and particularly maerl habitats, will recover in less than a century. Some soft bottom habitats in the southern North Sea may have undergone a profound ecosystem change, or regime shift, which might prove to be long-lasting. It is still difficult to distinguish clearly between ecological changes attributable to climatic factors and those resulting from anthropogenic impacts, but it is probable that continuing climate change and increasing human disturbance will result in even more marked change in the benthic ecosystems of the shallow seas of northwest Europe.

References and Further Reading

Alder, J. & Hancock, A. (1845–1855). *A Monograph of the British Nudibranchiate Mollusca*. Van Vooorst, London.

Allen, J. I. & Clarke, K. R. (2007). Effects of demersal trawling on ecosystem functioning in the North Sea: a modelling study. *Marine Ecology Progress Series*, **336**, 63–75.

Amezcua, F., Nash, R. D. M. & Veale, L. (2003). Feeding habits of the Order Pleuronectiformes and its relation to sediment type. *Journal of the Marine Biological Association of the United Kingdom*, **83**, 593–601.

Armonies, W. (2001). What an introduced species can tell us about the spatial extension of benthic populations. *Marine Ecology Progress Series*, **209**, 289–294.

Atkinson, R. J. A. & Taylor, A. C. (2005). Aspects of the physiology, biology and ecology of thalassinidean shrimps in relation to their burrow environment. *Oceanography and Marine Biology: an Annual Review*, **43**, 173–210.

Backeljau, T., Bouchet, P., Gofas, S. & de Bruyn, L. (1994). Genetic variation, systematics and distribution of the venerid clam *Chamelea gallina*. *Journal of the Marine Biological Association of the United Kingdom*, **74**, 211–223.

Ball, B. J., Costelloe, J., Könnecker, G. & Keegan, B. F. (1995). The rocky subtidal assemblages of Kinsale Harbour (south coast of Ireland). In: Eleftheriou, A., Ansell, A. D. & Smith, C. J. (eds), *Biology and Ecology of Shallow Coastal Waters: Proceedings of the 28th European Marine Biology Symposium*. Olsen & Olsen, Fredensborg, pp. 293–302.

Barnes, R. S. K. & Farnon Ellwood, M. D. (2011). Macrobenthic assemblage structure in a cool temperate intertidal dwarf eelgrass bed, in comparison with those from lower latitudes. *Biological Journal of the Linnean Society*, **104**, 527–540.

Barnes, R. S. K. & Hughes, R. N. (1982). *An Introduction to Marine Ecology*. Blackwell, Oxford.

Basford, D. J., Eleftheriou, A. & Raffaelli, D. (1989). The epifauna of the northern North Sea (56°–61° N). *Journal of the Marine Biological Association of the United Kingdom*, **69**, 387–407.

Bell, J. J., Barnes, D. K. A., Shaw, C., Heally, A. & Farrell, A. (2003). Seasonal 'fall out' of sessile macrofauna from submarine cliffs: quantification, causes and implications. *Journal of the Marine Biological Association of the United Kingdom*, **83**, 1199–1208.

Bergmann, M., Beare, D. J. & Moore, P. G. (2001). Damage sustained by epibenthic invertebrates discarded in the *Nephrops* fishery in the Clyde Sea area, Scotland. *Journal of Sea Research*, **45**, 105–118.

Bergmann, M., Wieczorek, S. K., Moore, P. G. & Atkinson, R. J. A. (2002). Utilisation of invertebrates discarded from the *Nephrops* fishery by variously selective benthic scavengers in the west of Scotland. *Marine Ecology Progress Series*, **233**, 185–198.

Beukema, J. J. & Cadée, G. C. (1991). Growth rates of the bivalve *Macoma balthica* in the Wadden Sea during a period of eutrophication: relationships with concentrations of pelagic diatoms and flagellates. *Marine Ecology Progress Series*, **68**, 249–256.

Beukema, J. J. & Dekker, R. (1995). Dynamics and growth of a recent invader into European coastal waters: the American Razor Clam, *Ensis directus*. *Journal of the Marine Biological Association of the United Kingdom*, **75**, 351–362.

BIOMAERL Team (1999). *Final Report, BIOMAERL Project* (Coordinator: P. G. Moore, UMBS Millport, Scotland). EC contract no. MAS3-CT95-0020.

Blight, A. J. & Thompson, R. C. (2008). Epibiont species richness varies between holdfasts of a northern and a southerly distributed kelp species. *Journal of the Marine Biological Association of the United Kingdom*, **88**, 469–475.

Boon, A. R., Duinveld, G. C. A., Berghuis, E. M. & van der Weele, J. A. (1998). Relationships between benthic activity and the annual phytopigment cycle in near-bottom water and sediments in the southern North Sea. *Estuarine, Coastal and Shelf Science*, **46**, 1–13.

Borja, Å., Franco, J. & Pérez, V. (2000). A marine biotic index to establish the ecological quality of soft-bottom benthos within European estuarine and coastal environments. *Marine Pollution Bulletin*, **40**, 1100–1114.

Borum, J., Kaas, H. & Wium-Andersen, S. (1984). Biomass variation and autotrophic production of an epiphyte-macrophyte community in a coastal Danish area: II. Epiphyte species composition, biomass and production. *Ophelia*, **23**, 165–179.

Borum, J., Pedersen, M., Krause-Jensen, D., Christensen, P. & Nielsen, K. (2002). Biomass, photosynthesis and growth of *Laminaria saccharina* in a high-arctic fjord, NE Greenland. *Marine Biology*, **141**, 11–19.

Bosselmann, A. (1989). Larval plankton and recruitment of macrofauna in a subtidal area in the German Bight. In: Ryland, J. S. & Tyler, P. A. (eds), *Reproduction, Genetics and Distributions of Marine Organisms. Proceedings of the 23rd European Marine Biology Symposium.* Olsen & Olsen, Fredensborg, pp. 43–54.

Boström, C. & Bonsdorff, E. (2000). Zoobenthic community establishment and habitat complexity–the importance of seagrass shoot density, morphology and physical disturbance for faunal recruitment. *Marine Ecology Progress Series*, **205**, 123–138.

Bowden, D. A., Rowden, A. A. & Attrill, M. (2001). Effect of patch size and in-patch location on the infaunal macroinvertebrate assemblages of *Zostera marina* seagrass beds. *Journal of*

Experimental Marine Biology and Ecology, **259**, 133–154.

Boyd, S. E., Limpenny, D. S., Ress, H. L. & Cooper, K. M. (2005).The effects of marine sand and gravel extraction on the macrobenthos at a commercial dredging site (results six years post dredging). *ICES Journal of Marine Research*, **62**, 145– 162.

Bradshaw, C., Veale, L. O. & Brand, A. R. (2002). The role of scallop-dredge disturbance in long-term changes in Irish Sea benthic communities: a re-analysis of an historical dataset. *Journal of Sea Research*, **47**, 161–184.

Buchanan, J. B. & Moore, J. J. (1986). A broad review of variability and persistence in the Northumberland benthic fauna: 1971–1985. *Journal of the Marine Biological Association of the United Kingdom*, **66**, 641–657.

Callaway, R., Alsvåg, J., de Boois, I. *et al.* (2002). Diversity and community structure of epibenthic invertebrates and fish in the North Sea. *ICES Journal of Marine Science*, **59**, 1199–1214.

Callaway, R., Engelhard, G. H., Dann, J., Cotter, J. & Rumohr, H. (2007). A century of North Sea epibenthos and trawling: comparison between 1902–1912, 1982–1985 and 2000. *Marine Ecology Progress Series*, **346**, 27–43.

Carroll, J., Gobler, C. J. & Peterson, B. J. (2008). Resource-restricted growth in New York estuaries: light limitation, and alleviation of nutrient stress by hard clam. *Marine Ecology Progress Series*, **369**, 51–62.

Cebrián, J., Duarte, C. M., Marbà, N. & Enríquez, S. (1997). Magnitude and fate of the production of four co-occurring Western Mediterranean seagrass species. *Marine Ecology Progress Series*, **155**, 29–44.

Chapman, N. D., Moore, C. G., Harries, D. B. & Lyndon, A. R. (2012). The community associated with biogenic reefs formed by the polychaete, *Serpula vermicularis*. *Journal of the Marine Biological Association of the United Kingdom*, **92**, 679–685.

Chauvaud, L., Jean, F., Raguenau, O. & Thouzeau, G. (2000). Long-term variation of the Bay of Brest ecosystem: benthic–pelagic coupling revisited. *Marine Ecology Progress Series*, **200**, 35–48.

Chia, F. S. (1989). Differential larval settlement of benthic marine invertebrates. Pages 3 – 12, In: Ryland, J. S. & Tyler, P. A. (eds), *Reproduction, Genetics and Distributions of Marine Organisms*. Olsen & Olsen, Fredensborg, pp. 3–12.

Chícharo, L., Chícharo, A., Gaspar, M., Alves, F. & Regala, J. (2002). Ecological characterization of dredged and non-dredged bivalve fishing areas off south Portugal. *Journal of the Marine Biological Association of the United Kingdom*, **82**, 41–50.

Christie, H., Jørgensen, N. M., Norderhaug, K. M., & Waage-Nielsen, E. (2003). Species distribution and habitat exploitation of fauna associated with kelp (*Laminaria hyperborea*) along the Norwegian coast. *Journal of the Marine Biological Association of the United Kingdom*, **83**, 687–699.

Clark, R. B. (2001). *Marine Pollution*, 5th edition. Oxford University Press, Oxford.

Connor, D. W., Dalkin, M. J., Hill, T. O., Holt. R. H. F. & Sanderson, W. G. (1997). *Marine Biotope Classification for Britain and Ireland. Volume 2. Sublittoral Biotopes*. JNCC Report 230.

Costello, M. J., Emblow, C. & White, R. (eds) (2001). European Register of Marine

Species: a Check-list of the Marine Species in Europe and a Bibliography of Guides to their Identification. Patrimoines naturels, **50**. Muséum National d'Histoire Naturelle, Paris.

Crisp, D. J. (ed) (1964). The effects of the severe winter of 1962–63 on marine life in Britain. *Journal of Animal Ecology*, **33**, 165–210.

Danovaro, R., Fraschetti, S., Belgrano, A. *et al.* (1995). The potential impact of meiofauna on the recruitment of macrobenthos in a subtidal coastal benthic community of the Ligurian Sea (north-western Mediterranean): a field result. In: Eleftheriou, A., Ansell, A. D. & Smith, C. J. (eds), *Biology and Ecology of Shallow Coastal Waters: Proceedings of the 28th European Marine Biology Symposium*. Olsen & Olsen, Fredensborg, pp. 115–122.

Dauwe, B., Herman, P. M. J. & Heip, C. H. R. (1998). Community structure and bioturbation potential of macrofauna at four North Sea stations with contrasting food supply. *Marine Ecology Progress Series*, **173**, 67–83.

Davoult, D. & Gounin, F. (1995). Suspension-feeding activity of a dense *Ophiothrix fragilis* (Abildgaard) population at the water-sediment interface: time coupling of food availability and feeding behaviour of the species. *Estuarine, Coastal and Shelf Science*, **41**, 567–577.

Day, F. (1880–1884). *The Fishes of Great Britain and Ireland*. Williams and Norgate, London.

de Kluijver, M. J. (1997). Sublittoral communities of North Sea hard substrata. Thesis, University of Amsterdam.

de Montaudouin, X., Labarraque, D., Giraud, K. & Bachelet, G. (2001). Why does the introduced gastropod *Crepidula fornicata* fail to invade Arcachon Bay (France)? *Journal of the Marine Biological Association of the United Kingdom*, **81**, 97–104.

Duineveld, G. C. A., Lavaleye, M. S. S. & Berghuis, E. M. (2004). Particle flux and food supply to a seamount cold-water coral community (Galicia Bank, NW Spain). *Marine Ecology Progress Series*, **277**, 13–23.

Duineveld, G. C. A., Bergman, M. J. N. & Lavaleye, M. S. S. (2007). Effects of an area closed to fisheries on the composition of the benthic fauna in the southern North Sea. *ICES Journal of Marine Science*, **64**, 899–908.

Dunstan, P. K. & Bax, N. J. (2007). How far can marine species go? Influence of population biology and larval movement on future range limits. *Marine Ecology Progress Series*, **344**, 15–28.

Eastwood, P. D., Mills, C. M., Aldridge, J. N., Houghton, C. A. & Rogers, S. I. (2007). Human activities in UK offshore waters: an assessment of direct, physical pressure on the seabed. *ICES Journal of Marine Science*, **64**, 453–463.

Ejdung, G. & Elmgren, R. (1998). Predation on newly settled bivalves by deposit-feeding amphipods: a Baltic Sea case study. *Marine Ecology Progress Series*, **168**, 87–94.

Eleftheriou, A. (ed) (2013). *Methods for the Study of Marine Benthos*, 4th edition. Wiley-Blackwell, Oxford.

Ellis, J. R. & Rogers, S. I. (2000). The distribution, relative abundance and diversity of echinoderms in the eastern English Channel, Bristol Channel and Irish Sea. *Journal of the Marine Biological Association of the United Kingdom*, **80**, 127–138.

Ellis, J. R., Martinez, I., Burt, G. J. & Scott, B. E. (2013). Epibenthic assemblages in the North Sea and associated with the Jones Bank. *Progress in Oceanography*, **117**, 76–88.

Engelhard, G. H., Pinnegar, J. H., Kell, L. T. & Rijnsdorp, A. D. (2011). Nine decades of North Sea sole and Plaice distribution. *ICES Journal of Marine Science*, **68**, 1090–1104.

Eno, N. C., Clark, R. A. & Sanderson, W. G. (1997). *Non-native Marine Species in British Waters: a Review and Directory*. JNCC, Peterborough.

Erftemeijer, P. L. A., van Beek, J. K. L., Ochieng, C. A., Jager, Z. & Los, H. L. (2008). Eelgrass seed dispersal via floating generative shoots in the Dutch Wadden Sea: a model approach. *Marine Ecology Progress Series*, **358**, 115–124.

Feller, R. J., Stancyk, S. E., Coull, B. C. & Edwards, D. G. (1992). Recruitment of polychaetes and bivalves: long-term assessment of predictability in a soft-bottom habitat. *Marine Ecology Progress Series*, **87**: 227–238.

Fosså, J. H., Mortensen, P. B. & Furevik, D. M. (2002). The deep-water coral *Lophelia pertusa* in Norwegian waters: distribution and fishery impacts. *Hydrobiologia*, **471**, 1–12.

Foster-Smith, J. (ed) (2000). *The Marine Fauna and Flora of the Cullercoats District*. Penshaw Press, Sunderland.

Fredriksen, S. (2003). Food web studies in a Norwegian kelp forest based on stable isotope ($\delta13C$ and $\delta15N$) analysis. *Marine Ecology Progress Series*, **260**, 71–81.

Freeman, R. B. & Wertheimer, D. (1980). *Philip Henry Gosse: a Bibliography*. William Dawson & Sons, Folkestone.

Freiwald, A., Henrich, R., Schäfer, P. & Willkomm, H. (1991). The significance of high-boreal to subarctic maerl deposits in northern Norway to reconstruct Holocene climatic changes and sea level oscillations. *Facies*, **25**, 315–339.

Frid, C. L. J., Buchanan, J. B. & Garwood, P. R. (1996). Variability and stability in benthos: twenty-two years of monitoring off Northumberland. *ICES Journal of Marine Science*, **53**, 978–980.

Frid, C. L. J., Garwood, P. R. & Robinson, L. A. (2009). The North Sea benthic system: a 36 year time–series. *Journal of the Marine Biological Association of the United Kingdom*, **89**, 1–10.

Gallardo, C. S. & Penchaszadeh, P. E. (2001). Hatching mode and latitude in marine gastropods: revisiting Thorson's paradigm in the southern hemisphere. *Marine Biology*, **138**, 547–552.

Gambi, M. C., Nowell, A. R. M. & Jumars, P. A. (1990). Flume observations on flow dynamics in *Zostera marina* (eelgrass) beds. *Marine Ecology Progress Series*, **61**, 159–169.

Garcia-Soto, C. & Pingree, R. D. (2012). Atlantic Multidecadal Oscillation (AMO) and sea surface temperature in the Bay of Biscay and adjacent regions. *Journal of the Marine Biological Association of the United Kingdom*, **92**, 213–234.

Gaspar, M. B., Castro, M. & Monteiro, C. C. (1999). Effect of tooth spacing and mesh size on the catch of the Portuguese clam and razor clam dredge. *ICES Journal of Marine Science*, **56**, 103–110.

Gaspar, M. B., Leitão, F., Santos, M. N. *et al*. (2003). A comparison of direct macrofaunal mortality using three types of clam dredges. *ICES Journal of Marine Science*, **60**, 733–742.

Gaspar, M. B., Carvalho, S., Constantino, R. *et al*. (2009). Can we infer dredge fishing effort from macrobenthic community structure? *ICES Journal of Marine Science*, **66**, 2121– 2132.

Gass, S. E. & Roberts, J. M. (2006). The occurrence of the cold-water coral *Lophelia pertusa* (Scleractinia) on oil and gas platforms in the North Sea: colony growth, recruitment and environmental controls on distribution. *Marine Pollution Bulletin*, **52**, 549–559.

Genner, M. J., Sims, D. W., Wearmouth, V. J. *et al.* (2004). Regional climatic warming drives long-term community changes of British marine fish. *Proceedings of the Royal Society of London B*, **271**, 655–661.

George, C. L. & Warwick, R. M. (1985). Annual macrofauna production in a hard-bottom reef community. *Journal of the Marine Biological Association of the United Kingdom*, **65**, 713–735.

Gevaert, F., Davoult, D., Creach, A. *et al.* (2001). Carbon and nitrogen content of *Laminaria saccharina* in the eastern English Channel: biometrics and seasonal variations. *Journal of the Marine Biological Association of the United Kingdom*, **81**, 727–734.

Gosse, P. H. (1858–1860). *Actinologica Britannica: a History of the British Sea-Anemones and Madrepores*. Van Voorst, London.

Grant, A. & Hayward, P. J. (1985). Bryozoan benthic assemblages in the English Channel. In: Nielsen, C. & Larwood, G. P. (eds), *Bryozoa: Ordovician to Recent*. Olsen & Olsen, Fredensborg, pp. 115–124.

Gray, J. S. & Elliott, M. (2009). *Ecology of Marine Sediments*, 2nd edition. Oxford University Press, Oxford.

Guillou, M. (1990). Biotic interactions between predators and super-predators in the Bay of Douarnenez, Brittany. In: Barnes, M. & Gibson, R. N. (eds), *Trophic Relationships in the Marine Environment. Proceedings of the 24th European Marine Biology Symposium*, pp. 141–156.

Guillou, M., Grall, J. & Connan, S. (2002). Can low sea urchin densities control macro-epiphytic biomass in a northeast Atlantic maerl bed ecosystem (Bay of Brest, Brittany, France)? *Journal of the Marine Biological Association of the United Kingdom*, **82**, 867–876.

Gulliksen, B. (1980). The macrobenthic rocky-bottom fauna of Borgenfjorden, North-Tröndelag, Norway. *Sarsia*, **65**, 115–138.

Günther, C.-P. (1992). Settlement and recruitment of *Mya arenaria* L. in the Wadden Sea. *Journal of Experimental Marine Biology and Ecology*, **159**, 203–215.

Hagen, N. T. (1995). Sea urchin outbreaks and epizootic disease as regulating mechanisms in coastal ecosystems. In: Eleftheriou, A., Ansell, A. D. & Smith, C. J. (eds), *Biology and Ecology of Shallow Coastal Waters: Proceedings of the 28th European Marine Biology Symposium*. Olsen & Olsen, Fredensborg, pp. 303–308.

Hall-Spencer, J. M. (1998). Conservation issues relating to maerl beds as habitats for molluscs. *Journal of Conchology, Special Publication*, **2**, 271–286.

Hall-Spencer, J. M. & Atkinson, R. J. A. (1999). *Upogebia deltaura* (Crustacea: Thalassinidea) in Clyde Sea maerl beds, Scotland. *Journal of the Marine Biological Association of the United Kingdom*, **79**, 871–880.

Hall-Spencer, J. M. & Moore, P. G. (2000a). *Limaria hians* (Mollusca: Limacea): a neglected reef-forming keystone species. *Aquatic conservation: marine and freshwater ecosystems*, **10**, 267–277.

Hall-Spencer, J. M. & Moore, P. G. (2000b). Impacts of scallop dredging on maerl grounds. In: Kaiser, M. J. & de Groot, S. J. (eds), *Effects of Fishing on Non-Target Species and Habitats: Biological, Conservation and Socio-economic Issues*. Blackwell, Oxford, pp. 105–117.

Hall-Spencer, J. M. & Moore, P. G. (2000c). Scallop dredging has profound, long-term impacts on maerl habitats. *ICES Journal of Marine Science*, **57**, 1407–1415.

Hall-Spencer, J. M., White, N., Gillespie, E., Gilham, K. & Foggo, A. (2006). Impact of fish farms on maerl beds in strongly tidal areas. *Marine Ecology Progress Series*, **326**, 1–9.

Hardy, A. (1956). *The Open Sea: its Natural History. Part 1. The World of Plankton*. New Naturalist 34. Collins, London.

Hartnoll, R. G. (1975). The annual cycle of *Alcyonium digitatum*. *Estuarine and Coastal Marine Science*, **3**, 71–78.

Hauton, C., Hall-Spencer, J. M. & Moore, P. G. (2003). An experimental study of the ecological impacts of hydraulic bivalve dredging on maerl. *ICES Journal of Marine Science*, **60**, 381–392.

Hauton, C., Howell, T. R. W., Atkinson, R. J. A. & Moore, P. G. (2007). Measures of hydraulic dredge efficiency and razor clam production, two aspects governing sustainability within the Scottish commercial fishery. *Journal of the Marine Biological Association of the United Kingdom*, **87**, 869–877.

Havenhand, J. N. & Svane, I. (1989). Larval behaviour, recruitment, and the role of adult attraction in *Ascidia mentula* O. F. Müller. In: Ryland, J. S. & Tyler, P. A. (eds), *Reproduction, Genetics and Distributions of Marine Organisms. Proceedings of the 23rd European Marine Biology Symposium*. Olsen & Olsen, Fredensborg, pp. 127–132.

Hayward, P. J. & Ryland, J. S. (eds) (1995). *Handbook of the Marine Fauna of North-West Europe*. Oxford University Press, Oxford.

Henderson, P. A. (2007). Discrete and continuous change in the fish community of the Bristol Channel in response to climate change. *Journal of the Marine Biological Association of the United Kingdom*, **87**, 589–598.

Henderson, P. A. & Seaby, R. M. (2005). The role of climate in determining the temporal variation in abundance, recruitment and growth in sole *Solea solea* in the Bristol Channel. *Journal of the Marine Biological Association of the United Kingdom*, **85**, 197–204.

Henderson, P. A., James, D. & Holmes, R. H. A. (1992). Trophic structure within the Bristol Channel: seasonality and stability in Bridgwater Bay. *Journal of the Marine Biological Association of the United Kingdom*, **72**, 675–690.

Henderson, P. A., Seaby, R. M. & Somes, J. R. (2006). A 25-year study of climatic and density-dependent population regulation of common shrimp *Crangon crangon* (Crustacea: Caridea) in the Bristol Channel. *Journal of the Marine Biological Association of the United Kingdom*, **86**, 287–298.

Henrich, R., Freiwald, A., Betzler, C. *et al.* (1995). Controls on modern carbonate sedimentation on warm-temperate to Arctic coasts, shelves and seamounts in the northern hemisphere: implications for fossil counterparts. *Facies*, **32**, 71–108.

Herbert, R. J. H., Hawkins, S. J., Sheader, M. & Southward, A. J. (2003). Range extension and reproduction of the barnacle *Balanus perforatus* in the eastern English Channel. *Journal of the Marine Biological Association of the United Kingdom*, **83**, 73–82.

Heslenfeld, P. & Enserink, E. L. (2008). OSPAR Ecological Quality Objectives:

the utility of health indicators for the North Sea. *ICES Journal of Marine Science*, **65**, 1392–1397.

Hewitt, G. M. (1996). Some genetic consequences of ice ages, and their role in divergence and speciation. *Biological Journal of the Linnean Society*, **58**, 247–276.

Hincks, T. (1880). *A History of the British Marine Polyzoa*. Van Voorst, London.

Hiscock, K. (ed) (1998). Marine Nature Conservation Review. Benthic Marine Ecosystems of Great Britain and the North-east Atlantic. JNCC, Peterborough.

Hiscock, K., Sharrock, S., Highfield, J. & Snelling, D. (2010). Colonization of an artificial reef in south-west England– ex-HMS 'Scylla'. *Journal of the Marine Biological Association of the United Kingdom*, **90**, 69–94.

Hollertz, K. (2002). Feeding biology and carbon budget of the sediment-burrowing heart urchin *Brissopsis lyrifera* (Echinoidea: Spatangoida). *Marine Biology*, **140**, 959–969.

Holme, N. A. (1961). The bottom fauna of the English Channel. *Journal of the Marine Biological Association of the United Kingdom*, **41**, 397–461.

Holme, N. A. (1966). The bottom fauna of the English Channel. Part II. *Journal of the Marine Biological Association of the United Kingdom*, **46**, 401–493.

Howe, R. L., Rees, A. P. & Widdicombe, S. (2004). The impact of two species of bioturbating shrimp (*Callianassa subterranea* and *Upogebia deltaura*) on sediment denitrification. *Journal of the Marine Biological Association of the United Kingdom*, **84**, 629–632.

Howson, C. M. & Picton, B. E. (eds) (1997). *The Species Directory of the Marine Fauna and Flora of the British Isles and Surrounding Seas*. Ulster Museum and the Marine Conservation Society, Belfast and Ross-on-Wye.

Hughes, D. J. & Atkinson, R. J. A. (1997). A towed video survey of megafaunal bioturbation in the northeastern Irish Sea. *Journal of the Marine Biological Association of the United Kingdom*, **77**, 635–653.

Hughes, D. J., Ansell, A. D. & Atkinson, R. J. A. (1996a). Distribution, ecology and life cycle of *Maxmuelleria lankesteri* (Echiura: Bonellidae): a review with notes on field identification. *Journal of the Marine Biological Association of the United Kingdom*, **76**, 897– 908.

Hughes, D. J., Ansell, A. D. & Atkinson, R. J. A. (1996b). Sediment bioturbation by the echiuran worm *Maxmuelleria lankesteri* (Herdman) and its consequences for radionuclide dispersal in Irish Sea sediments. *Journal of Experimental Marine Biology and Ecology*, **195**, 203–220.

Hughes, R. G. (1983). The life history of *Tubularia indivisa* (Hydrozoa: Tubulariidae) with observations on the status of *T. ceratogyne*. *Journal of the Marine Biological Association of the United Kingdom*, **63**, 467–479.

Humphreys, J., Caldow, R. W. G., McGrorty, S., West, A. D. & Jensen, A. C. (2007). Population dynamics of naturalised Manila clams *Ruditapes phillipinarum* in British coastal waters. *Marine Biology*, **151**, 2255–2270.

Hunt, H. L. & Scheibling, R. E. (1997). Role of early post-settlement mortality in recruitment of benthic marine invertebrates. *Marine Ecology Progress Series*, **155**, 269–301.

Huston, M. A. (1994). *Biological Diversity*. Cambridge University Press, Cambridge.

Irvine, L. M. & Chamberlain, Y. M. (1994). *Seaweeds of the British Isles*. Vol. 1, *Rhodophyta*. Part 2B, *Corallinales, Hildenbrandiales*. HMSO, London.

Jackson, J. B. C. (2001). What was natural in the coastal oceans? *Proceedings of the National Academy of Science*, **98**, 5411–5418.

Jamieson, G. S., Grosholz, E. D., Armstrong, D. A. & Elner, R. W. (1998). Potential ecological implications from the introduction of the European green crab, *Carcinus maenas* (Linnaeus), to British Columbia, Canada, and Washington, USA. *Journal of Natural History*, **32**, 1587–1598.

Jaschinski, S., Brepoh, D. C. & Sommer, U. (2008). Carbon sources and trophic structure in an eelgrass *Zostera marina* bed, based on stable isotope and fatty acid analyses. *Marine Ecology Progress Series*, **358**, 103–114.

Jeffreys, J. G. (1862–1869). *British Conchology: or an Account of the Mollusca Which Now Inhabit the British Isles and the Surrounding Seas*. Van Voorst, London.

Jenkins, S. R., Beukers-Stewart, B. D. & Brand, A. R. (2001). Impact of scallop dredging on benthic megafauna: a comparison of damage levels in captured and non-captured organisms. *Marine Ecology Progress Series*, **215**, 297–301.

Jennings, S. & Kaiser, M. J. (1998). The effects of fishing on marine ecosystems. *Advances in Marine Biology*, **34**, 201–351.

Jennings, S., Lancaster, J. E., Woolmer, A. & Cotter, J. (1999). Distribution, diversity and abundance of epibenthic fauna in the North Sea. *Journal of the Marine Biological Association of the United Kingdom*, **79**, 385–399.

Jennings, S., Pinnegar, J. K., Polunin, N. V. C. & Warr, K. J. (2001a). Impacts of trawling disturbance on the trophic structure of benthic invertebrate communities. *Marine Ecology Progress Series*, **213**, 127–142.

Jennings, S., Dinmore, T. A., Duplisea, D. E., Warr, K. J. & Lancaster, J. E. (2001b). Trawling disturbance can modify benthic production processes. *Journal of Animal Ecology*, **70**, 459–475.

Jennings, S., Nicholson, M. D., Dinmore, T. A. & Lancaster, J. E. (2002). Effects of chronic trawling disturbance on the production of infaunal communities. *Marine Ecology Progress Series*, **243**, 251–260.

Jensen, A. & Frederiksen, R. (1992). The fauna associated with the bank-forming deep-water coral *Lophelia pertusa* (Scleractinaria) On the Faroe Shelf. *Sarsia*, **77**, 53–69.

Jensen, A. C., Humphreys, J., Caldow, R. W. G., Grisley, C. & Dyrynda, P. E. J. (2004). Naturalization of the Manila Clam (*Tapes philippinarum*), an alien species, and establishment of a clam fishery within Poole Harbour, Dorset. *Journal of the Marine Biological Association of the United Kingdom*, **84**, 1069–1073.

Jensen, K. T. & Jensen, J. N. (1985). The importance of some epibenthic predators on the density of juvenile benthic macrofauna in the Danish Wadden Sea. *Journal of Experimental Marine Biology and Ecology*, **89**, 157–174.

Jephson, T., Nyström, P., Moksnes, P.-O. & Baden, S. (2008). Trophic interactions in *Zostera marina* beds along the Swedish coast. *Marine Ecology Progress Series*, **369**, 63–76.

Jernakoff, P., Brearley, A. & Nielsen, J. (1996). Factors affecting grazer-epiphyte interactions in temperate seagrass meadows. *Oceanography and*

Marine Biology: an Annual Review, **34**, 109–162.

John, D. M. (1969). An ecological study on *Laminaria ochroleuca*. *Journal of the Marine Biological Association of the United Kingdom*, **49**, 175–187.

Johnson, D. (2008). Environmental indicators: their utility in meeting the OSPAR Convention's regulatory needs. *ICES Journal of Marine Science*, **65**, 1387–1391.

Jonsson, L. G., Nilsson, P. G., Floruta, F and Lundälv, T. (2004). Distributional patterns of macro- and megafauna associated with a reef of the cold-water coral *Lophelia pertusa* on the Swedish west coast. *Marine Ecology Progress Series*, **284**, 163–171.

Jorgensen, P., Ibarro-Obando, S. E. & Carriquiry, J. D. (2007). Top-down and bottom-up stabilizing mechanisms in eelgrass meadows differentially affected by coastal upwelling. *Marine Ecology Progress Series*, **333**, 81–93.

Josefson, A. B. & Conley, D. J. (1997). Benthic response to a pelagic front. *Marine Ecology Progress Series*, **147**, 49–62.

Kain, J. M. (1979). A view of the genus *Laminaria*. *Oceanography and Marine Biology: an Annual Review*, **17**, 101–161.

Kaiser, M. J. & Spence, F. E. (2002). Inconsistent temporal changes in the megabenthos of the English Channel. *Marine Biology*, **141**, 321–331.

Kaiser, M. J. & Spencer, B. E. (1996). The effects of beam-trawl disturbance on infaunal communities in different habitats. *Journal of Animal Ecology*, **65**, 348–358.

Kaiser, M. J., Attrill, M. J., Jennings, S. *et al.* (2005). *Marine Ecology: Processes, Systems, and Impacts*. Oxford University Press, Oxford.

Kamenos, N. A., Moore, P. G. & Hall-Spencer, J. M. (2003). Substratum heterogeneity of dredged vs undredged maerl grounds. *Journal of the Marine Biological Association of the United Kingdom*, **83**, 411–413.

Kamenos, N. A., Moore, P. G. & Hall-Spencer, J. M. (2004). Small-scale distribution of juvenile gadoids in shallow inshore waters: what role does maerl play? *ICES Journal of Marine Science*, **61**, 422–429.

Klitgaard, A. B. (1995). The fauna associated with outer shelf and upper slope sponges at the Faroe Islands, northeastern Atlantic. *Sarsia*, **80**, 1–22.

Klitgaard, A. B. & Tendal, O. S. (2001). 'Ostur'– 'cheese bottoms'– sponge dominated areas in the Faroese shelf and slope areas. In: Bruntse, G. & Tendal, O. S. (eds), *Marine Biological Investigations and Assemblages of Benthic Invertebrates from the Faroe Islands*. Kaldbak Marine Biological Laboratory, pp. 9–21.

Knudsen, M. F., Seidenkrantz, M.-S., Jacobsen, B. H. & Kuijpers, A. (2011). Tracking the Atlantic Multidecadal Oscillation through the last 8,000 years. *Nature Communications*, **2**, article number 178. doi: 10.1038/ncomms1186.

Kröncke, I. & Reiss, H. (2010). Influence of macrofauna long-term variability on benthic indices used in ecological quality assessment. *Marine Pollution Bulletin*, **60**, 58–68.

Kube, S., Postel, L., Honnef, C. & Augustin, C. B. (2007). *Mnemiopsis leidyi* in the Baltic Sea–distribution and overwintering between autumn 2006 and spring 2007. *Aquatic Invasions*, **2**, 137–145.

Künitzer, A. (1989). Factors affecting the population dynamics of *Amphiura filiformis* (Echinodermata: Ophiuroidea) and *Mysella bidentata* (Bivalvia: Galeommatacea) in the North Sea. In: Ryland, J. S. & Tyler, P. A. (eds), *Reproduction, Genetics and Distributions of Marine Organisms. Proceedings of the 23rd European Marine Biology Symposium.* Olsen & Olsen, Fredensborg, pp. 395–406.

Künitzer, A. (1992). Does settlement influence population dynamics of macrobenthos? A case study in the central North Sea. In: Colombo, G., Ferrari, I., Ceccherelli, V. U. & Rossi, R. (eds), *Marine Eutrophication and Population Dynamics. Proceedings of the 25th European Marine Biology Symposium.* Olsen & Olsen, Fredensborg, pp. 285–291.

Lambert, W. J., Levin, P. S. & Berman, J. (1992). Changes in the structure of a New England (USA) kelp bed: the effects of an introduced species? *Marine Ecology Progress Series*, **88**, 303–307.

Lasota, R., Hummel, H. & Wolawicz, M. (2004). Genetic diversity of European populations of the invasive soft-shell clam, *Mya arenaria* (Bivalvia). *Journal of the Marine Biological Association of the United Kingdom*, **84**, 1051–1056.

Lee, A. J. & Ramster, J. W. (eds) (1981). *Atlas of the Seas around the British Isles.* Ministry of Agriculture, Fisheries and Food, London.

Lefebvre, A. & Davoult, D. (2000). Larval distribution of *Ophiothrix fragilis* (Echinodermata: Ophiuroidea) in a macrotidal area, the Dover Strait (eastern English Channel, France). *Journal of the Marine Biological Association of the United Kingdom*, **80**, 567–568.

Lehtonen, K. K. & Andersin, A.-B. (1998). Population dynamics, response to sedimentation and role in benthic metabolism of the amphipod *Monoporeia affinis* in an open-sea area of the northern Baltic Sea. *Marine Ecology Progress Series*, **168**, 71–85.

Levin, L. A. & Bridges, T. S. (1995). Pattern and diversity in reproduction and development. In: McEdward, L. (ed), *Ecology of Marine Invertebrate Larvae.* CRC Press, New York, pp. 1–77.

Levinton, J. S. (2011). *Marine Biology: Function, Biodiversity, Ecology*, 3rd edition. Oxford University Press, Oxford.

Levinton, J. S., Ward, J. E. & Shumway, S. E. (2002). Feeding responses of the bivalves *Crassostrea gigas* and *Mytilus trossulus* to chemical composition of fresh and aged kelp detritus. *Marine Biology*, **141**, 367–376.

Lindenbaum, C., Bennell, J. D., Rees, E. I. S. *et al.* (2008). Small-scale variation within a *Modiolus modiolus* (Mollusca: Bivalvia) reef in the Irish Sea. I. Seabed mapping and reef morphology. *Journal of the Marine Biological Association of the United Kingdom*, **88**, 133–141.

Magurran, A. E. (1996). *Ecological Diversity and its Measurement*, 2nd edition. Chapman & Hall, London.

Maier, C., Hegeman, J., Weinbauer, M. G. & Gattuso, J.-P. (2009). Calcification of the cold-water coral *Lophelia pertusa* under ambient and reduced pH. *Biogeosciences*, **6**, 1671–1680.

Malatesta, A. & Zarlenga, F. (1986). Northern guests in the Pleistocene Mediterranean Sea. *Geologica Roma*, **25**, 91–154.

Marin, M. G., Moschino, V., Pampanin, D. M. *et al.* (2003). Effects of hydraulic dredging on target species *Chamelea gallina* from the northern Adriatic Sea. *Journal of the Marine Biological*

Association of the United Kingdom, **83**, 1281–1285.

Marine Biological Association (1957). *The Plymouth Marine Fauna Online.* http://www.mba.ac.uk/pmf (accessed 21 August 2015).

Martin, S., Thouzeau, G., Richard, M., Chauvaud, L., Jean, F. & Clavier, J. (2007). Benthic community respiration in areas impacted by the invasive mollusk *Crepidula fornicata*. *Marine Ecology Progress Series*, **347**, 51–60.

McEdward, L. (ed) (1995). *Ecology of Marine Invertebrate Larvae.* CRC Press, New York.

Mieszkowska, N., Hawkins, S. J., Burrows, M. T. & Kendall, M. A. (2007). Long-term changes in the geographic distribution and population structures of *Osilinus lineatus* (Gastropoda: Trochidae) in Britain and Ireland. *Journal of the Marine Biological Association of the United Kingdom*, **87**, 537–545.

Migné, A., Davoult, D. & Gattuso, J.-P. (1999). Calcium carbonate production of a dense population of the brittle-star *Ophiothrix fragilis* (Echinodermata: Ophiuroidea): role in the carbon cycle of a temperate coastal ecosystem. *Marine Ecology Progress Series*, **173**, 305–308.

Mineur, F., Cook, E. J., Minchin, D. *et al.* (2012). Changing coasts: marine aliens and artificial structures. *Oceanography and Marine Biology: an Annual Review*, **50**, 189–234.

Möller, P., Pihl, L. & Rosenberg, R. (1985). Benthic faunal energy flow and biological interaction in some shallow marine soft bottom habitats. *Marine Ecology Progress Series*, **27**, 109–121.

Morrow, C. C., Thorpe, J. P. & Picton, B. E. (1992). Genetic divergence and cryptic speciation in two morphs of the common subtidal nudibranch *Doto coronata* (Ospisthobranchia: Dendronotacea: Dotoidae) from the northern Irish Sea. *Marine Ecology Progress Series*, **84**, 53–61.

Mortensen, P. B. (2001). Aquarium observations on the deep-water coral *Lophelia pertusa* (L., 1758) (Scleractinia) and selected associated invertebrates. *Ophelia*, **54**, 83–104.

Mortensen, P. B., Roberts, J. M. & Sundt, R. C. (2000). Video-assisted grabbing: a minimally destructive method of sampling azooxanthellate coral banks. *Journal of the Marine Biological Association of the United Kingdom*, **80**, 365–366.

Mortensen, P. B., Hovland, M. T., Fosså, J. H. & Furevik, D. M. (2001). Distribution, abundance and size of *Lophelia pertusa* coral reefs in mid-Norway in relation to seabed characteristics. *Journal of the Marine Biological Association of the United Kingdom*, **81**, 581–597.

Morton, B. & Dinesen, G. E. (2011). The biology and functional morphology of *Modiolarca subpicta* (Bivalvia: Mytilidae: Musculinae), epizoically symbiotic with *Ascidiella aspersa* (Urochordata: Ascidiacea), from the Kattegat, northern Jutland, Denmark. *Journal of the Marine Biological Association of the United Kingdom*, **91**, 1637–1649.

Muxika, I., Borja, Å. & Bald, J. (2007). Using historical data, expert judgment and multivariate analysis in assessing reference conditions and benthic ecological status according to the European Water Framework Directive. *Marine Pollution Bulletin*, **55** (1–6), 16–29.

Navarro, J. M. & Thompson, R. J. (1996). Physiological energetics of the horse mussel *Modiolus modiolus* in a cold ocean environment. *Marine Ecology Progress Series*, **138**, 135–148.

Navarro, J. M. & Thompson, R. J. (1997). Biodeposition by the horse mussel *Modiolus modiolus* (Dillwyn) during the spring diatom bloom. *Journal of Experimental Marine Biology and Ecology*, **209**, 1–13.

Neulinger, S. C., Järnegren, J., Ludvigsen, M., Lochte, K. & Dullo, W.-C. (2008). Phenotypic-specific bacterial communities in the cold-water coral *Lophelia pertusa* (Scleractinia) and their implications for the coral's nutrition, health and distribution. *Applied and Environmental Microbiology* **74**, 7272–7285.

Neumann, H., Reiss, H., Rakers, S., Ehrich, S. & Kröncke, I. (2009). Temporal variability in southern North Sea epifauna communities after the cold winter of 1995/1196. *ICES Journal of Marine Science*, **66**, 2233–2243.

Newell, G. E. & Newell, R. C. (1963). *Marine Plankton: A Practical Guide*. Hutchinson, London.

Newell, R. C., Seiderer, L. J. & Hitchcock, D. R. (1998). The impact of dredging works in coastal waters: a review of the sensitivity to disturbance and subsequent recovery of biological resources on the sea bed. *Oceanography and Marine Biology: an Annual Review*, **36**, 127–178.

Newell, R. C., Seiderer, L. J. & Robinson, J. E. (2001). Animal:sediment relationships in coastal deposits of the eastern English Channel. *Journal of the Marine Biological Association of the United Kingdom*, **81**, 1–9.

Nickell, L. A. & Sayer, M. D. J. (1998). The occurrence and activity of mobile macrofauna on a sublittoral reef: diel and seasonal variation. *Journal of the Marine Biological Association of the United Kingdom*, **78**, 1061–1082.

Norderhaug, K. M., Christie, H. & Rinde, E. (2002). Colonisation of kelp imitations by epiphyte and holdfast fauna; a study of mobility patterns. *Marine Biology*, **141**, 965–973.

Norderhaug, K. M., Fredriksen, S. & Nygeard, K. (2003). Trophic importance of *Laminaria hyperborea* to kelp forest consumers and the importance of bacterial degradation to food quality. *Marine Ecology* Progress *Series*, **255**, 135–144.

Norderhaug, K. M., Christie, J. H., Fosså, J. H. & Fredriksen, S. (2005). Fish–macrofauna interactions in a kelp (*Laminaria hyperborea*) forest. *Journal of the Marine Biological Association of the United Kingdom*, **85**, 1279–1286.

Orejas, C., Gori, A. & Gili, J. M. (2008). Growth rates of live *Lophelia pertusa* and *Madrepora oculata* from the Mediterranean Sea maintained in aquaria. *Coral Reefs*, **27**, 255.

OSPAR Commission (2013). Report of the EIHA Common indicator Workshop. OSPAR Biodiversity Series, 606/2013.

Ottersen, G., Stenseth, N. C. & Hurrell, J. W. (2004). Climatic fluctuations and marine systems: a general introduction to the ecological effects. In: Stenseth, N. C., Ottersen, G., Hurrell, J. W. & Belgrano, A. (eds), *Marine Ecosystems and Climate Variation*. Oxford: Oxford University Press, Oxford, pp. 3–14.

Palmer, D. W. (2004). Growth of the razor clam *Ensis directus*, an alien species in the Wash on the east coast of England. *Journal of the Marine Biological Association of the United Kingdom*, **84**, 1075–1076.

Pathmanaban, O. N., Porter, J. S. & White, R. (2005). Dogger Bank itch in the eastern English Channel: a newly described geographical distribution of an old problem. *Clinical and Experimental Dermatology*, **30**, 622–626.

Pätzold, J., Ristedt, H. & Wefer, G. (1987). Rate of growth and longevity of a large colony of *Pentapora foliacea* (Bryozoa) recorded in their oxygen isotope profiles. *Marine Biology*, **96**, 535–538.

Pearson, T. H. (2001). Functional group ecology in soft sediment marine benthos: the role of bioturbation. *Oceanography and Marine Biology: an Annual Review*, **39**, 233–267.

Peralta, G., Brun, F. G., Pérez-Lloréns, J. L. & Bouma, T. J. (2006). Direct effects of current velocity on the growth, morphometry and architecture of seagrasses: a case study on *Zostera noltii*. *Marine Ecology Progress Series*, **327**, 135–142.

Persson, M., Andersson, S., Baden, S. & Moksnes, P.-O. (2008). Trophic role of the omnivorous grass shrimp *Palaemon elegans* in a Swedish eelgrass system. *Marine Ecology Progress Series*, **371**, 203–212.

Peters, L., König, G. M., Wright, A. D. *et al.* (2003). Secondary metabolites of *Flustra foliacea* and their influence on bacteria. *Applied Environmental Microbiology*, **69**, 3469–3475.

Peterson, C. H., Luettich, R. A., Micheli, F. & Skilleter, G. A. (2004). Attenuation of water flow inside seagrass canopies of differing structure. *Marine Ecology Progress Series*, **268**, 81–92.

Picton, B. E. & Morrow, C. C. (1994). *A Field Guide to the Nudibranchs of the British Isles*. Immel, London.

Piet, G. J., Rijnsdorp, A. D., Bergman, M. J. N. *et al.* (2000). A quantitative evaluation of the impact of beam trawling on benthic fauna in the southern North Sea. *ICES Journal of Marine Science*, **57**, 1332–1339.

Pihl Baden, S. (1990). The cryptofauna of *Zostera marina* (L.): abundance, biomass and population dynamics. *Netherlands Journal of Sea Research*, **27**, 81–92.

Pihl Baden, S. & Pihl, L. (1984). Abundance, biomass and production of mobile epibenthic fauna in *Zostera marina* (L.) meadows, western Sweden. *Ophelia*, **23**, 65–90.

Pingree, R. (2005). North Atlantic and North Sea climate change: curl up, shut down, NAO and ocean colour. *Journal of the Marine Biological Association of the United Kingdom*, **85**, 1301–1315.

Pinn, E. H. & Atkinson, R. J. A. (2010). Burrow development, nutrient fluxes, carnivory and caching behaviour by *Calocaris macandreae* (Crustacea: Decapoda: Thalassinidea). *Journal of the Marine Biological Association of the United Kingdom*, **90**, 247–253.

Polte, P., Schanz, A. & Asmus, H. (2005). The contribution of seagrass beds (*Zostera noltii*) to the function of tidal flats as a juvenile habitat for dominant, mobile epibenthos in the Wadden Sea. *Marine Biology*, **147**, 813–822.

Ragaini, L., Cantalamessa, G., Di Celma, C. *et al.* (2007). First Emilian record of the boreal-affinity bivalve *Portlandia impressa* Perri, 1975 from Montefiore dell'Aso (Marche, Italy). *Bollettino della Società Paleontologica Italiana*, **45**, 227–234.

Ramsay, K., Kaiser, M. J. & Hughes, R. N. (1996). Changes in hermit crab feeding patterns in response to trawling disturbance. *Marine Ecology Progress Series*, **144**, 63–72.

Rasmussen, E. (1977). The wasting disease of eelgrass (*Zostera marina*) and its effects on environmental factors and fauna. In: McRoy, C. P. & Helfferich, C. (eds), *Seagrass Ecosystems, a Scientific Perspective*. Marcel Dekker, New York, pp. 1–51.

Rees, E. I. S., Sanderson, W. G., Mackie, A. S. Y. & Holt, R. H. F. (2008). Small-scale variation within a *Modiolus modiolus* (Mollusca: Bivalvia) reef in the Irish Sea. III. Crevice, sediment infauna and epifauna from targeted cores. *Journal of the Marine Biological Association of the United Kingdom*, **88**, 151–156.

Rees, H. L. (1984). A note on mesh selection and sampling efficiency in benthic studies. *Marine Pollution Bulletin*, **15**, 225–229.

Rees, H. L., Pendle, M. A., Waldock, R., Limpenny, D. S. & Boyd, S. E. (1998). A comparison of benthic biodiversity in the North Sea, English Channel, and Celtic Sea. *ICES Journal of Marine Science*, **56**, 228–246.

Reiss, H., Degraer, S., Duinveld, G. C. A. *et al.* (2010). Spatial patterns of infauna, epifauna, and demersal fish communities in the North Sea. *ICES Journal of Marine Science*, **67**, 278–293.

Roberts, J. M. (2005). Reef-aggregating behaviour by symbiotic eunicid polychaetes from cold-water corals: do worms assemble reefs? *Journal of the Marine Biological Association of the United Kingdom*, **85**, 813–819.

Robertson, A. I. & Mann, K. H. (1980). The role of isopods and amphipods in the initial fragmentation of eelgrass detritus in Nova Scotia, Canada. *Marine Biology*, **59**, 63–69.

Rogers, A. D. (1999). The biology of *Lophelia pertusa* (Linnaeus, 1758) and other deep-water reef-forming corals and impacts from human activities. *International Review of Hydrobiology*, **84**, 315–406.

Romanelli, M., Cordisco, A. A. & Giovanardi, O. (2009). The long-term decline of the *Chamelea gallina* L. (Bivalvia: Veneridae) clam fishery in the Adriatic Sea: is a synthesis possible? *Acta Adriatica*, **50** (2), 171–205.

Rosenberg, R., Nilsson, H. C., Hollertz, K. & Hellman, B. (1997). Density-dependent migration in an *Amphiura filiformis* (Amphiuridae, Echinodermata) infaunal population. *Marine Ecology Progress Series*, **159**, 121–131.

Rowden, A. A. & Jones, M. B. (1994). A contribution to the biology of the burrowing mud shrimp, *Callianassa subterranea* (Decapoda: Thalassinidea). *Journal of the Marine Biological Association of the United Kingdom*, **74**, 623–635.

Rumohr, H. & Kujawski, T. (2000). The impact of trawl fishery on the epifauna of the southern North Sea. *ICES Journal of Marine Science*, **57**, 1389–1394.

Sanderson, W. G., Holt, R. H. F., Kay, L. *et al.* (2008). Small-scale variation within a *Modiolus modiolus* (Mollusca: Bivalvia) reef in the Irish Sea. II. Epifauna recorded by divers and cameras. *Journal of the Marine Biological Association of the United Kingdom* **88**, 143–149.

Saunders, J. E., Attrill, M. J., Shaw, S. M. & Rowden, A. (2003). Spatial variability in the epiphytic algal assemblages of *Zostera marina* seagrass beds. *Marine Ecology Progress Series*, **249**, 107–115.

Schaffelke, B. & Lüning, K. (1994). A circa-annual rhythm controls seasonal growth in kelps *Laminaria hyperborea* and *L. digitata* from Helgoland (North Sea). *European Journal of Phycology*, **29**, 49–56.

Schanz, A. & Asmus, H. (2003). Impact of hydrodynamics on development and morphology of intertidal seagrasses in the Wadden Sea. *Marine Ecology Progress Series*, **261**, 123–134.

Schanz, A., Polte, P. & Asmus, H. (2002). Cascading effects of hydrodynamics on an epiphyte–grazer system in intertidal seagrass beds of the Wadden Sea. *Marine Biology*, **141**, 287–297.

Schratzberger, M. & Jennings, S. (2002). Impacts of chronic trawling disturbance on meiofaunal communities. *Marine Biology*, **141**, 991–1000.

Schratzberger, M., Dinmore, T. A. & Jennings, S. (2002). Impacts of trawling on the diversity, biomass and structure of meiofauna assemblages. *Marine Biology*, **140**, 83–93.

Sharp, J. H., Winson, M. K., Wade, S. *et al.* (2008). Differential microbial fouling on the marine bryozoan *Pentapora fascialis*. *Journal of the Marine Biological Association of the United Kingdom*, **88**, 705–710.

Singh-Renton, S. & Bromley, P. J. (1999). Feeding of small whiting (*Merlangius merlangus*) in the central and southern North Sea. *Journal of the Marine Biological Association of the United Kingdom*, **79**, 957–960.

Smith, S. H. (1995). Trends among echinoderms in a coastal area of the Kattegat, Sweden. In: Eleftheriou, A., Ansell, A. D. & Smith, C. J. (eds), *Biology and Ecology of Shallow Coastal Waters: Proceedings of the 28th European Marine Biology Symposium*. Olsen & Olsen, Fredensborg, pp. 323–331.

Southward, A. J. (1991). Forty years of changes in species composition and population density of barnacles on a rocky shore near Plymouth. *Journal of the Marine Biological Association of the United Kingdom*, **71**, 495–513.

Southward, A. J. & Southward, E. C. (1977). Distribution and ecology of the hermit crab *Clibanarius erythropus* in the western Channel. *Journal of the Marine Biological Association of the United Kingdom*, **57**, 441–452.

Southward, A. J. & Southward, E. C. (1988). Disappearance of the warm-water hermit crab *Erythropus clibanarius* from southwest Britain. *Journal of the Marine Biological Association of the United Kingdom*, **68**, 409–412.

Southward, A. J., Hiscock, K., Kerckhof, F., Moyse, J. & Elfimov, A. S. (2004). Habitat and distribution of the warm-water barnacle *Solidibalanus fallax* (Crustacea: Cirripedia). *Journal of the Marine Biological Association of the United Kingdom*, **84**, 1169–1177.

Southward, A. J., Longmead, O., Hardman-Mountford, N. J. *et al.* (2005). Long-term oceanographic and ecological research in the western English Channel. *Advances in Marine Biology*, **47**, 1–105.

Stebbing, A. R. D. (1971). The epizoic fauna of *Flustra foliacea* [Bryozoa]. *Journal of the Marine Biological Association of the United Kingdom*, **51**, 283–300.

Stebbing, A. R. D., Turk, S. M. T., Wheeler, A. & Clarke, K. R. (2002). Immigration of southern fish species to south-west England linked to warming of the North Atlantic (1960–2001). *Journal of the Marine Biological Association of the United Kingdom*, **82**, 177–180.

Streftaris, N., Zenetos, A. & Papathanassiou, E. (2005). Globalisation in marine ecosystems: the story of non-indigenous marine species across European seas. *Oceanography and Marine Biology: an Annual Review*, **43**, 419–453.

Swaby, S. E. & Potts, G. W. (1999). The sailfin dory, a first British record. *Journal of Fish Biology*, **54**, 1338–1340.

Teixeira, H., Weisberg, S. B., Borja, Å. et al. (2012). Calibration and validation of the AZTI's Marine Biotic Index (AMBI) for southern California marine bays. *Ecological Indicators*, **12** (1), 84–95.

ter Hofstede, R. & Rijnsdorp, A. D. (2011). Comparing demersal fish assemblages between periods of contrasting climate and fishing pressure. *ICES Journal of Marine Science*, **68**, 1189–1198.

Thiébaut, E., Dauvin, J.-C. & Lagadeuc, Y. (1994). Horizontal distribution and retention of *Owenia fusiformis* larvae (Annelida: Polychaeta) in the Bay of Seine. *Journal of the Marine Biological Association of the United Kingdom*, **74**, 129–142.

Thiébaut, E., Dauvin, J.-C. & Wang, Z. (1996). Tidal transport of *Pectinaria koreni* postlarvae (Annelida: Polychaeta) in the Bay of Seine (eastern English Channel). *Marine Ecology Progress Series*, **138**, 63–70.

Thorson, G. (1946). *Reproduction and Larval Development of Danish Marine Bottom Invertebrates*. C. A. Reitzels, Copenhagen.

Thorson, G. (1950). Reproductive and larval ecology of marine bottom invertebrates. *Biological Reviews*, **138**, 547–552.

Toth, G. B. & Pavia, H. (2002a). Intraplant habitat and feeding preference of two gastropod herbivores inhabiting the kelp *Laminaria hyperborea*. *Journal of the Marine Biological Association of the United Kingdom*, **82**, 243–247.

Toth, G. B. & Pavia, H. (2002b). Lack of phlorotannin induction in the kelp *Laminaria hyperborea* in response to grazing by two gastropod herbivores. *Marine Biology*, **140**, 403–409.

Trigg, C., Harries, D., Lyndon, A. & Moore, C. G. (2011). Community composition and diversity of two *Limaria hians* (Mollusca: Limacea) beds on the west coast of Scotland. *Journal of the Marine Biological Association of the United Kingdom*, **91**, 1402–1412.

Tuck, I. D., Hall, S. J., Robertson, M. R., Armstrong, E. & Basford, D. J. (1998). Effects of physical trawling disturbance in a previously unfished sheltered Scottish sea loch. *Marine Ecology Progress Series*, **162**, 227–242.

van Katwijk, M. M. & Hermus, D. C. R. (2000). Effects of water dynamics on *Zostera marina*: transplantation experiments in the intertidal Dutch Wadden Sea. *Marine Ecology Progress Series*, **208**, 107–118.

van Lent, F., Verschuure, J. M. & van Veghel, M. L. J. (1995). Comparative study on populations of *Zostera marina* L. (eelgrass): in situ nitrogen enrichment and light manipulation. *Journal of Experimental Marine Biology and Ecology*, **185**, 55–76.

van Nes, E. H., Amaro, T., Scheffer, M. & Duineveld, G. C. A. (2007). Possible mechanisms for a marine benthic regime shift in the North Sea. *Marine Ecology Progress Series*, **330**, 39–47.

Veale, L. O., Hills, A. S., Hawkins, S. J. & Brand, A. R. (2001). Distribution and damage to the by-catch assemblages of the northern Irish Sea scallop dredge fisheries. *Journal of the Marine Biological Association of the United Kingdom*, **81**, 85–96.

Verdier-Bonnet, C., Carlotti, F., Rey, C. & Bhaud, M. (1997). A model of larval dispersion coupling wind-driven currents and vertical larval behaviour: application to the recruitment of the annelid *Owenia fusiformis* in Banyuls

Bay, France. *Marine Ecology Progress Series*, **160**, 217–231.

Vorberg, R. (2000). Effects of shrimp fisheries on reefs of *Sabellaria spinulosa* (Polychaeta). *ICES Journal of Marine Science*, **57**, 1416–1420.

Wall, C. C., Peterson, B. J. & Gobler, C. J. (2008). Facilitation of *Zostera marina* productivity by suspension-feeding bivalves. *Marine Ecology Progress Series*, **357**, 165–174.

Warner, G. F. (1971). On the ecology of a dense bed of the brittle star *Ophiothrix fragilis*. *Journal of the Marine Biological Association of the United Kingdom*, **51**, 267–282.

Warwick, R. M. & Clarke, K. R. (2001). Practical measures of marine biodiversity based on relatedness of species. *Oceanography and Marine Biology: an Annual Review*, **39**, 207–231.

Warwick, R. M. & Davies, J. R. (1977). The distribution of sublittoral macrofauna communities in the Bristol Channel in relation to the substrate. *Estuarine and Coastal Marine Science*, **5**, 267–288.

Wenngren, J. & Ólafsson, E. (2002). Intraspecific competition for food within and between year classes in the deposit-feeding amphipod *Monoporeia affinis*: the cause of population fluctuations? *Marine Ecology Progress Series*, **240**, 205–213.

Whittick, A. (1983). Spatial and temporal distributions of dominant epiphytes on the stipes of *Laminaria hyperborea* (Gunn.) Fosl, (Phaeophyta: Laminariales) in S. E. Scotland. *Journal of Experimental Marine Biology and Ecology*, **73**, 1–10.

Whomersley, P. & Picken, G. B. (2003). Long-term dynamics of fouling communities found on offshore installations in the North Sea. *Journal of the Marine Biological Association of the United Kingdom*, **83**, 897–901.

Whomersley, P., Huxham, M., Schratzberger, M. & Bolam, S. (2009). Differential response of meio- and macrofauna to *in situ* burial. *Journal of the Marine Biological Association of the United Kingdom*, **89**, 1091–1098.

Widdicombe, S. & Austen, M. C. (1999). Mesocosm investigation into the effects of bioturbation on the diversity and structure of a subtidal macrobenthic community. *Marine Ecology Progress Series*, **189**, 181–193.

Widdicombe, S., Austen, M. C., Kendall, M. A. *et al.* (2004). Importance of bioturbators for biodiversity maintenance: indirect effect of fishing disturbance. *Marine Ecology Progress Series*, **275**, 1–10.

Widdows, J., Pope, N. D., Brinsley, M. D., Asmus, H. & Asmus, R. M. (2008). Effects of seagrass beds (*Zostera noltii* and *Z. marina*) on near-bed hydrodynamics and sediment resuspension. *Marine Ecology Progress Series*, **358**, 125–136.

Willems, K. A., Vanosmael, C., Claeys, D., Vincx, M. & Heip, C. (1982). Benthos of a sublittoral sandbank in the southern bight of the North Sea: general considerations. *Journal of the Marine Biological Association of the United Kingdom*, **62**, 549–557.

Williams, M., Haywood, A. M., Harper, E. M. *et al.* (2009). Pliocene climate and seasonality in North Atlantic shelf seas. *Philosophical Transactions of the Royal Society A*, **367**, 85–108.

Wilson, J. B. (1979). Patch development of the deep-water coral *Lophelia pertusa* (L.) on Rockall Bank. *Journal of the Marine Biological Association of the United Kingdom*, **59**, 165–177.

Witbard, R., Duineveld, G. C. A., Amaro, T. & Bergman, M. J. N. (2005). Growth trends in three bivalve species indicate climate forcing on the benthic ecosystem in the southeastern North Sea. *Climate Research*, **30**, 29–38.

Yonge, C. M. (1949). *The Sea Shore*. New Naturalist 12. Collins, London.

Yonge, C. M. & Thompson, T. E. (1976). *Living Marine Molluscs*. Collins, London.

Zintzen, V., Norro, A., Massin, C. & Mellefet, J. (2008). Temporal variation of *Tubularia indivisa* (Cnidaria, Tubulariidae) and associated epizoites on artificial habitat communities in the North Sea. *Marine Biology*, **153**, 405–420.

Zühlke, R., Alsvåg, J., de Boois, I. *et al.* (2001). Epibenthic diversity in the North Sea. *Senckenbergiana Maritima*, **31**, 269–281.

Index

SPECIES INDEX

Page numbers in **bold** include illustrations.

Abra alba 35, 37, 87, 92, 224, 242, 263, 264, 275
  A. nitida 93, 275, 353
Acanthocardia echinata 284
Actinauge richardi 116
Aequipecten opercularis (Queen Scallop) 142, 241, 248, 260, **261**, 280, 282, 297, 298, 328, 333, 334–5
Alcyonidium diaphonum 94, **95**, 99, 165
Alcyonium digitatum 94, 100, **149**, 170, 179, 261, 271
Alvania 242
Ammodytes 58, 143
Amphiblestrum flemingii 164
Amphiura filiformis 35, 92, 98, 119–22, **120**, 126, 127, 243, 276, 336
  A. chiajei 35, 124
Anapagurus hyndmanni **272**
  A. laevis 100, 104, 108, 114
Anemone, Plumose (*Metridium senile*) 100, **147**, 157, 177, 179
Anseropoda placenta (Goosefoot Starfish) 98, 114

Antedon bifida (Feather Star) 98, 114
  A. petasus 278
Aphrodita aculeata (Sea Mouse) 37, 86, **87**, 273, 351
Aplysia punctata (sea hare) **199**, 200
Aporrhais 350
  A. pespelecani (Pelican's Foot Shell) 36
  A. serresianus 275
Arctica islandica 280, **281**, 284, 350, 353
Arenicola marina 35
Argopecten irradians (Bay Scallops) 230
Arnoglossus laterna (Scaldfish) 61
Ascidia mentula 152, **153**, 158, **159**, 169
Ascidiella scabra 61, 169
  A. aspersa 158, **159**, 160, 169, 177, 261, 278
Aspitrigla cuculus (Gurnard) 332
Astarte sulcata 35, 87, **88**
Asterias rubens 37, **42**, 94, 95, **97**, 98, 99, 100, 101, 105, 107, 114, 116, 123, 137, 140, 142, 167, 180, 228, 327, 328, 329, 330, 331, 332, 334, 336

SPECIES INDEX · 383

Astropecten irregularis 37, 95, 98, 99, 101, 105, 107, 108, 114, 123, **138**, 142, 327, 329
Atelecyclus rotundatus (Round Crab) 262, **263**
Aurelia aurita (Moon Jellyfish) 117

Balanoglossus **43**
Balanophyllia regia 264
Balanus crenatus (barnacle) 175, 297
Balistes carolinensis 295
Banded Chink Shell (*Lacuna vincta*) 193, **194**, 195, 198, 200, 202
barnacle (*Balanus crenatus*) 175, 297
Bass (*Dicentrarchus labrax*) 299
Beroe 315
Bittium reticulatum 243
Botryllus schlosseri **159**
Branchiostoma lanceolatum 356
Branta bernicla (Brent Goose) 215, 290
Brill (*Scophthalmus rhombus*) 348
Brissopsis lyrifera 35, **36**, 37, 82, 86, 87, 124, 358
Buccinum undatum (Common Whelk) 39, 244, **275**, 284, 328, 334, 336, 350
Buglossidium luteum (Solenette) 61, 108, 143, 144
Bugula flabellata 161–2, **163**
Bull Rout (*Myoxocephalus scorpius*) 167, 168, 201, 202

Calanus finmarchicus 201
Callianassa subterranea 35, 37, 80, 81, 92, 121, 122, 353, 356
Callionymus lyra (Dragonet) 109, **111**, 143
Calliostoma 275
Callopora dumerilii 164
Calocaris macandreae 35, **80**, 81

Cancer pagurus (Edible Crab) **33**, 200, 278, 292, 329, 334
Caprella tuberculata 178
Carcinus maenas (Green Crab) **42**, 136, 180, 224, **225**, 229, 315–16, 329, 330
Caryophyllia smithii (Devonshire Cup Coral) 114, 116, 264
Ceramium 218
Cerianthus lloydii 37, 82, **84**, 240
Chaetopterus variopedatus (Parchment Worm) 48, 84, 92, 240, 273
Chamelea striatula (Striped Venus) 35, 69, **97**, 140, 285, **286**, 287, 338, 341, 350
C. gallina (Mediterranean Striped Venus) 69, 285–6, 338, 342–5, 356
Chelonia mydas (Florida Green Turtle) 288–9
Chlamys varia 241, 260, **261**
Chthalamus 287
C. montagui **42**
Ciona intestinalis **155**, 156–7, 176, 177, 261, 263, 278
Cladophora 218
Clam, Hard (*Mercenaria mercenaria*) 230, 231, 306, 309
Manila (*Ruditapes philippinarum*) 69, 314, 338
Sand Gaper (*Mya arenaria*) 53, 136, 224, 307
Clausinella fasciata 241, 341
Clibanarius erythropus 291–2
Clupea harengus (Herring) 30, 109, 142, 287–8
Cocconeis scutellum 216
Cockle (*Cerastoderma edule*) 134, 136, 140, 224, 284, 339, 340
Dog (*Glycymeris glycymeris*) **25**, 36, 75, 112, 240, 241

Cod (*Gadus morhua*) 143, 200, 201, 202, 218, 220, 228, 244, 348
  Poor (*Trisopterus minutus*) 144, 299
Colpomenia peregrina 310
Colus gracilis (Slender Whelk) 108, **110**, 244, 350
Conger conger (Conger Eel) 292
Coral, Devonshire Cup (*Caryophyllia smithii*) 114, 116, 264
  Ross (*Pentapora foliacea*) **157**, 160, 162–5, 253
  Stone (*Lophelia pertusa*) 180, 181, 233, 264–77, **265**, 325, 326
Corallina officinalis 197
Cormorant (*Phalacrocorax carbo*) 200
Corophium 256
  C. sextonae 308
  C. volutator 222, **223**
Corynactis viridis **252**
Corystes cassivelaunus 107
Crab, Angular (*Goneplax rhomboides*) 62, **79**
  Edible (*Cancer pagurus*) **33**, 200, 278, 292, 329, 334
  Green (*Carcinus maenas*) **42**, 136, 180, 224, **225**, 229, 315–16, 329, 330
  Round (*Atelecyclus rotundatus*) 262, **263**
  Sponge (*Dromia personata*) **295**
  Stone (*Lithodes maja*) 278
Crangon allmani 101, **102**, 107, 116, 328, 329
  C. crangon (Brown Shrimp) 107, 134–6, **135**, 142, 143, 144, 224, 229, 300, 301, 329
Crassostrea gigas (Pacific Oyster) 203, 316
  C. virginica (Eastern Oyster) 230, 232, 311

Crepidula fornicata (American Slipper Limpet) 130, **131**, 308, 309, 311–12, 316–17
Crisia 336
  C. eburnea 161
Crossaster papposus (Common Sun Star) 276, **277**
Cultellus pellucidus 353
Cushion Star (*Porania pulvillus*) 114, **115**, 276
Cuspidaria rostrata **275**
Cymodoce nodosa 203

Dab (*Limanda limanda*) 109, 143, 144, 299, 329, 332, 358
  Long Rough (*Hippoglossoides platessoides*) 109, 143, 144, 329
Delesseria sanguinea 197, 214, 228
Dendrobeania 161
Dendrodoa grossularia 155, 170
Dendronotus frondosus 170, **171**, 178
Dicentrarchus labrax (Bass) 299
Dinophysis 130
Donax trunculus 35, 338, 341, 356
  D. vittatus 338
Dory, Sailfin (*Zenopsis conchifer*) 294
Dosinia exoleta **244**, 341, 347, 350
  D. lupinus 35, 353
Doto 171–3, 243, 251
  D. fragilis **172**
Dragonet (*Callionymus lyra*) 109, **111**, 143
Dromia personata (Sponge Crab) **295**
Dyctyosyphon foeniculaceus 216

Ebalia 274
  E. tuberosa 114, 262, **274**
Echinocardium cordatum (Sea Potato) 35, **36**, **42**, 50, 82, 90, 92, 98
  E. flavescens 35, 276

E. pennatifida 240
Echinocyamus pusillus 246, 276
Echinus 98, 105, 276
   E. esculentus (Edible Sea Urchin) 167, **168**, 193, 199, 200, 334
Ecrobia ventrosa 223
Ectocarpus 218
   E. siliculosa 216
Ectopleura larynx 170, **172**, 176
Edwardsia claparedii 244
Eel, Conger (*Conger conger*) 292
Eel Grass, Common (*Zostera marina*) **204**, 205–15, 216, 218, 219, 220, 221, 223, 224, 226, 227, 228, 229, 230, 231, 232, 290
   Dwarf (*Nanozostera noltii*) 204, 205, 207, 209, 210, 214, 215, 221
Eider Duck (*Somateria mollissima*) 200
Electra pilosa 160, **161**
Eledone cirrhosa (Curled Octopus) **334**
Emarginula 275
Ennucula tenuis 86, 87
Ensis 35, 274, 338–9, 345
   E. arcuatus 241, 341, 345, **346**, 350
   E. directus (American Razor Clam) 92, **312**, 313, 338
   E. siliqua 35, 341, 345, **346**, 350, 356
Epizoanthus incrustans 272
Erichthonius difformis 222, **223**
Eubranchus tricolor **243**
Euclymene robusta 273
Eulalia 273
Eumida 273
Eunice norvegica 273
Eunicella verrucosa 298
Eurynome aspersa 114
Euspira montagui 244
   E. nitida 244

Eutrigla gurnardus (Grey Gurnard) 109, 329, 332

Feather Star (*Antedon bifida*) 98, 114
Ficopotamus enigmaticus **174**, 310
File Shell, Gaping (*Limaria hians*) **248**, 249, 274, 338
Flounder (*Platichthys flesus*) 144
Flustra foliacea (Hornwrack) 94, **96**, 99, 112, 147, 157, 160–2, 164, 165, 251
Fucus serratus 169, 190, 193, 223
   F. vesiculosus 223
Fulmarus glacialis (Fulmar) 58
Funiculina quadrangularis 82

Gadus morhua (Cod) 143, 200, 201, 202, 218, 220, 228, 244, 348
Galathea 274
   G. squamifera **202**
Gammarus locusta 218, 220, 223, 224
   G. oceanus 229
Gari tellinella 242
Gasterosteus aculeatus 225
Geodia 271
   G. barretti 270
Gibbula 242
Glycymeris glycymeris (Dog Cockle) **25**, 35, 75, 112, 140, 241
Gobiusculus flavescens (Two Spot Goby) **226**
Goby, Sand (*Pomatoschistus minutus*) 61, 143, 225
   Two Spot (*Gobiusculus flavescens*) **226**
Golfingia vulgaris 256
Goneplax rhomboides (Angular Crab) 62, **79**
Goniodoris castanea 170
   G. nodosa 170
Goose, Brent (*Branta bernicla*) 215, 290

Guillemot, Common (*Uria aalge*) 56, 58
Gurnard (*Aspitrigla cuculus*) 332
 Grey (*Eutrigla gurnardus*) 109, 143, 329, 332
Gymnodinium 130

Haddock (*Melanogrammus aeglefinus*) 109, 143, 348
Hagfish (*Myxine glutinosa*) 109, 143
Halecium 251
Halophila decipiens 203
 H. stipulacea 203
Harmothoe 273
Hathrometra sarsi 278
Heart Urchin, Purple (*Spatangus purpureus*) 35, 36, 76, 98, 334, **335**
Hediste diversicolor 134
Henricia 276
 H. oculata 98
 H. sanguinolenta 350
Herring (*Clupea harengus*) 30, 109, 142, 287–8
Hesionura augeneri 76
Heteranomia squamula 274
Heteromastus filiformis 134
Hiatella arctica 200, **201**, 256
Hippasteria phryngiana 276
Hippoglossoides platessoides (Long Rough Dab) 109, 143, 144, 329
Hippolyte 225
Hornwrack (*Flustra foliacea*) 94, **96**, 99, 112, 147, 156, 160–2, 164, 165, 251
Hormathia digitata 108
Hyalinoecia tubicula 108
Hyas araneus 274
 H. coarctatus 99, **101**, 108, **274**
Hydractinia echinata 100, 101, **102**
Hydrobia 223

Hydroides norvegica 272

Idotea balthica 218, 220, 223, 227, 228
 I. chelipes 223
 I. granulosa 223
Inachus 274

Jassa 177
 J. herdmani 178
Jaxea nocturna 79
Jellyfish, Moon (*Aurelia aurita*) 117

Kelp, Sugar (*Laminaria saccharina*) 183, **184**, 185, 186, 187, 188, 190, 192
Kirchenpaueria pinnata 251
Kittiwake (*Rissa tridactyla*) 58
Kurtiella bidentata 37, 92, 119, **120**, 121, 125, 126, 127, 243

Labrus bergylta (Wrasse, Ballan) 201, 202
Lacuna vincta (Banded Chink Shell) 193, **194**, 195, 198, 200, 202
Lafoea dumosa 337
Lagis koreni 84, 118, 333
Lamellaria 275
Laminaria 32, 181–97, 200, 216
 L. digitata 183, **184**, 185, 187, 188, 190, **191**, 192, **194**, **195**, 197, **198**
 L. hyperborea 155, **168**, 183, **184**, 186, 188, **189**, 190, 192, **196**, 197, 200–1
 L. ochroleuca 183, 189, 190
 L. saccharina (Sugar Kelp) 183, **184**, 185, 186, 187, 188, 190, 192
Lanice conchilega 253
Leptasterias muelleri 98, 350
Leucothoe spinicarpa 158, **158**
Limacia clavigera 243
Limanda limanda (Dab) 109, 143, 144, 299, 329, 332, 358

SPECIES INDEX · 387

Limaria hians (Gaping File Shell) **248**, 249, 274, 338
  L. loscombi 274
Limaropsis aurita 274
Limatula subauriculata 274
Limpet, Blue–rayed (*Patella pellucida*) 183, 193, **194**, **195**, 200
  American Slipper (*Crepidula fornicata*) 130, **131**, 308, 311–12, 316–17, 309
Liocarcinus 143, 262, 292, 327, 335
  L. depurator 99, 114, 328, 329, **330**
  L. holsatus 61, 99, 101, 114, 116
Liparis liparis (Sea Snail) 299, 301
  L. montagui 200, 244
Lithodes maja (Stone Crab) 274, 278
Lithophyllum racemus 235
Lithothamnion corallioides 235, 238, 239, 247
  L. glaciale 235, 239
  L. incrustans 237
  L. lophiforme 235
Littorina littorea 228
Lobster, Norway (*Nephrops norvegicus*) 35, **78**, 79, 116, 292, 326–8, 329, 332
Lophelia pertusa (Stone Coral) 180, 181, 233, 264–77, **265**, 325, 326
Luidia ciliaris 98, 123, 167, 276, 328
  L. sarsi 98, 123, **139**, 140, 142, 167, 276, 351
Lutraria angustior 241, 341

Mackerel (*Scomber scombrus*) 142, 201
Macoma balthica (Baltic Tellin) 35, **36**, 54, 132, 136, 224, 315, 340
Macropipus tuberculatus 114
Macropodia tenuirostris 114
Madrepora oculata 270, 271
Maja brachydactyla 292, 298

Marshallora adversa 243
Marthasterias glacialis (Spiny Starfish) 98, 137, **139**, 140, 141, 142, 276, 328
Maxmuelleria lankesteri 37, 85–6, 89, 359
Melanogrammus aeglefinus (Haddock) 109, 143, 348
Membranipora **43**, 190–2
  M. membranacea 167, **169**, 190, **191**, 192,
Membraniporella nitida 164, **169**
Membranoptera alata 196, 197
Mercenaria mercenaria (Hard Clam) 230, 231, 306, 309
Metridium senile (Plumose Anemone) 100, **147**, 157, 177, 179
Microdeutopus gryllotalpa 223
Microstomus kitt (Lemon Sole) 144
Mnemiopsis leidyi 315
Modiolarca subpicta 158–60, 169
Modiolus modiolus (Horse Mussel) 31, 35, 181, 233, 248, 254, 256–64, **257**, **262**, 283
Monocorophium insidiosum 222, 223
Monoporeia affinis 37, 54, 126–9, 136
Monostroma grevillea 216
Munida 274
  M. rugosa 278, 329
Munidopsis serricornis 278
Mussel, Blue (*Mytilus edulis*) 136, 179, 224, 228, 230, 256, 283–4
  Horse (*Modiolus modiolus*) 31, 35, 181, 233, 248, 254, 256–64, **257**, **262**, 283
Mya arenaria (Sand Gaper Clam) **53**, 136, 224, 307
  M. truncata 240, 241
Mytilus edulis (Blue Mussel) 136, 179, 224, 228, 256, 283–4
  M. trossulus 203

Myxine glutinosa (Hagfish) 109, 143
Myoxocephalus scorpius (Bull Rout) 167, 168, 201, 202

Nanozostera noltii (Dwarf Eelgrass) 204, 205, 207, 209–10, 214, 215, 221
Nassarius 275
    N. reticulatus **40**
Neanthes 273
Necora puber 330
Nemertesia antennina **172**, 251, **254**
    N. ramosa 251
Neopentadactyla mixta 240
Neotromatella monostromatica 216
Nephrops norvegicus (Norway Lobster, Dublin Bay Prawn) 35, **78**, 79, 116, 292, 326–8, 329, 332
Nephtys 144, 325
    N. hombergii 35, 134
Neptunea antiqua (Red Whelk) 108, **110**, 244, 328, 336, 350
    N. contraria **280**
Nereis 273
Nerophis ophidion (Straight-nosed Pipefish) 225
Notopterophorus papilio 158,
Nucula 36, 275
    N. nitidosa 35, 37, 92, 116
    N. nucleus 35, 241
Nuculana 275

Obelia dichotoma 175
Octopus, Common (*Octopus vulgaris*) 291
    Curled (*Eledone cirrhosa*) **334**
Oenopota 275
Onchidoris 170
    O. bilamellata 243
    O. muricata 243

Onoba 242
    O. vitrea 127
Ophiocomina nigra 98, 112, **137**, 165, 244–5, 260, 262, 336
Ophiopholis aculeata 336
Ophiothrix fragilis 35, **42**, 94, 95, 98, 99, 112, 118, **119**, 137, 155, 165–7, **257**, 261, 263, 264, 276, 336
    O. luetkeni 116
Ophiura albida 61, 98, 105, 107, 336
    O. ophiura 35, 61, 98, 99, 101, 105, 107, 114, 116, 137, **138**, 140, 141, 144, 327, 328, 329, 331, 332, 336
Owenia fusiformis 37, 92–3, 117–18, 240, 244, 253
Oyster, Eastern (*Crassostrea virginica*) 230, 232, 311
    Pacific (*Crassostrea gigas*) 203, 316
Oyster Drill, American (*Urosalpinx cinerea*) 306, **307**
oysters 130, 148, 203, 230, 274, 306, 289, 290, 306, 309, 310, 311, 316, 338, 347

Pagurus 143
    P. bernhardus 61, 100, 101, **102**, 104, 105, 107, 298, 328, 329, 330, 331, 332, 333, 336
    P. prideaux 104, 105, 114, 116, 332, 333
    P. pubescens 100, 104, 108
    P. variabilis 116
Palaemon 224
    P. adspersus 218, 220, 229
    P. elegans 218, 220
Palliolum tigerinum 241, **242**
Palmaria palmata 196, 197, **198**
Pandalus 143, 278, 329
    P. montagui 101
Paracentrotus lividus 245, 246
Paragorgia arborea 271

Paramuricea placomus 271
Parchment Worm (*Chaetopterus
  variopedatus*) 48, 84, 92, 240, 273
Patella pellucida (Blue Rayed
  Limpet) 183, 193, **194, 195**, 200
Pecten jacobeus 285
  P. maximus (Great Scallop) 37, **241**,
    247, 285, 298, 333, 334, 335
Pelican's Foot Shell (*Aporrhais
  pespelecani*) 36
Pennatula phosphorea 82, **83**, 116
Pentapora foliacea (Ross Coral) **157**, 160,
  162–5, 253
Perforatus perforatus **296**, 297
Peringia ulvae 221
Perophora japonica 309
Phaeocystis pouchetti 132
Phalacrocorax carbo (Cormorant) 200
Pharus legumen 341
pilchard 287, 288, 291
Philine 243
Philoceras trispinosus 107
Phorcus lineatus 292, **293**
Phtisica marina 178
Phycodrys rubens 196, 197
Phymatolithon calcareum 234–5, 239, 247
Pilaiella littoralis 216
Pipefish, Deep-nosed (*Syngnathus
  typhle*) 225, **226**
  Straight-nosed (*Nerophis
    ophidion*) 225
Pisidia longicornis 160, 164, 178, 262
Plaice (*Pleuronectes platessa*) 109, 134,
  144, 201, 303–5, 329, 348, 358
Platichthys flesus (Flounder) 144
Pollack (*Pollachius pollachius*) 244
Pollachius virens (Saithe) 201, 244
Polycarpa pomaria 278
Polycera quadrilineata 191, 243

Polyclinum aurantium **159**
Polysiphonia 218
Pomatoceros triqueter 155, 175
Pomatoschistus 134, **135**
  P. minutus (Sand Goby) 61, 143, 225
  P. microps 225
  P. pictus 225
Pondweed, Beaked Tassel (*Ruppia
  maritima*) 204
Pontophilus spinosus 114
Porania pulvillus (Cushion Star) 114,
  115, 276
Posidonia oceanica 204
Potamogeton 204
Pout (*Trisopterus luscus*) 144, 299
  Norway (*Trisopterus esmarkii*) 109
Protanthea simplex 278, 327, 329
Psammechinus miliaris 94, 98, 107, 160,
  245, 246
Pseudamussium peslutrae **284**
Pseudoprotella phasma 177
Ptilota gunneri 196, 197
Pungitius pungitius 225
Pycnogonum littorale 116

Radix baltica 223
Raniceps raninus (Tadpole Fish) 168
Razor Clam, American (*Ensis
  directus*) 92, **312**, 313, 338
razor shells 35, 274, 338–9, 341, 345–7,
  346, 356
Retusa truncata 243
Rhizostoma octopus 117
Rissa tridactyla (Kittiwake) 58
Rissoa 198, 223, 242
  R. membranacea **199**, 228
  R. parva 202, 203
Rout, Bull (*Myxocephalus scorpius*) 167,
  168, 201, 202

Ruditapes philippinarum (Manila Clam) 69, 314, 338
Ruppia maritima (Beaked Tassel Pondweed) 204

Sabellaria alveolata 254–6
S. *spinulosa* 31, 147, 254–6, 261, 264
Saccorhiza polyschides 182, 183, 190, 192
Saithe (*Pollachius virens*) 201, 244
Sarcodictyon roseum **271**
Scad (*Trachurus trachurus*) 109, 143
Scaldfish (*Arnoglossus laterna*) 61
Scallop, Bay (*Argopecten irradians*) 230
  Great (*Pecten maximus*) 37, **241**, 247, 285, 298, 333, 334, 335
  Queen (*Aequipecten opercularis*) 142, 241, 247, 260, **261**, 280, 282, 297, 298, 328, 333, 334–5
Scalpellum scalpellum 116
Scoloplos armiger 61, 134
Scomber scombrus (Mackerel) 142, 201
Scopelocheirus hopei 331
Scophthalmus maximus (Turbot) 348
  S. *rhombus* (Brill) 348
Scrobicularia plana 315
Scrupocellaria **162**, 336
  S. *reptans* 161
Sea Hare (*Aplysia punctata*) **199**, 200
Sea Mouse (*Aphrodita aculeata*) 37, 86, **87**, 273, 351
Sea Potato (*Echinocardium cordatum*) 35, **38**, **42**, 50, 82, 90, 92, 98
Sea Snail (*Liparis liparis*) 299, 301
Sea Urchin, Edible (*Echinus esculentus*) 167, **168**, 193, 199, 200, 334
Semibalanus balanoides 287
Serpula vermicularis 253, 254, 272
Sertularia 336
  S. *cupressina* 251
Shrimp, Brown (*Crangon crangon*) 107, 134–6, **135**, 142, 143, 144, 224, 229, 300, 301, 329
Smittoidea reticulata 164
Sole, Dover (*Solea solea*) 134, 292, 299, 302–5, 348, 358
  Lemon (*Microstomus kitt*) 144
Solenette (*Buglossidium luteum*) 61, 108, 143, 144
Solidobalanus fallax 297, **298**, 309
Somateria mollissima (Eider Duck) 200
Spadella cephaloptera 74
Spatangus purpureus (Purple Heart Urchin) 36, 76, 334, **335**
Sphaerechinus granularis 245
Spiophanes bombyx 61
Spinachia spinachia 225
Spisula elliptica 35, 284, 350
  S. *solida* 35, 76, 284, 338, 341, 356
  S. *subtruncata* 92, 353
Sprat (*Sprattus sprattus*) 142, 144
Starfish, Goosefoot (*Anseropoda placenta*) 98, 114
Starfish, Spiny (*Marthasterias glacialis*) 98, 137, **139**, 140, 141, 142, 276, 328
Strongylocentrotus droebachiensis 192–3
Sun Star, Common (*Crossaster papposus*) 276, **277**
Syngnathus typhle (Deepnosed Pipefish) 225, **226**

Tadpole Fish (*Raniceps raninus*) 168
Talochlamys pusio 274
Tapes rhomboides 241, **242**, 336, 341
Taurulus bubalis **167**
Tectura virginea 242

SPECIES INDEX · 391

Tellin, Baltic (*Macoma balthica*) 35, **36**, 54, 132, 136, 224, 315, 340
　Thin (*Tellina tenuis*) 35, **36**, 134, 340
Terebratula maxima **279**
Teredo navalis 307
Testudinalia testudinalis 242
Theodoxus fluviatilis 223
Thracia convexa 353
Thyasira flexuosa 93
Timoclea ovata 337
Trisopterus minutus (Poor Cod) 144, 299
Tritonia hombergii **170**, 171
Trivia arctica 243, 275, **276**
　T. monacha 243
Topknot (*Zeugopterus punctatus*) 168
Trophon 275
Tubularia indivisa 170, **176**, 177–9
Turbot (*Scophthalmus maximus*) 348
Turtle, Florida Green (*Chelonia mydas*) 288–9
Turritella communis (Auger Shell) 35, 37, 114

Ulothrix 216
Ulva 218
Undaria pinnatifida 182
Upogebia deltaura 35, 80, 81, 240, 243, 246–7, 338, 353, 356
　U. stellata 80, 81, 92, 356
Uria aalge (Common Guillemot) 58
Urosalpinx cinerea (American Oyster Drill) 306, **307**

Velutina velutina 275

Venerupis senegalensis 256
Venus, Mediterranean Striped (*Chamelea gallina*) 69, 285–6, 338, 342–5, 356
　Striped (*Chamelea striatula*) 35, 69, **97**, 140, 285, **286**, 287, 341, 350
Venus 264
　V. casina 35, 241, **280**
　V. fasciatus 76
　V. verrucosa 337
Victorella pavida 307–8
Virgularia mirabilis 82

Watersipora aterrima 309
Whelk, Common (*Buccinum undatum*) **39**, 244, **275**, 284, 328, 334, 336, 350
　Red (*Neptunea antiqua*) 108, **110**, 244, 328, 336, 350
　Slender (*Colus gracilis*) 108, **110**, 244, 350
Whiting (*Merlangius merlangus*) 109, 143, 144, 299, 301, 332, 329, 348
Wigeon (*Anas penelope*) 215, 290
Wrasse, Ballan (*Labrus bergylta*) 201, 202

Zenopsis conchifer (Sailfin Dory) 294
Zeugopterus punctatus (Topknot) 168
Zostera 204–32
　Z. angustifolia 207
　Z. marina (Common Eelgrass) **204**, 205–215, 216, 218, 219, 220, 221, 223, 224, 226, 227, 228, 229, 230, 231, 232, 290

GENERAL INDEX

Page numbers in **bold** include illustrations.

acoustic surveys 14–15
actaeplanic larvae 47
Adriatic Sea 341–5
algae 175, 176, 186, 192, 196, 197, 206, 216, 219, 222, 238, 249, 250, 261, 269, 309
   brown 216
   coralline 32, 145, 148–9, 181, **189**, 192, 213, 233, **234**, **235**, 236–7
   green 179, 216, 224
   red 158, 188, 200, 214, 228
Algarve 356–7
alien species 306–17
   critical conditions 311
   via aquaculture 309–10
   via human agency 311–12
   via shipping 307, 309, 310
Ampharetidae 93
amphipods 54, 65, 90, 92, 126, 128, 129, 136, 144, 150, 155, 158, 160, 164, 169, 177, 178–9, 198–203, 215, 218, 223–4, 225, 226, 228–9, 253, 256, 263, 273, 308, 325, 331, 332
anchiplanic larvae 47
aplanic larvae 47
Argentine Shelf 44
Aristotle 4
Atlantic Multidecadal Oscillation (AMO) 20–2, 287
Atlantic Subarctic Upper Water 24
Atlantic, northwest 44, 92, 192, 214, 288–90, 314

Baltic Sea 54, 124, 126, 134, 136, 183, 204, 206, 214, 218, 225, 254, 309, 315

barnacles 152, 168, 173, 175, 263, 297, **298**, 299
bathyal zone 2
Belgian Coast 75–8, 179, 279, 313
'benthic', definition 2
   development of study of 4–11
   zones and characteristics of 2–3
bentho–pelagic coupling 124ff
Bill Bailey's Bank 17
biodiversity indices 62–9
   AZTI Marine Biotic Index (AMBI) 68–9
   Shannon–Wiener index 61–2, 64
   Simpson Index 61–2
   taxonomic diversity and distinctness indices 65–8
biogenic reefs 145, 231, 251–78
   locations 251
biotopes 38
bioturbation 86–93
Biscay, Bay of 256, 283–5, 307
bivalves 25, 32, 33, 38, 54, 67, 68, 75, 80, 86, **90**, **91**, 92, 93, 95, 117, 123, 125, 132, 136, 137, 140, 141–3, 144, 146, 154, 164, 190, 198, 200, 224, 229–32, 241, 243, 244, 250, 259, 263, 274, 280, 283–4, 290, 292
Black Sea 8, 286, 308, 315
blennies 168
Bothnia, Gulf of 126, 129
box corers 13, 76
brachiopod **279**
Brent Spar 180
Brest, Bay of 130–1, 245, 311, 316–7
Bridgwater Bay 26, 135, 144, 299–302

GENERAL INDEX · 393

Bristol Channel 26, 94–5, 98–99, 114, 134, 165, 256, 263, 264, 297, 300, 320
Brittany Coast 7, 75, 114, 130, 137, 154, 183, 185, 234–6, 238–9, 240, 245, 283, 285, 291–2, 298, 311
brittlestars 30, 32, 41, 61, 92, 94, 99, 105, 112, 116, 118–22, **119**, **120**, 123–4, 126, 136, **137**, **138**, 144, 155, 165–7, 200, 243–5, **257**, 260–2, 276, 327, 329, 336, 350
bryozoans 4, 25, 32, **33**, 38, 41, **43**, 44, 46–7, 94, **95**, **96**, 99, 101, 112, **113**, 146–7, **148**, 150, 152, 154, 156–8, 160–1, **162**, **163**, 164–5, 167, 168, **169**, 170, 173, 178, 190–2, 195, 197, 200, 213, 233, 243, 249, 251, **252**, **262**, 272, 283, 307, 336
Busk, Rev. G. 6
bycatch/discards 56, 303, 311, 327–37, 339, 341–2

calcium carbonate production 236–7, 238
Canada 7, 69, 182, 192–3, 239, 315
Canary Isles 44
carboniferous limestone 175
Celtic Seas Ecoregion 8, 17
Central Electricity Generating Board (CEGB) 7
Centre for Environment, Fisheries and Aquaculture Science (CEFAS) 7, 94, 99
chaetognaths (arrow worms) 30, 74
Challenger, HMS 3
Chesapeake Bay 289–90
Chile 44
chlorophyll 93, 124–6, 129, 130–1, 230–1, 260

clams 36, 53, 69, 76, **97**, 130, 136, 179, 224, 230–2, 240, 241, 248, 280, 290, 306, 307, 310, 312–13, 314, 316, 323, 339, 314, 315, 338–9, 340, 341, 342, 343–5, **346**, 347, 356, 357
dredging 338–9, 341–5, 357
climate change 293–306
previous 287
clonal animals, definition 146
Clyde Sea 6, 16, 78, 79, 240–1, 243, 244, 245, 246, 249, 326, 327, 329
cnidarians 32, 82, 94, 100, **102**, 146, 149, 150, 154, 179, 195, 261, 263, 271, 272, 274, 278
cockles **25**, 36, 75, 112, 134, 136, 140, 240, 274, 284, 338, 339, 340
dredging 338, 340
collecting 5–6
colonial animals, definition 146
comb jellies (ctenophores) 74, 315
Copenhagen 7
copepods 28, 75, 76, 117, 158, 200, 201, 225, 269
coral banks 264–78
biological profile 267–9
creation 265–6
distribution 266–7
studies of 177–8
coral communities 270–8
bryozoa 270
cnidarians 269–70, 278
crustaceans 273–4
echinoderms 276–7
molluscs 274–6
polychaetes 273
sponges 270
coral, cup 114, 264
Ross **157**, 162, 164
soft 94, 101, **149**, 150, 179, 263

Stone 180, 181, 233, 264–78, 325–6
Coralline Crag 279, 281–2
Cornwall 112, 163, 175, 237, 291, 294, 297, 307
cottid fish 167
crabs 62, 93, 94, 114, 134, 169, 218, 249, 258, 274, 316, 333
   Angular 62, **79**
   Edible **33**, 200, 278, 292, 329, **334**
   Green 42, 136, 224, **225**, 229, 258, 315, 329
   hermit 61, 99, 100, **102**, 104–6, 114–16, 143, 144, **272**, 291–2, 298, 330–33
   nut 262, **274**, 291
   porcelain 160, 178, 262
   Round 262, **263**
   spider 99, 101, 169, 258, **274**, 298, 336
   Sponge **295**
   Stone 274
   swimming 61, 99, 114, 143, 262, 327, 328–9, **330**, 332, 337, 340, 345, 346
Creran, Loch 249, 253–4
crinoids 41, 278
ctenophores (comb jellies) 74, 315

Danish Coast 34
decapods 6, 32, 41, 48, 78–9, 83, 107–8, 134, 135, 136, 154, 164, 167, 250, 254, 262, 278, 299, 316, 320, 327–9, 338, 351
Denmark 7, 17, 23, 105, 125, 126, 154, 204, 313
Department of Environment, Food and Rural Affairs (DEFRA) 7
Devon 185, 287, 291
diatoms 26–30, 124, 129, 131–2, 149, 152, 172, 213, 214, 216, 228–9, 259, 308, 314, 316
Dogger Bank 51, 120, 121, 165

Dohrn, Anton 6
Dove Laboratory, Tynemouth 6, 7, 124
Dover Straits 19, 111, 118, 119, 166
dredged samples 4, 8, 13, 112, 153, 258, 297
dredging, clam 338–9, 341–5, 357
   cockle 338, 340
   scallop 8, 246, 333–9, 348
dredging and trawling, damage by 153, 319–58
   by aggregate extraction 319–23
   by fishing 311, 323–58
Dunstaffnage 7

East North Atlantic Central Water 24
echinoderms 32, 74, 90–1, 94–5, 98, 99, 101, 107, 112, 114, 127, 137, 140–2, 146, 167, 192, 222, 272, 276, 320, 327–8, 334, 351
echinoids 41, 123, 124, 152, 160, 169, 276, 336
echinopluteus 41, **42**
echiurans (spoon worms) 74, 85, 89
eelgrass communities 215–32
   amphipods 222–4, 225
   bivalves 230–2
   crabs 224
   diatoms 213, 215, 216, 228, 239, 259
   energy flow 228–32
   fish 225
   isopods 222–3, 225–6
   sediment–dwellers 227
eelgrasses 203–32
   anthropogenic pressures on 229–30
   degradation of 289–90
   distribution 203–4, 205–7
   growth rates 212

habitat 211–13
hydrodynamic factors 208–10
morphology 205
resilience 229
species 203–4
Ellis, John 4
energy transfers 123–4
English Channel 23, 26, 62, 94, 98, 99, 112–13, 114, 144, 164, 166, 188, 287, 288, 291, 292, 293, 294, 296, 300–2, 313, 336–7
epifauna, definition 32
euphotic zone 2–3, 28, 30, 264
European Union Habitats Directive 247
European Water Framework Directive 98
eutrophication 54–7, 130–2, 212, 217–19, 229–30, 289, 314, 351, 357

Fal, River 237
Faroe Islands 17, 271, 277, 284
feeding methods, grouping by 37–8
Field Studies Council (FSC) 7
fish communities
  changes in 300–6
  damage to 323–58
  overfishing effects 317–18
fish farming 249–50
fisheries science 3, 94, 317–19
fishing, commercial 93, 287, 303, 306, 317–19
  damage to habitat and fauna 323–58
  development of 348
Flamborough Head 17, 111
flatworms 132
foraminiferans 283
Forbes, Edward 4, 5
fossils, evidence from 283–6

'fouling communities' 174–5, 179, 306, 309–10, 314, 318
Frisian Front 61, 65, 69, 91, 92, 121–2
Fyne, Loch 249

Gareloch, Loch 324
gastropods 32, 40, 41, 44, 47, 48, 67, 75, 93, 167, 190, **194**, 198–9, 201, 202, 218, 220, 222, 228, 253, 274–6, 283, 316, 344, 351
gastrotrichs 75
German Bight 61, 89, 91, 92, 93, 110, 303
Glyceridae 85
gobies 61, 134, 135, 143, 144, 168, 218, 225, **226**
Gosse, Philip Henry 5
grabs 11–13, 93
  Day 11
  Hamon 12–13
  Petersen 11
  van Veen 11, **12**, 76
Greater North Sea Ecoregion 8
Gullmarsfjord 223, 225, 229

habitat disturbance and loss 54–5, 317–61
  data collection of 348–57
  by aggregate extraction 319–23
  by fishing 311, 323–58
habitat engineers see biogenic reefs
hard grounds 145–80
  artificial surfaces **174**, 175–6
  communities, defining physical factors 148–80
  ascidians 155, 156, 176–7
  bryozoans 147–8, 160–2, **162–3**
  corals 149–50, 162–5, 180
  echinoderms 165–8

hard grounds, communities, defining physical factors *continued*
  fish 168–9
  'fouling communities' 174–5, 179, 306, 309–10, 314, 318
  hydroids 147–8, **176**, 177–9
  larval behaviours 150–2
  molluscs 158–60, 169
  sea slugs 149, **170–2**, 173
  monitoring 153–6
  populations 156–166
  settlement 150–2, 173–80
  spatial competition 148–50, 173
heart urchins 35, **36**, 42, 50, 82, 86, 90, 240, 246, 276, **335**, 348, 352–5, 358, 359
Hebrides 17, 164, 254
hemichordates (acorn worms) 41, **43**, 74
Hincks, Rev. T. 6
holoplankton 40, 50
holothurians (sea cucumbers) 41, 277
Humber 105, 106, 114, 207, 313
hydroids 4, 100, 145, 146, 149, 154, 156, 158, 160, 166, 170, 171, **172**, 173, 175, **176**, 177, 179, 192, 197, 200, 213, 243, 248, 251, **252**, 271, 272, 297, 336
hydrozoans 74, 76, 271
Hyne, Lough 168

Iceland 18, 19, 44, 182, 183, 284
infauna 75–93
  definition 32
International Bottom Trawl Survey (IBTS) 100–1
International Council for the Exploration of the Sea (ICES – CIEM) 7–9, 16, 55
  ICES rectangles 8–9, 100, 104–8, 303, 351–3

invertebrate assemblages 34–7
Irish Sea 4, 19, 24, 29, 79, 85, 86, 89, 94, 95, 98, 99, 114, 143, 163, 165, 235, 256, 346
Isle of Man 6, 16, 85, 163, 256, 334–7
Isle of Wight 185, 293, 297
isopods 126, 154, 199, 201, 215, 218, 222–3, 227, 229, 253, 273

Jeffreys, J.G. 6
jellyfish 117

K-dominance curves 58–60, 62, 250–1, 326
kelps 32, 154, 156, 160, 165, 167, 182–203, 221–2
  associated fauna 188–200
  canopy habitat 190–5
  distribution 182–5
  energy flow 200–3
  epiphytic seaweeds in 196–8
  fish in 200–2
  growth 184–8
  holdfast habitat 198–200
  morphology 183
  reproduction 183–4
  species 182–5
  stipe habitat 195–8
Kiel Fjord 214, 228, 350
kinorhynchs 75
Kwinte Bank 75–7

larval stages 39–54
  classification systems 43–8
  densities 49–54, 116–22
  recruitment 48–9
  reproductive and mortality rates 48–9, 116
  settlement 48–9, 150–2

lecithotrophs 40, 43, 45–7, 116, 150–2, 159
limpets 130–2, 183, 192–5, 242, 275, 306, 308, 311–12, 316–17
Lister, Martin 4
Liverpool 6, 7, 308, 325
lobsters **78**, 169, 193, **202**, 222, 274, 326, 327, 329, 336
Long Island 230–1
loriciferans 75
Lowestoft 7
lugworms 35

macroalgae 2, 26, 30, 31, 154, 172, 182, 186, 195, 202, 213, 216–17, 218, 221, 227, 228, 232, 237, 240, 245–6, 315
maerl beds 233–51
  biodiversity of 247
  communities:
    hydroids 251–3
    molluscs 241–4, 248–9
    polychaetes 251, 253–6
    sea urchins 245–6
    shrimps 246
  damage by dredging 337–8
  description 233–4
  development and structure 236–40
  locations 234–5
  recovery 247
  vulnerability 247
Maldanidae 93
Malta 238
marine aggregates, extracting 319–23
marine biology study, development of 4–11
marine food webs 123–44
Marine Strategy Framework Directive, EU (MSFD) 58
megafauna, definition 33
meiofauna, definition 33

Menai Bridge, Anglesey 7
meroplankton 40–1, 48, 50, 117
microfauna, definition 32–4
Millport Marine Biology Station 6, 7
mineral extraction 55
mineralisation 88
Ministry of Agriculture, Fisheries and Food (MAF) 7
molluscs 12, 15, 25, 30, 32, 41, 47, 48, 51, 74, 112, 146, 154, 155, 169, 179, 228, 241, 243, 249, 253–4, 272, 275, 320, 324–5, 326, 327, 328, 333, 335, 338
  larval stage 41, 46, 47, 48, 51
mussels 31, 130, 136, 145, 158–60, 164, 179–80, 181, 203, 224, 230, 232, 233, 275, 283, 290, 338, 347
mussel beds 233, 248, 254, 256–64
mussel fossils 283–4

Natural Environment Research Council (NERC) 7
nauplius 41, **42**
nematodes 75, 76, 77, 132–3, 134, 200, 360
Nephtyidae 85
Newfoundland 259–60
Newhaven dredge 333, 335, 337, 339
nitrates 26, 27, 88, 89, 130, 186, 212, 359
Norfolk coast 221, 288, 312
Norman, Canon A.M. 6
North Atlantic Current (NAC) 18–19, 20
North Atlantic Oscillation Index (NAO) 19–20, 287, 299–300, 302
North Sea 1, 8, 17, 19, 20, 22–4, 26, 28, 49–51, 55, 56–7, 58, 60–1, 62, 65, 81, 89, 91, 94, 99–110, 113–16, 119, 121, 123, 124, 134, 142–3, 154, 178–80, 204, 225, 268, 284, 292, 298, 299, 303–5, 309, 313, 315, 323, 348, 349, 351–61

Northumberland coast 16, 124, 154, 196
Norwegian coast 7, 18, 17, 125, 154, 180, 183, 185, 186, 188–90, 193, 196, 200, 202, 225, 234–6, 238–9, 256, 266–8, 278, 283, 284–5
　West Norwegian Shelf 325–6
Norwich Crag 279
nutrients, cycle of 28–31

oil rigs 174, 179–80, 268
ophiopluteus 41, **42**
Oslo Convention 1972 54–5
Oslo/Paris Commission (OSPAR) 55–8
osmotrophy 46, 47
ostracod 76, 154, 200
overfishing 55
oysters 130, 148, 203, 230, 274, 306, 289, 290, 306, 309, 310, 311, 316, 338, 347

Paris Convention 1974 55
particulate organic matter (POM) 30, 31, 183
pelagic environment 2–3
Petersen, C.G.J. 11, 34
phlorotannins 202–3
phoronids (horseshoe worms) 41, **43**
phosphates 26, 27, 28, 57, 88, 157, 159, 358
photosynthesis 26–7
phytoplankton 26–30, 49, 57, 71, 77, 82, 88–9, 121, 123, 124, 126, 130–1, 166, 200–1, 204, 213–14, 217, 218, 228–30, 232, 242, 260, 268–9, 289, 291, 299, 309, 316–17, 358–9
'pioneer' species 176
pipefish 143, 218, 221, 225, **226**
plankton, *see* larval stages, phytoplankton, zooplankton

planktotrophs 39, 43–7, 116, 150–1, 159
Pleistocene Epoch 282–6
Pliocene Epoch 281–2, 285
Plymouth Marine Laboratory 6, 7, 189, 222, 297, 308
Plymouth Routines in Multivariate Ecological Research (PRIMER) 70
polychaetes 12, 33, 34, 40–1, 46, 48, 51, 52, 61, 65, 68, 75–7, 80, 83–4, **85**, 86, 90–3, 108, 117–18, 126, 132–5, 143, 144, 146, 149–50, 154–5, 158, 169, 178, 190, 199, 224, 228–9, 233, 240, 243, 251, 253–6, 263, 269, 272–3, 308, 316, 321, 324–5, 333, 348, 352–6, 358
Poole Harbour 177, 314–15
population density and fluctuation 116–22
Port Erin, IOM 6, 336
predator feeding 133
printed/electronic information, coming of 15–16

ragworm 273
Ray, John 4
razor shells 35, 274, 338–9, 341, 345–7, **346**, 356
　American 92, **312**, 313, 338
Red Crag 279, 280, 281
rias 237
Rinne 17, 19
Rissoidae 190
Roscoff 7, 5, 76, 77
ROVs 13–14
Russell Cycles 287, 292

salinity values 21–2
salmon farming 249–50

scallops 142, 164, 230, 240, **241**, **242**, 247, 260, 261, 274, 282, **284**, 285, 297, 298, 323, 328
  dredging 8, 333–9, 347, 348
scavengers 80, 85, 95, 132, 134, 167, 168, 173, 240, 244, 249, 250, 262, 273, 274, 319, 320, 323, 329–33, 336, 338, 351, 356
Scilly, Isles of 183, 206, 227
Scotland, coast of 18–19, 26, 78, 87, 106, 186, 193, 204, 207, 234, 237, 284, 303, 305, 345, 346
scuba diving 13, 15
Scylla, HMS 175–7
sea anemones 5, 94, 116, 144, 146, 148, 149, 154, 157, 173, 179, 244, 272, 278, 292, 328
sea cucumbers 41, 277
sea floor damage
  by aggregate extraction 319–22
  by fishing 323–4, 333, 335–6, 337–8, 341
  by pollution 317–19
sea lochs 82, 85–6, 237, 249, 253–4, 261, 324
sea pens 82, 83, 116, 328
sea slugs 149, **170–2**, 173, 178, 191, 192, 199, **243**, 251, 274
sea squirts 61, 74, 145, 146, 152, **153**, 169, 243, 263, 309
sea temperature variation 23–4, 291–3
sea urchins 93–4, 105, 167, 168, 192–4, 199, 245–6, 276, 290
seabirds 58
seagrasses 81, 203–4, 289 *see also* eelgrass
seaweeds 6, 26, 32, 148, 168, 182–3, 188, **189**, 195, 204, 213, 217, 220, 223, 240, **241**, 245–6, 290

epiphytic 195–7
see also kelp
sediment grade 33–4
sedimentary environment 71–144
  categories 74
  classification 72–5
  community:
    Cnidaria 82–3
    crustaceans 78, 100, 112, 114
    echinoderms 94–5, **97**, 98–101, 105–6, 107, 112, 114, 116, 118–22, 123, 126, 136, **137–8**, 140–2
    fish 108–9, 142
    heart urchins 82, 86
    molluscs 75, 108, 112, 130, **131**, 134, 136
    shrimps 79–81, 107, 114, 134–6
    worms 83–6, 117–18, 126–30, 132–4
  composition 71–5
  epifauna 93–116
  infauna 75–93
  populations 116–44
  sampling 75–6
  Wentworth classification scale 72
Seine, Bay of 112, 117–18
Shetland Is 4, 17, 19, 20, 106, 111, 164, 180, 234, 250, 285, 286, 303
shipworm 307
shrimps 101, **102**, 106, 114, 134–6, **135**, 142, 144, 224–5, 229, 300–1, 336
  grass 218, 220
  mud 36–7, 79–81, 92, 121–2, 240, 243, 246–7, 338, 353, 356, 359
silica 27, 28, 71, 131, 316, 359
sipunculans 37, 45, 47, 48, 74, 256
Skagerrak 17, 22, 89, 90, 93, 125, 127, 313
sleds, used for research 13, **14**

soft sediment epifauna, surveys 93–106
Solander, Daniel 4
sonar equipment 15
Southampton 7, 311
species' distribution patterns, origins of 282–6
  historic changes in 287–93
sponges 4, 32, 46, 74, 149–50, 151, 154, 156, 158, 169, 170, 195, 197, 200, 248, 249, 260, 262, 263, 265, 266, 270, 271, 272, 274, 295
St Andrews 6, 7
St Malo, Gulf of 112, 291, 300, 336
starfish 32, 41, 42, 74, 93, 94, **97**, 99–100, 114, **115**, 123, 136–7, **138**, 141–2, 167, 228, 229, 250, 258, 262, 276, **277**, 290, 327, 328, 329, 331, 350, 351
sticklebacks 218, 225
Strangford Lough 258, 60
sublittoral zone 2
submersibles 3, 13–14
Sussex, West 163, 204
Swedish Coast 7, 121, 125, 126, 154, 206, 218–9, 220, 222–3, 224, 229, 278, 307
synchronous spawning 48

taxonomic diversity and distinctness indices 65–8
teleplanic larvae 47–8
Thomson, Charles Wyville 5, 17
Thorson, Gunnar 43–4, 47
top shells 242, 275, 292, **293**
Torrey Canyon 54–5, 292, 317
toxins 54–5, 317–18
trawling, *see* dredging and trawling
  Agassiz trawl 8, **10**

beam trawler, damage from 122, 142, 303, 322–7, 332–40, 347, 351–2, 357–8
  sampling with 8, **10**, 94, 99–109, 114
trigger fish 295
trochophore **40**, 41
trophic group analysis 37–8
tube anemones 82, **84**, 240
tubeworms 146, 147, 148, 152, 155, 173, **174**, 175, 179, 192, 244, 253, 310
tunicates 61, 74, 145, 146, 152, **153**, 169, 243, 263, 309

unitary animals, definition 146

veliger **40**, 41, 47

Wadden Sea 53, 87, 132, 206, 207, 208–9, 212, 214, 228, 313
water temperature, significance 3, 13, 15, 20–4, 34–5, 44, 49, 48, 51, 61, 103, 112, 116, 121, 126, 136, 147, 174, 177, 180, 188, 205, 206, 213, 216, 237, 238, 257, 266–7, 282, 285, 287–8, 290–2, 294, 297, 300–6, 309, 311, 315, 327, 345, 359
Waters, Rev. A.W. 6
Welsh coast, north 259, 262, 292
  south 114, 134, 160–1, 163, 237, 295–7, 310, 320
Weymouth 7
whelks 38, **39**, 44, 108, **110**, 167, 244, 249, 250, 258, 275, 280, 284, 328, 331, 334, 338, 350
Whitesand Bay reef 175–7
worms 12, 74, 85, 92–3, 117–18, 190, 224, 228, 240, 272, 273
  acorn 41, **43**, 74
  arrow 30, 74

bamboo 84, 272
honeycomb 31, 254–6, **255**
horseshoe 41, **43**
nematodes 75–6
paddle 273
peanut 74
scale 273
spoon 74, 85, 89

tubeworms 146, 147, 148, 152, 155, 173, **174**, 175, 179, 192, 244, 253, 310
wrecks 54, 168, 178–9, 353

Yorkshire coast 17, 256, 311, 352

zooplankton 28–31, 40, 41, 88, 268, 287, 291, 299, 309, 315

The New Naturalist Library

1. *Butterflies* — E. B. Ford
2. *British Game* — B. Vesey-Fitzgerald
3. *London's Natural History* — R. S. R. Fitter
4. *Britain's Structure and Scenery* — L. Dudley Stamp
5. *Wild Flowers* — J. Gilmour & M. Walters
6. *The Highlands & Islands* — F. Fraser Darling & J. M. Boyd
7. *Mushrooms & Toadstools* — J. Ramsbottom
8. *Insect Natural History* — A. D. Imms
9. *A Country Parish* — A. W. Boyd
10. *British Plant Life* — W. B. Turrill
11. *Mountains & Moorlands* — W. H. Pearsall
12. *The Sea Shore* — C. M. Yonge
13. *Snowdonia* — F. J. North, B. Campbell & R. Scott
14. *The Art of Botanical Illustration* — W. Blunt
15. *Life in Lakes & Rivers* — T. T. Macan & E. B. Worthington
16. *Wild Flowers of Chalk & Limestone* — J. E. Lousley
17. *Birds & Men* — E. M. Nicholson
18. *A Natural History of Man in Britain* — H. J. Fleure & M. Davies
19. *Wild Orchids of Britain* — V. S. Summerhayes
20. *The British Amphibians & Reptiles* — M. Smith
21. *British Mammals* — L. Harrison Matthews
22. *Climate and the British Scene* — G. Manley
23. *An Angler's Entomology* — J. R. Harris
24. *Flowers of the Coast* — I. Hepburn
25. *The Sea Coast* — J. A. Steers
26. *The Weald* — S. W. Wooldridge & F. Goldring
27. *Dartmoor* — L. A. Harvey & D. St Leger Gordon
28. *Sea Birds* — J. Fisher & R. M. Lockley
29. *The World of the Honeybee* — C. G. Butler
30. *Moths* — E. B. Ford
31. *Man and the Land* — L. Dudley Stamp
32. *Trees, Woods and Man* — H. L. Edlin
33. *Mountain Flowers* — J. Raven & M. Walters
34. *The Open Sea: I. The World of Plankton* — A. Hardy
35. *The World of the Soil* — E. J. Russell
36. *Insect Migration* — C. B. Williams
37. *The Open Sea: II. Fish & Fisheries* — A. Hardy
38. *The World of Spiders* — W. S. Bristowe
39. *The Folklore of Birds* — E. A. Armstrong
40. *Bumblebees* — J. B. Free & C. G. Butler
41. *Dragonflies* — P. S. Corbet, C. Longfield & N. W. Moore
42. *Fossils* — H. H. Swinnerton
43. *Weeds & Aliens* — E. Salisbury
44. *The Peak District* — K. C. Edwards
45. *The Common Lands of England & Wales* — L. Dudley Stamp & W. G. Hoskins
46. *The Broads* — E. A. Ellis
47. *The Snowdonia National Park* — W. M. Condry
48. *Grass and Grasslands* — I. Moore
49. *Nature Conservation in Britain* — L. Dudley Stamp
50. *Pesticides and Pollution* — K. Mellanby
51. *Man & Birds* — R. K. Murton
52. *Woodland Birds* — E. Simms
53. *The Lake District* — W. H. Pearsall & W. Pennington
54. *The Pollination of Flowers* — M. Proctor & P. Yeo
55. *Finches* — I. Newton
56. *Pedigree: Words from Nature* — S. Potter & L. Sargent
57. *British Seals* — H. R. Hewer
58. *Hedges* — E. Pollard, M. D. Hooper & N. W. Moore
59. *Ants* — M. V. Brian
60. *British Birds of Prey* — L. Brown
61. *Inheritance and Natural History* — R. J. Berry
62. *British Tits* — C. Perrins
63. *British Thrushes* — E. Simms
64. *The Natural History of Shetland* — R. J. Berry & J. L. Johnston

65. *Waders* — W. G. Hale
66. *The Natural History of Wales* — W. M. Condry
67. *Farming and Wildlife* — K. Mellanby
68. *Mammals in the British Isles* — L. Harrison Matthews
69. *Reptiles and Amphibians in Britain* — D. Frazer
70. *The Natural History of Orkney* — R. J. Berry
71. *British Warblers* — E. Simms
72. *Heathlands* — N. R. Webb
73. *The New Forest* — C. R. Tubbs
74. *Ferns* — C. N. Page
75. *Freshwater Fish* — P. S. Maitland & R. N. Campbell
76. *The Hebrides* — J. M. Boyd & I. L. Boyd
77. *The Soil* — B. Davis, N. Walker, D. Ball & A. Fitter
78. *British Larks, Pipits & Wagtails* — E. Simms
79. *Caves & Cave Life* — P. Chapman
80. *Wild & Garden Plants* — M. Walters
81. *Ladybirds* — M. E. N. Majerus
82. *The New Naturalists* — P. Marren
83. *The Natural History of Pollination* — M. Proctor, P. Yeo & A. Lack
84. *Ireland: A Natural History* — D. Cabot
85. *Plant Disease* — D. Ingram & N. Robertson
86. *Lichens* — Oliver Gilbert
87. *Amphibians and Reptiles* — T. Beebee & R. Griffiths
88. *Loch Lomondside* — J. Mitchell
89. *The Broads* — B. Moss
90. *Moths* — M. Majerus
91. *Nature Conservation* — P. Marren
92. *Lakeland* — D. Ratcliffe
93. *British Bats* — John Altringham
94. *Seashore* — Peter Hayward
95. *Northumberland* — Angus Lunn
96. *Fungi* — Brian Spooner & Peter Roberts
97. *Mosses & Liverworts* — Nick Hodgetts & Ron Porley
98. *Bumblebees* — Ted Benton
99. *Gower* — Jonathan Mullard
100. *Woodlands* — Oliver Rackham
101. *Galloway and the Borders* — Derek Ratcliffe
102. *Garden Natural History* — Stefan Buczacki
103. *The Isles of Scilly* — Rosemary Parslow
104. *A History of Ornithology* — Peter Bircham
105. *Wye Valley* — George Peterken
106. *Dragonflies* — Philip Corbet & Stephen Brooks
107. *Grouse* — Adam Watson & Robert Moss
108. *Southern England* — Peter Friend
109. *Islands* — R. J. Berry
110. *Wildfowl* — David Cabot
111. *Dartmoor* — Ian Mercer
112. *Books and Naturalists* — David E. Allen
113. *Bird Migration* — Ian Newton
114. *Badger* — Timothy J. Roper
115. *Climate and Weather* — John Kington
116. *Plant Pests* — David V. Alford
117. *Plant Galls* — Margaret Redfern
118. *Marches* — Andrew Allott
119. *Scotland* — Peter Friend
120. *Grasshoppers & Crickets* — Ted Benton
121. *Partridges* — G. R. (Dick) Potts
122. *Vegetation of Britain & Ireland* — Michael Proctor
123. *Terns* — David Cabot & Ian Nisbet
124. *Bird Populations* — Ian Newton
125. *Owls* — Mike Toms
126. *Brecon Beacons* — Jonathan Mullard
127. *Nature in Towns and Cities* — David Goode
128. *Lakes, Loughs and Lochs* — Brian Moss
129. *Alien Plants* — Clive A. Stace and Michael J. Crawley
130. *Yorkshire Dales* — John Lee